Statistical Applications in the Spatial Sciences

Statistical Applications in the Spatial Sciences

Editor N Wrigley

10/1980
geog.

 Pion Limited, 207 Brondesbury Park, London NW2 5JN

Printed in Great Britain by Page Bros (Norwich) Limited.

Preface

The Quantitative Methods Study Group of the Institute of British Geographers was formed in 1964 [see Gregory (1976)] in the initial phases of the so-called 'quantitative revolution' in geography. In its early years it was a relatively small group, providing a forum for the few British geographers who used statistical techniques in research and who were pioneering the introduction of statistics teaching into undergraduate degree courses in geography. Today, after fifteen years of increasing activity and strength (Unwin, 1978), it is a group of 450 geographers whose interests cover the whole spectrum of possible interrelationships between statistics, mathematics, and computing, and the various subdisciplines within human and physical geography. In addition to conferences held at regular intervals[1] it has its own publications[2], a role in postgraduate research training (Hepple and Wrigley, 1979), and is developing increasingly important links with related organisations in other disciplines and with sister groups in other countries (Unwin, 1979).

Since its foundation, one of the Group's major objectives has been to develop links, both formal and informal, between statisticians and geographers. In its early years the Group benefited considerably from the participation of certain eminent statisticians in its activities, and informal contacts on an individual basis have continued. It was not until the 1970s, however, with the increasing maturity of quantitative geography, that the first joint conferences and publications with statistical societies took place. These included: a joint conference with the Applied Stochastic Processes Study Group in 1975 (ASPSG, 1976), which attracted a grant from the Statistics Committee of the Social Science Research Council; a review paper by Cliff and Ord (1975) to the Royal Statistical Society Research Section; and a special issue of *The Statistician* (1974). In all of these ventures quantitative geographers began to introduce statisticians to the techniques which they had been using and developing over the previous decade for analysing spatial data. They also demonstrated how the origins of such work could be traced to the papers of statisticians such as Student (1907; 1914), Yule (1926), Kendall (1939), Stephan (1934), Neprash (1934), Moran (1948), and Geary (1954). Although these first formal contacts were invaluable in developing a dialogue between geographers and statisticians, a number of statisticians clearly would have preferred the meetings which were held to have concentrated upon the discussion of more detailed examples of statistical applications in geographical research. To this end, a joint conference of the Group with the Royal Statistical Society General Applications Section was held at the University of Bristol in September 1977

---

[1] Summaries of these are published regularly in the IBG journal *Area*.
[2] The Group publishes the CATMOG—*Concepts and Techniques in Modern Geography*—monograph series available from Geo Abstracts Ltd, University of East Anglia, Norwich NR4 7TJ, and books of collected conference papers or review essays (Martin et al, 1978; Bennett and Wrigley, 1980).

on the theme of *Statistical applications in the spatial sciences*. This book consists of thirteen of the sixteen papers presented at that conference. The majority of the papers are by geographers but the subject matter discussed ranges from archaeology through climatology, economics, epidemiology, entomology, and hydrology, to plant science and soil science. Despite the wide range of subject matter the papers confront common problems and share similar approaches. Together they form a partial, but it is hoped nonetheless valuable, view on the applied spatial analysis scene in the latter half of the 1970s.

Neil Wrigley
Chairman, IBG Quantitative Methods Study Group

References
ASPG, 1976 "Patterns and processes in the plane" conference report by the Applied Stochastic Processes Study Group, *Advances in Applied Probability* **8** 651-658
Bennett R J, Wrigley N (Eds), 1980 *Quantitative Geography in Britain: Retrospect and Prospect* (Routledge and Kegan Paul, London)
Cliff A D, Ord J K, 1975 "Model building and the analysis of spatial pattern in human geography" *Journal of the Royal Statistical Society* **37B** 297-348
Geary R C, 1954 "The contiguity ratio and statistical mapping" *Incorporated Statistician* **5** 115-145
Gregory S, 1976 "On geographical myths and statistical fables" *Transactions Institute of British Geographers, New Series* **1** 386-389
Hepple L W, Wrigley N, 1979 "SSRC research training course" *Area* **11** 261-262
Kendall M G, 1939 "The geographical distribution of crop productivity in England" *Journal of the Royal Statistical Society* **102** 21-48
Martin R L, Thrift N J, Bennett R J (Eds), 1978 *Towards the Dynamic Analysis of Spatial Systems* (Pion, London)
Moran P A P, 1948 "The interpretation of statistical maps" *Journal of the Royal Statistical Society* **10B** 243-251
Neprash J, 1934 "Some problems in the correlation of spatially distributed variables" *Journal of the American Statistical Association* **29** Supplement 167-168
The Statistician, 1974 **23**(3, 4)
Stephan F F, 1934 "Sampling errors and the interpretation of social data ordered in time and space" *Journal of the American Statistical Association* **29** 165-166
Student, 1907 "On the error of counting with a haemacytometer" *Biometrika* **5** 351-360
Student, 1914 "The elimination of spurious correlation due to position in time and space" *Biometrika* **10** 179-180
Unwin D J, 1978 "Quantitative and theoretical geography in the United Kingdom" *Area* **10** 337-344
Unwin D J, 1979 "Theoretical and quantitative geography in north-west Europe" *Area* **11** 164-166
Yule G U, 1926 "Why do we sometimes get nonsense-correlations between time-series?—a study in sampling and the nature of time-series" *Journal of the Royal Statistical Society* **89** 1-69

Contributors

M G Anderson
Department of Geography, University of Bristol, University Road, Bristol BS8 1SS

R J Bennett
Department of Geography, University of Cambridge, Downing Place, Cambridge CB2 3EN

A D Cliff
Department of Geography, University of Cambridge, Downing Place, Cambridge CB2 3EN

I S Evans
Department of Geography, University of Durham, Science Laboratories, South Road, Durham DH1 3LE

R I Ferguson
Earth and Environmental Science, University of Stirling, Stirling FK9 4LA

A C Gatrell
Department of Geography, University of Salford, Salford M5 4WT

P Haggett
Department of Geography, University of Bristol, University Road, Bristol BS8 1SS

L W Hepple
Department of Geography, University of Bristol, University Road, Bristol BS8 1SS

I Hodder
Department of Archaeology, University of Cambridge, Downing Street, Cambridge CB2 3DZ

R L Martin
Department of Geography, University of Cambridge, Downing Place, Cambridge CB2 3EN

F H C Marriott
Department of Biomathematics, University of Oxford, Pusey Street, Oxford OX1 2JZ

S Openshaw
Department of Town and Country Planning, University of Newcastle upon Tyne, Newcastle upon Tyne NE1 7RU

K S Richards
Department of Geography, The University, Hull HU6 7RX

P J Taylor
Department of Geography, University of Newcastle upon Tyne, Newcastle upon Tyne NE1 7RU

R Webster
Soil Survey of England and Wales, ARC Weed Research Organization, Sandy Lane, Yarnton, Oxford OX5 1PF

I Woiwod
Department of Entomology, Rothamsted Experimental Station, Harpenden, Herts AL5 2JQ

Contents

Part 1

Social science applications

Introduction

In the early 1960s, when the IBG Quantitative Methods Group was formed, British quantitative geographers were engaged in applying conventional statistical methods to geographical problems and in diffusing knowledge of such methods to the wider geographical community. By the late 1960s the use of such statistical methods had become commonplace in geographical research, and quantitative geographers began to turn their attention to an examination of those properties of geographical data that make the use of conventional statistical methods in geographical research more difficult than had appeared likely in the early 1960s. Attention was drawn to the problem of spatial dependence. It was noted that whereas most conventional statistical methods assume independent observations, geographical data typically exhibit systematic ordering over space, thus breaking the assumption of independence and giving rise to potentially serious problems when geographers attempt to apply standard statistical procedures to spatially located data. Attempts were made to develop measures of the degree of spatial dependence or spatial autocorrelation exhibited in geographical data sets (Cliff and Ord, 1973) and to demonstrate the implications of failing to meet the independence assumption in the case of certain standard statistical models and tests of inference. A deeper understanding of the difference between the 'chance set-up' perspective of classical probability and statistics and the more appropriate chance process or stochastic process perspective, given the spatial dependence apparent in most geographical data, emerged (Hepple, 1974), and it was realised that the approach to stochastic process inference adopted in econometrics was particularly relevant in geographical research since there were many similarities between the time-series problems faced by econometricians and the spatial-series problems faced by geographers. In addition to the problem of spatial dependence, attention was also drawn to the modifiable areal unit problem. Quantitative geographers became increasingly sensitive to the dependence of statistical measures upon the sizes and shapes of the areal units for which data are collected.

During the 1970s the problems which were identified in the late 1960s have been considered in greater depth. Attempts have been made to develop techniques to cope with spatial dependence. This has involved modification of statistical tests (Cliff and Ord, 1975) and development of statistical models that incorporate spatial dependence, or the transformation of data in such a way that spatial autocorrelation is removed, thus allowing the use of conventional models and tests [for example, the spatial differencing employed by Martin (1974)]. A linkage of the econometrician's concern with temporal dependence and the geographer's concern with spatial dependence has occurred in the development of a variety of statistical

models for space–time series, and many of these have been used in a
forecasting context (Cliff et al, 1975; Haggett et al, 1977; Martin et al,
1978). This in turn has resulted in the development of models that
incorporate parameters which vary through time and/or over space, and
with a concern with appropriate identification and estimation procedures
(Martin and Oeppen, 1975; Bennett, 1975). Attempts have also been
made to investigate and provide solutions to aspects of the modifiable
areal unit problem (Openshaw, 1977), and in addition to consider other
problems faced in the statistical analysis of geographical data such as its
nonstationarity and nonnormality, and the prevalence of categorical
(Wrigley, 1979) and closed-number system observations.

Although the 1970s have also seen continuing progress in numerous
other areas of spatial analysis such as point-pattern analysis, regionalisation,
surface mapping, Markov process modelling, and multidimensional scaling,
the studies presented in this first section of the book reflect only the
major developments discussed above. The first four chapters, by Cliff and
Haggett, Hepple, Martin, and Gatrell, relate to the space–time dependence
theme. Cliff and Haggett discuss attempts to describe and model the
diffusion of epidemics in closed communities by using the spatial autocorrelation measures and space–time forecasting models discussed above.
They also employ explicitly process-based models such as the Hamer–Soper
and chain binomial models. Hepple presents an analysis of regional responses
to national unemployment fluctuations in Great Britain which employs
time-varying parameter models and recursive least-squares techniques.
Martin discusses some of the specification and identification problems in
constructing models of the spatial transmission of wage inflation through
US labour markets. Gatrell explores the spatial autocorrelation structure
of central-place populations in Southern Germany, treating Christaller's
population map as a realisation of a spatial stochastic process and examining
it with the use of two-dimensional autocorrelation functions.

These four chapters on the subject of space–time dependence are
followed by two which relate to the theme of the modifiable areal unit.
Openshaw and Taylor provide a useful introduction to the problem and
follow it with a discussion of three experiments on variations in the
correlation coefficient under different spatial and statistical conditions.
Evans then discusses correlations between census variables at a given
scale—the 1 km aggregate level. His study based upon 1971 Census 1 km
grid square data represents the first such study at a national scale and
provides a datum against which previous correlation analyses at other
scales can be compared.

The section is concluded with a chapter by Hodder, an archaeologist,
who demonstrates how many of the statistical techniques used or developed
by quantitative geographers have been adopted in archaeology. Many of
the problems encountered in the statistical analysis of archaeological data
are those already recognised by quantitative geographers, but there are also

additional problems, for example the differential rates of the survival of archaeological sites, which present major challenges to the spatial analyst in archaeology.

References

Bennett R J, 1975 "The representation and identification of spatio-temporal systems: an example of population diffusion in North-West England" *Transactions, Institute of British Geographers* **66** 73–94

Cliff A D, Haggett P, Ord J K, Bassett K, Davies R B, 1975 *Elements of Spatial Structure: A Quantitative Approach* (Cambridge University Press, London)

Cliff A D, Ord J K, 1973 *Spatial Autocorrelation* (Pion, London)

Cliff A D, Ord J K, 1975 "The comparison of means when samples consist of spatially autocorrelated observations" *Environment and Planning A* **7** 725–734

Haggett P, Cliff A D, Frey A, 1977 *Locational Analysis in Human Geography, 2nd edition, volume 2: Locational Methods* (Edward Arnold, London)

Hepple L W, 1974 "The impact of stochastic process theory upon spatial analysis in human geography" *Progress in Geography* **6** 91–142

Martin R L, 1974 "On autocorrelation, bias and the use of first spatial differences in regression analysis" *Area* **6** 185–194

Martin R L, Oeppen J E, 1975 "The identification of regional forecasting models using space–time correlation functions" *Transactions, Institute of British Geographers* **66** 95–118

Martin R L, Thrift N J, Bennett R J (Eds), 1978 *Towards the Dynamic Analysis of Spatial Systems* (Pion, London)

Openshaw S, 1977 "A geographical solution to scale and aggregation problems in region-building, partitioning and spatial modelling" *Transactions, Institute of British Geographers, New Series* **2** 459–472

Wrigley N, 1979 "Developments in the statistical analysis of categorical data" *Progress in Human Geography* **3** (forthcoming)

1

Geographical aspects of epidemic diffusion in closed communities

A D Cliff, P Haggett

1.1 Introduction

The historical record suggests that, at least as far back as the Black Death, inhabitants of one geographical area have seen the approach of an epidemic from another area as a wave about to engulf their own population. By the late nineteenth century, the pressures for some kind of early-warning system for the approach of such waves had become, at least in the most industrialised countries, very strong. Dr J Tatham argued in 1888 for a compulsory notification system for infectious diseases so that "... the sanitary authorities of one notification town may be warned in time of the approach of any infectious disease from other towns" (Benjamin, 1968, page 142). Legislation, which was first adopted locally by individual towns, later made standard at the national level, and still later enacted at the international level, ensures that a wealth of data, highly variable in quality but of fine geographical mesh, pours forth each week and month on the detailed pattern of disease distribution around the world.

Modern research on the etiology of infectious diseases at the clinical and laboratory level has produced falls both in incidence and mortality levels and has tended to reduce the importance of mass data gathering. But simple mechanistic models of spread in terms of epidemic waves continue to be used to describe the large-scale movements of diseases through human populations. From time to time, the wave patterns formed by infectious diseases attract researchers from other fields to try to sieve through the records to find those persistencies in spatial or temporal behaviour which might throw light on the progress of a disease, or which might allow some early-warning system to be developed. Frequently the results have been disappointing, but always with a sufficient hint of regularity to attract a later generation of workers back to the same problem. The perennial questions remain: does the distribution map throw light on the origin or progress of an epidemic? Does the epidemic spread in recognisable ways? Can we learn anything from the maps that would help to moderate or to delay its impact?

In this chapter we review some of our own work in this tradition. The work forms part of an ongoing project, financed by the Social Science Research Council, entitled "Geographical modelling and forecasting for epidemiological and population data". We look first at the nature of the epidemic data and the selection of test areas (section 1.2). This leads on to a consideration of the models that can economically describe (section 1.3) and, hopefully, forecast (section 1.4) some of the evolving epidemic maps.

Finally we look at some future lines of enquiry, and try to answer the question of whether, despite its many limitations, work at this geographical scale is capable of yielding information not apparent at the detailed clinical level.

1.2 Sources of information

1.2.1 Selection of epidemic disease

In the work reported here, attention is focussed on the geographical patterns formed by one particular infectious disease—measles—and this is used as an indicator for epidemic spread. The selection of measles notifications from the group of infectious diseases for which spatial data are available was determined by a number of factors:

(1) Measles is a notifiable disease in many countries of the world. World Health Organization records suggest that around 128 of its 210 reporting countries collect measles records with some degree of accuracy. For England and Wales, the major published source of epidemiological data is the Registrar General's *Weekly Return*. This includes notifications of several infectious diseases, in addition to measles, as supplied by the Medical Office of Health for each local authority for the week (ending on a Friday). Detailed clinical work suggests that notifications may seriously underestimate the actual incidence of some diseases. Apart from variations in diagnosis, the practice of individual medical practitioners in notifying cases to local Medical Officers of Health is known to vary widely. Not enough information is available about regional variations to allow the application of individual correction factors, but these might be expected to run at around one-and-a-half to two times the notifications actually recorded in the case of measles (Stocks, 1949).

(2) The high rate of incidence of measles in unvaccinated populations yields a large number of notifications, so that statistical estimates of critical parameters can be made with reasonable precision. For the main area studied, Southwest England, the number of notifications averaged over 250 a week throughout the study period. For Iceland the corresponding figure was thirty-three a week.

(3) Transmission of the disease is directly from person to person. Absence of an intermediate host (as in malaria) allows simpler models to be employed, and enables demographic data to be incorporated directly into the model.

Typical infection chains for a measles outbreak are set out in figure 1.1. For an individual, the first indication of a measles attack occurs with the onset of a high temperature some eight to twelve days after infection. This *latent period* is followed by a seven- to nine-day *infectious period* during which the virus can be passed on to other individuals. Three or four days after onset symptoms, characteristic rashes usually appear, and isolation is common in the latter part of the infectious period [figure 1.1(a)].

Given the variable length of the latent and infectious periods, and the chance timing of contacts between infected and susceptible individuals, a measles chain may have links from as short as eight days [figure 1.1(b)] to as long as three weeks [figure 1.1(c)]. As figure 1.1(d) shows, a two-week period (the *serial interval*)[1] is a reasonable estimate for chain length and this is adopted here as in earlier statistical studies of measles epidemics (Bailey, 1975, pages 273–275).

(4) Measles is a highly infectious disease. Up to 99% of susceptibles appear to contract the disease after first contact with an infective, and a large number (30%–40%) of cases are children of preschool age. The strong age bias towards a section of the population with low spatial mobility suggests that propinquity may play a major part in guiding the geographical pattern of outbreaks, and lends added validity to the search for spatial models.

(5) Measles outbreaks show a characteristic recurrence peak every two to four years, with clear interepidemic phases (see figure 1.2).

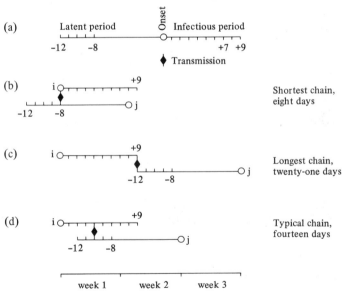

Figure 1.1. The infection process as a chain structure. Schematic model for measles contacts between an infective (i) and a susceptible (j). Open circles denote onset of infection. In the shortest chain, (b), the infective makes his contact on day 1 of his infectious period and the susceptible is 'latent' for as short a time as possible. The longest chain arises when i transmits on his last possible day of infection and j is latent for as long as possible (c).

[1] Bailey (1957, pages 15–16) defines the serial interval as "the period from the observation of symptoms in a second case directly infected from the first ... . Thus the serial interval is the observable epidemiological unit".

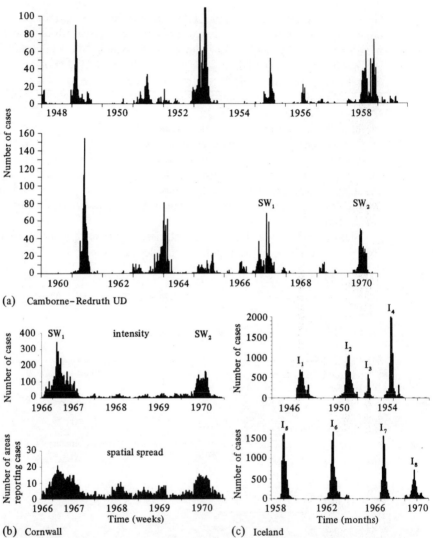

Figure 1.2. Recurrence intervals for major measles outbreaks in Southwest England and Iceland. (a) Sample Cornish GRO for twenty-three-year period, 1948–1970, showing two-year recurrence interval typical of open urban systems with a high density of susceptible individuals in which measles is endemic. (b) Waves chosen for modelling in Cornwall (SW_1 and SW_2). (c) Icelandic data for twenty-six-year period, 1945–1970, which show four-year recurrence interval typical of isolated communities in which measles is not endemic. The four-year gap is related to the buildup of the susceptible population (up to 40% of cases occur in the under-five age group in unvaccinated populations). Waves are labelled I_1 to I_8 for identification in modelling. In (b) note the continuous record of infection, both in terms of intensity and number of areas reporting cases, typical of areas in which measles is endemic. 1970 was chosen as the stopping point because vaccination programmes were introduced in both countries during the preceding decade.

This recurrence permits the testing of models developed on one set of waves with later outbreaks. For this reason measles has played a central role in the development of quantitative epidemiological theory, and a number of classic deterministic and stochastic models of epidemic spread were first derived from a consideration of measles returns (Stocks and Karn, 1928; Soper, 1929; Bailey, 1957).

1.2.2 Selection of geographical areas and time periods

The characteristics of the geographical areas and time periods analysed to date in the project are summarised in table 1.1. Areas were chosen largely to emphasise the degree of closure in the outbreaks studied both in a spatial and temporal sense.

(1) *Southwest England.* Study of the spatial characteristics of outbreaks is being conducted for the whole of Southwest England. This is conventionally defined as the six geographical counties of Cornwall, Devon, Dorset, Gloucester, Somerset, and Wiltshire. It covers an area of some 9000 square miles, which is slightly larger than that of Massachusetts, and in 1971 it had a civilian population of around $3 \cdot 7$ millions. For statistical purposes the main units used by the General Register Office (GRO) for data collection during the time period considered in the present study were the 'old' (pre-1970) local authority administrative areas: the county boroughs (CB), municipal boroughs (MB), urban districts (UD), and rural districts (RD). The Southwest was divided into 179 such units. If we ignore the Scilly Isles, the remaining 178 units constituted a contiguous network of areas.

For other parts of the analysis the county of Cornwall, the combined counties of Devon and Cornwall, and the Bristol area have been separately analysed (see figure 1.3). Cornwall, with its twenty-seven GRO areas,

Table 1.1. Data sets used in the study of the spatial spread of measles outbreaks.

Area	Time periods	Number of spatial units	Waves	Models used later in paper
Southwest England:				
Bristol	222 weeks	10	2	Hamer–Soper (stochastic)
Devon and Cornwall	222 weeks	72	2	spectral and cross-spectral analyses
Southwest Cornwall	1600 weeks	11	12	–
Cornwall	222 weeks	27	2	autoregressive, exponential smoothing
Southwest England	222 weeks	178	2	autoregressive
Iceland:				
Reykjavik	312 months	1	8–12	autoregressive, exponential smoothing, chain binomial, Kalman filter
Iceland	312 months	50	8–12	–

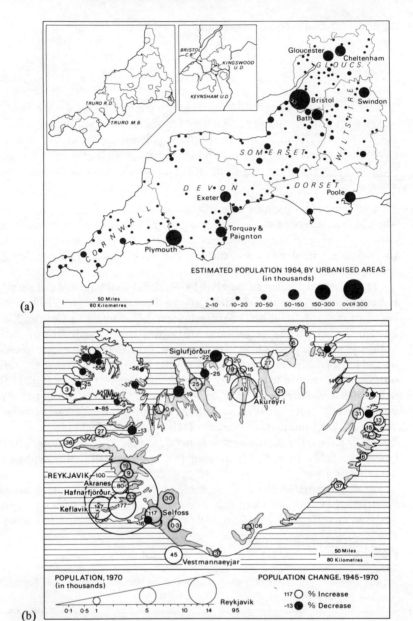

Figure 1.3. Population structure of the two main sample areas. (a) Southwest
England, with insets for Cornwall and the Bristol region on the same scale.
(b) Iceland, where circles give 1970 population and the figures give population change.
Stipple indicates settled areas. The circles are on different scales on each map
(Reykjavik is about the same size as Bath). Iceland is over twice the size of the
Southwest, but its 1970 population density was *circa* 5 persons per square mile,
compared with *circa* 410 persons per square mile for the Southwest. In the Southwest,
Bristol and Plymouth are the only centres in which measles is truly endemic; Iceland
has no centre big enough to sustain endemicity. In (b) the rapid population increase
in the Reykjavik area may be contrasted with the depopulation of towns in the
northwest, north, and east. The population of Iceland increased from *circa* 120000
to *circa* 204000 during 1945–1970; half the current total population is in the
Reykjavik area.

forms a particularly interesting test region because of (a) its isolation from the major population centres of the United Kingdom and (b) its linear and peninsular form which presents fewer feasible paths of infection for epidemic outbreaks. It has, in consequence, been the focus of a number of diffusion investigations (Scott, 1971). In the case of Devon and Cornwall combined (with seventy-two GRO areas in total), the same advantages accrue in only slightly diminished form.

In the time domain a 222-week record has been investigated, which begins at the end of September 1966 and runs through to the last week of 1970. The series includes two major measles outbreaks in the Southwest, the first of which peaked in late February 1967, and the second three-and-a-half years later in July/August 1970. It should be emphasised that these 'peaking dates' are based on total notifications for the region as a whole. For most subareas examined, the choice of study period meant that the data record began and ended with a string of zero readings relating to 'interepidemic' phases. For eleven GRO areas in western Cornwall, findings have been checked by examining notifications over an 1199-week period back to and including 1948. The twelve peaks identified there are shown in figure 1.2(a) for Camborne–Redruth UD.

(2) *Iceland*. For Southwest England, with its large total population, measles cases were reported somewhere within the area in every week of the time series analysed. To allow a clearer definition of separate epidemic waves, a search was made to find a more isolated geographical area which would have the following characteristics: (a) a population small enough to break the continuity of disease transmission. Data reported from British and American cities by Bartlett (1957; 1960) suggested that 4000 to 5000 cases of measles each year were needed to prevent such breaks; this implies that study of populations of less than 250 000 to 300 000 inhabitants might be appropriate. (b) A long enough record of public health administration to give a notification series showing both multiple waves and a satisfactory reporting rate. (c) Sufficient external isolation to permit treatment of the whole area as a closed population, coupled with enough internal isolation of settlements to allow each community to be regarded as a separate epidemiological cell within the study area.

To approach condition (c) in a twentieth-century world it was decided to extend the study from a peninsular to an island setting. Study of measles cases for nineteen island communities over periods of up to fifteen years by Black (1966) had shown breaks in the continuity of measles transmission for all island communities of less than around 500 000 inhabitants.

Of the islands whose measles records were studied by Black, Iceland, with a 1970 population of 204 000, came closest to meeting the necessary design criteria. Under Danish rule, its medical records for measles extend back to 1896, and its reporting rates are high by international standards.

Some indication of the relative reporting rate can be obtained from a
comparison of the total notifications for Iceland with that for one of the
other areas studied. For Cornwall, a total of 8398 cases was recorded
over a 222-week period; for Iceland the total was 45388 over a 312-
month period. On weighting the figures for resident population, the
Cornish rate is around $4 \cdot 27$ notified cases per 1000 population per year,
whereas the Icelandic rate is $10 \cdot 45$ notified cases per 1000 population
per year (with population averaged over the 1945/1970 period). The
differences in the rates may partly reflect age composition, since the
Icelandic population is faster growing and has a younger age structure
than its English counterpart [35% under fifteen years of age, compared
with 23%; Keyfitz and Flieger (1968)]. However, it is more likely to
reflect differences in reporting practice between the two areas and
confirms the generally high standard of the Icelandic data. In addition,
Iceland is a country which showed major changes in regional demographic
structure and interregional transport connections in the postwar period
(Preusser, 1976; Thorarinsson, 1961). The main difficulty with the
Icelandic data is that it is on a monthly basis whereas, as discussed in
section 1.2.1, the serial interval for measles is about two weeks.

The period chosen for analysis was the twenty-six years from 1946 to
1970 inclusive, with notifications aggregated for fifty communities on a
comparable basis. As figure 1.2(c) indicates, this allowed study of eight
major epidemic peaks.

1.2.3 Problems of closure

Comparison of the degree of 'closure' of the two main study areas is
difficult and only preliminary estimates are possible. Southwest England,
with a resident population of $3 \cdot 7$ millions in 1970, has a passenger flow
across its boundaries from the rest of the country in excess of twenty
millions per year; this includes daily commuting traffic as well as short-
term visitors. Iceland, with a resident population of $0 \cdot 2$ millions in 1970,
has a passenger inflow of 6000 by sea and 360000 by air. Much of the
latter is made up of (a) employees at the Keflavik air base and their
families (about one third of the air traffic) and (b) short-term transit
passengers en route between Europe and North America. Even with
allowances for these crude figures, the ratio of movements to resident
population in 1970 is greater than $5 : 1$ for the English study area and less
than $2 : 1$ for Iceland. In earlier periods the disparity in isolation would
have been much greater, with airflows into Iceland of only 6000 in 1945
and 65000 in 1960. By contrast, the western parts of Southwest England
have traditionally had large influxes of temporary visitors in the holiday
season with the peak summer population commonly one-and-a-half times
the resident population.

Lack of closure of the populations under study makes it difficult to
obtain very realistic estimates of susceptible populations, or to model a

diffusion process wholly in terms of local conditions. A highly mobile population will clearly mean that infectives may move over substantial geographical distances and trigger outbreaks at widely separate locations.

1.3 Description of the diffusion process
A simple model of a diffusion process for measles epidemics in a regional context is given in Cliff et al (1975, page 173). These authors postulated initial outbreaks of measles in major urban areas in the early stages of an epidemic, followed by contagious spread from the towns into surrounding rural areas, and hierarchical spread from large to small towns. The purpose of this section is to see in what ways this model is supported and can be refined by applying various analytical techniques to the data sets described in section 1.2. The spread of measles outbreaks in the study areas may be examined by techniques of increasing sophistication. We choose to begin with the most straightforward and work upwards.

1.3.1 Epidemic maps
For each time period (week or month), a map was constructed to show the number of cases reported at each geographical location. Since administrative boundaries may change over time, it was necessary, in the case of the longer time series, to edit the records by combining areas to ensure spatial comparability. An atlas of standardised distribution maps allows some generalisations to be made simply on the basis of 'eyeballing' map sequences.

For example, figure 1.4 shows a sequence of eight monthly maps for the wave of measles I_1. Note the way in which the epidemic starts in the largest urban area, Reykjavik. The epidemic spreads rapidly to, and intensifies in, the Reykjavik region and in the towns of the southwest corner of Iceland. Wholesale spread to the north, northwest, and east coast settlements follows later. This same pattern is repeated for other epidemic waves in Iceland. To illustrate this we plot in figure 1.5 one useful summary measure of the spatial progression of the various waves, namely the successive locations of the weighted centroid of outbreaks. Figure 1.5 shows the trajectories of the centroid for the course of each of the eight Icelandic waves. Apart from the wave I_3, which follows rapidly after wave I_2 and may be correspondingly atypical, outbreaks seem either to start in the Reykjavik area or to reach it very quickly. A general northeasterly drift tends to occur after the seventh month of each wave and, with the exception of wave I_7, the trajectory terminates at the end of each epidemic in the north or northwest corner of Iceland.

The general picture outlined above is confirmed by figure 1.6, which shows, for the eight epidemic waves that hit Iceland during the study period, the average lag time (in months) from the start of an epidemic before each settlement was reached. The diagram clearly demonstrates that larger places and places nearer to Reykjavik are generally hit first;

smaller settlements and settlements which are isolated from Reykjavik are
reached later. While this evidence may, on the surface, seem to support
the idea of a simple spread model for epidemics *down* the hierarchy of
urban size and the contagious spread of epidemic waves *out* from the
initial centres of introduction, the interpretation is complicated by the
high inverse correlation between population size of settlement and distance
from Reykjavik. The smaller settlements are generally distant from the
source areas, and so it is difficult, even with partial correlation analysis, to
disentangle the size and distance effects. We therefore turn to the
Southwest England data for further evidence.

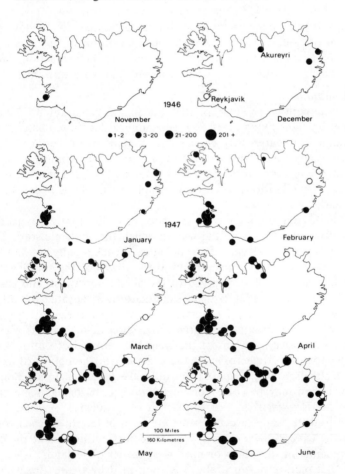

Figure 1.4. Monthly sequence of spread for the first Icelandic wave, I_1. Open circles
denote break in the record of cases reported. By June 80% of all communities had
been infected. In general, epidemic waves in Iceland peak during the winter months.

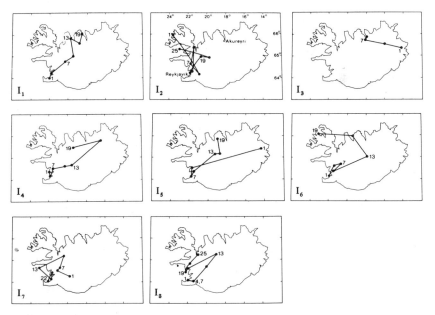

Figure 1.5. Trajectory of geographical centroids for waves I_1 to I_8. Centroids are a measure of the spatial mean of reported cases in each month. Numbers refer to timing within the wave in months (for example, 1 = month 1 of the outbreak, 7 = month 7). I_1 is the best behaved with a simple spread from Reykjavik to Akureyri. Note the contrast of I_2 and I_3. I_2 (6645 cases) is concentrated in the western part of Iceland whereas I_3 (1872 cases) was a brief epidemic following rapidly after I_2 and was confined to the eastern half of the island. I_3 exhausted the susceptible population missed by I_2.

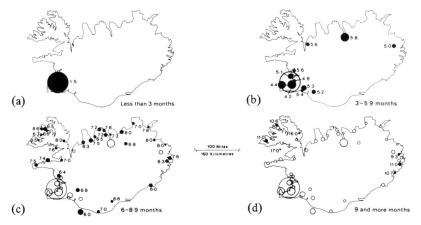

Figure 1.6. Average lag time in months from the start of an epidemic before each settlement was reached. Circles refer to population size on a proportionate scale as in figure 1.3. New centres on each map are coded black. The Reykjavik area was the starting point for most waves.

Mapping of the returns for each week on choropleth maps for Southwest England allowed an atlas of spatial spread to be built up. Figure 1.7 illustrates a typical four-week sequence. This shows that epidemics were recurrent over the region with a distinctive 'waxing and waning' or 'swash and backwash' pattern based on five major centres, namely Greater Bristol, Plymouth, West Cornwall, Southeast Dorset and South Wiltshire, and the Gloucester–Cheltenham area. At the lowest point of the 222-week period only twenty-seven GRO areas in the entire Southwest had outbreaks (about one area in seven), whereas at the maximum over 130 were infected (about two areas in three) [see Haggett (1972, page 312) for details]. Thus it appears that the relevance of our initial simple model for measles diffusion is much easier to establish for the Southwest than for Iceland, and so we now turn to other approaches in order to elaborate the model for both areas.

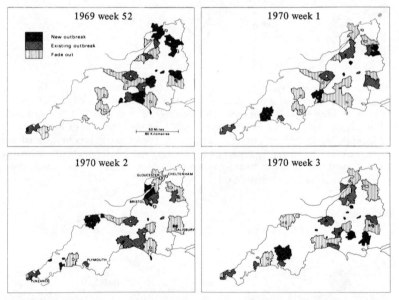

Figure 1.7. Changing distribution of GRO areas that reported measles cases in Southwest England for a four-week sequence, 1969–1970. Note the consistent abutting new to existing outbreak areas, which implies spatial contagion of outbreaks. Over the Southwest as a whole, 1966–1970, of the GRO areas which were at risk and which reported measles cases in a given week, one in eight was contiguous to a GRO that had reported cases in the preceding week. The corresponding figures for GRO areas two and three steps away from existing outbreaks were one in twenty-one and one in thirty-one.

1.3.2 Location of 'new' outbreak areas
The location of 'new' outbreaks with respect to existing outbreaks is of special interest if we assume that (1) such outbreaks represent the reinfection of temporarily free areas from source areas (either inside or

outside the study region where the epidemic is being maintained) and
(2) *ceteris paribus* the probability of population mixing is higher for
adjacent areas than for very distant areas. In the case of the Southwest
the definition of an outbreak in any GRO area used here is based on the
recorded notifications in the *Weekly Returns*. An outbreak was taken to
have ended when a GRO area failed to report new cases for three or more
consecutive weeks, and to have begun again with the next reported case;
that is, when one or more notifications occurred in any one week. It
should be stressed that this definition of 'new' outbreaks is entirely
dependent on the GRO notifications, and that, because of underreporting,
outbreaks may be erroneously defined as new which in reality are the
continuation of an existing outbreak. Other definitions of an epidemic
outbreak were considered. The first was to follow Bartlett (1957) and to
define an epidemic outbreak in terms of a threshold number of four cases
per thousand persons. However, with small areas there are substantial
differences in demographic structure. For example, rural districts
bordering large urban centres frequently contain substantial suburban
overspill composed of young families with a high proportion of children
of primary school age. Conversely coastal retirement areas usually have a
relatively high proportion of their population aged sixty years and over.
Thus any attempt to define a threshold rate would require a careful
estimate to be made of the number of susceptibles in each GRO area.
A second approach might be to define epidemic outbreaks in terms of an
excess above the mean notification ratio for a given location, with the
deviation measured as a standard z score. Both of these definitions are
highly conservative and the sample of 'new' outbreaks available for study
is therefore correspondingly restricted.

(1) *Distance decay effects*. By making use of the original definition,
the distribution of new outbreaks in the Southwest was mapped over the
sixty-week period spanning the (July/August 1970) major epidemic, SW_2
(see figure 1.7 for representative patterns). During this period the 179
Southwest areas recorded some 532 outbreaks, a gross average of $2 \cdot 97$
outbreaks per area. However, an examination of the returns for a 'lead in'
period of a fortnight preceding the first week of the study period
established that forty of the GRO areas were recording outbreak conditions
at the beginning of the period. This left 416 new outbreaks available for
study. Table 1.2 summarises the characteristics of these outbreaks. Half
the outbreaks were less than four weeks in duration and involved four or
less notifications. The pattern of new outbreaks formed a distinctly
contagious distribution, with three-quarters of them occurring in areas
contiguous to existing outbreaks. By three links distance from an existing
outbreak, the proportion of new outbreaks had fallen to less than 5%, and
no new outbreaks were recorded four or more links from an existing
outbreak. Again the strong contagious element in the spread process is
confirmed.

(2) *Estimated contact fields.* The method of defining contact probabilities, c_{ij}, between a reference subarea i and each other subarea $j = 1, 2, ..., i-1, i+1, ..., n$ in a study region composed of n subareas is shown in figure 1.8. The square schematically indicates 'the rest of the world'. The method consists of five basic steps.

Step 1:   Identify all new outbreaks l occurring within i.

Step 2:   For the first outbreak in i, identify all the potential sources among the $(n-1)$ other subareas in the study region. A potential source area is defined as one with an existing outbreak in the previous week, and which is therefore a possible subarea from which the infection might have spread. Such subareas are shaded in figure 1.8.

Step 3:   Assign the score $1/(k+1)$ to each source area and to the rest of the world, where k is the number of inferred source areas *within* the region for any one outbreak in i. The $+1$ in the denominator is used to allow for unidentified infection sources located *outside* the region in the rest of the world.

Step 4:   Repeat steps 2 and 3 for all new outbreaks l in i.

Table 1.2. Pattern of new outbreaks with distance from existing outbreaks, Southwest England, 1966–1970.

Distance from existing outbreaks	GRO areas outside existing outbreaks		New outbreaks		Ratio of column 4 to column 2
	number	percent	number	percent	
One link	2603	52·74	316	75·96	1/8
Two links	1749	35·43	83	19·95	1/21
Three links	529	10·72	17	4·09	1/31
Four links	55	1·11	–	–	
Total	4936	100	416	100	1/12

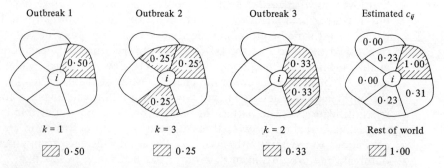

Figure 1.8. Estimation of contact probabilities, c_{ij}, between a reference subarea, i, and other subareas, j; k is the number of inferred source areas within the study region.

Step 5: Compute the quantity $c_j = \Sigma[1/(k+1)]$ for each subarea $j \neq i$ by summing over the l scores assigned to subarea j, and define the estimated contact probability, $\hat{c}_{ij}$, between i and j as

$$\hat{c}_{ij} = \frac{c_j}{\max_{j \neq i} c_j}. \qquad (1.1)$$

The denominator term will be equal to the c_j value for the rest of the world. Evidently $0 \leqslant \hat{c}_{ij} \leqslant 1$. If the jth subarea has no record of infectious disease and has therefore played no part in starting the new outbreak at i, then $\hat{c}_{ij} = 0$. Conversely, if it has a continuous record of infection its potential role is high and $\hat{c}_{ij} = 1$.

To proceed with the analysis, the map of the Cornish GRO areas was first reduced to a planar graph as in Cliff et al (1975, page 96) by regarding each GRO area as a vertex on such a graph and each common boundary between any two contiguous GRO areas as an edge between the corresponding vertices. Table 1.3 gives the estimated contact values for the Cornish GRO areas, with areas which had fewer than six new outbreaks over the study period omitted. Values range from zero up to a maximum, excluding the rest of the world, of $\hat{c}_{ij} = 0.896$, with an average value of 0.393. The pattern of values appears to be a complex one, and regression analyses were used to determine whether the $\{\hat{c}_{ij}\}$ values reflect the urban hierarchy/contagious spread model for measles epidemics described earlier. Where the $\hat{c}_{ij}$ were not too close to zero, spatial interaction models of the form

$$\ln \hat{c}_{ij} = \ln\beta_0 + \beta_1 \ln P_j - \beta_2 \ln d_{ij} \qquad (1.2)$$

were fitted, where
P_j is the population size of subarea j, and
d_{ij} is the distance on the graph between subareas i and j.
The analysis suggested a variable relationship between $\hat{c}_{ij}$ and size and distance. R^2 values ranged between 0.10 and 0.54 for the individual GRO areas. The results of t tests on the regression coefficients consistently implied that the distance term (that is, local neighbourhood effects) was more important than size effects.

However, central-place diffusion is not totally absent. The relationship between average $\hat{c}_{ij}$ terms for urban and rural GROs is shown in table 1.4. The initial contact values show little significant difference between the interurban, interrural, and urban–rural links; but when adjustments are made for the different average distances separating areas of the three types, the contacts between the high-density urban areas are emphasised.

The evidence therefore seems to suggest that the $\hat{c}_{ij}$ values in table 1.3 might profitably be modified to allow for intervening opportunity and distance effects. As discussed in section 1.3.2, the great bulk of new outbreaks in Southwest England occurred within two steps of an existing outbreak. There would therefore seem to be a case *either* (a) for editing

Table 1.3. Estimated contact probabilities, $\hat{c}_{ij}$, for spatial units within the study area ($\times$ 100).

GRO areas[a]	GRO identity number																						
	1	2	3	4	5	6	7	8	9	10	11	12	13	14	15	16	17	18	19	20	21	22	23
1 Bodmin MB	–	43	35	53	53	65	39	50	64	24	48	66	31	62	36	49	32	48	46	71	41	31	37
2 Bude–Stratton UD	26	–	38	11	21	21	13	60	24	24	12	43	06	44	21	20	18	34	26	22	06	35	14
3 Camborne–Redruth UD	34	68	–	53	41	72	48	72	62	19	30	62	52	53	54	50	39	28	52	62	29	51	48
4 Falmouth MB	18	64	33	–	44	68	49	55	48	42	17	45	31	40	20	53	46	56	43	72	36	25	29
5 Helston MB	53	44	58	78	–	90	46	90	70	48	40	86	53	63	43	69	18	28	58	65	51	59	78
6 Launceston MB	04	35	17	30	19	–	11	02	22	19	17	20	28	27	26	18	18	29	28	32	00	21	15
7 Newquay UD	41	52	32	40	56	61	–	55	30	51	52	37	79	69	38	30	59	76	45	40	74	39	51
8 Penryn MB	15	24	00	00	08	30	27	–	36	04	16	13	06	12	00	24	18	20	07	37	06	08	14
9 Penzance MB	38	37	20	41	76	56	64	45	–	65	75	45	35	44	41	42	61	47	52	49	33	40	65
10 St Austell MB	36	32	28	86	32	47	28	40	45	–	47	37	33	60	69	31	44	43	51	60	41	55	66
11 St Just UD	44	15	07	17	29	19	39	29	37	48	–	46	41	43	27	20	33	51	50	21	54	55	16
12 Saltash MB	26	44	54	49	25	51	60	60	57	19	21	–	76	48	10	66	31	20	35	53	48	28	47
13 Torpoint UD	20	32	16	33	19	21	19	19	10	32	31	19	–	26	32	07	25	47	33	34	50	27	24
14 Truro MB	21	52	72	09	24	60	44	67	15	24	11	45	42	–	00	34	21	55	26	29	16	18	35
15 Kerrier RD	24	61	24	71	47	79	29	60	57	37	41	69	41	64	–	69	39	45	33	42	42	39	59
16 Launceston RD	19	28	47	09	28	61	35	28	39	27	16	38	21	45	27	–	21	15	40	47	09	08	10
17 Liskeard RD	45	69	25	46	56	46	50	38	35	62	26	43	58	43	46	33	–	55	35	65	60	50	42
18 St Austell RD	17	36	28	62	30	88	27	59	62	28	22	44	13	48	42	43	58	–	35	50	20	34	54
19 St Germans RD	51	52	54	40	45	53	53	60	37	42	38	66	55	44	32	48	65	34	–	70	37	19	53
20 Stratton RD	31	64	09	11	27	60	30	45	16	09	07	35	00	45	21	32	42	34	09	–	09	25	19
21 Truro RD	43	81	62	58	52	79	47	45	59	33	31	58	19	59	49	64	52	57	44	64	–	48	37
22 Wadebridge RD	22	22	35	40	45	47	29	74	51	04	22	41	27	45	36	39	29	21	34	57	32	–	49
23 West Penwith RD	62	51	46	50	75	68	58	36	73	34	75	74	31	81	88	79	50	34	73	70	51	45	–

[a] GRO areas with less than six new outbreaks during the study period have been excluded.

table 1.3 to set all existing values in which $d_{ij} > 2$ steps on the graph formed by the Southwest GRO areas to $\hat{c}_{ij} = 0$, or (b) for recomputing the contact probabilities on the reduced basis for all pairs of GRO areas in which $d_{ij} \leqslant 2$ steps. Whichever solution is adopted, the contact values gained will provide some guide to *relative* magnitudes only. Again, however, the relevance of the effects of population size and the distance from source areas in the spread of measles epidemics is brought out.

Table 1.4. Contact probabilities for Cornish GRO areas[a].

Type of spatial interaction	Number of pairs	Mean separation distance (d_{ij})	Average contact values, $\hat{c}_{ij}$	
			unweighted	weighted
Urban – urban	272	4·18	0·440	0·519
Urban – rural	340	3·30	0·459	0·427
Rural – rural	90	2·49	0·478	0·335

[a] Average $\hat{c}_{ij}$ calculated from all twenty-seven GRO areas. Weighted $\hat{c}_{ij}$s are standardised to allow for separation distances.

1.3.3 Spread through graphs
In the preceding section, the joins or edges between the GRO areas on the graph of Cornwall were determined solely by the presence or absence of a common boundary. Thus for Cornwall the twenty-seven areas yield thirty-four joins (if external links to the contiguous county are ignored). Nearness on this planar graph can clearly be measured in terms of a shortest-path matrix. But it is evident that modern transport allows long-distance movements of infectives, so that contacts may occur between GRO areas *without* intervening areas being placed at risk. This is now particularly true of Iceland, where domestic air transport forms an increasingly important means of passenger movement. Theoretically therefore, each area may be potentially linked to all others in the form of a nonplanar graph: for Cornwall such a graph would have 351 joins, ten times the number of joins of the planar graph, if we ignore direction of linkage.

Haggett (1976) made use of different combinations of joins from the nonplanar Cornish data matrix to construct different graphs, each of which was designed to correspond as closely as possible with a hypothetical diffusion process. Seven different graphs were tried.

G_1 Local contagion with assumption of spread only between contiguous GRO areas. Thirty-four joins. Planar graph.

G_2 Wave contagion with assumption of spread by shortest paths from an endemic reservoir area (Plymouth). Twenty-eight joins. Planar graph.

G_3 Regional contagion with the assumption that spread is locally contagious and not on a county-wide basis, but rather within two separate regional subsystems (east and west). Thirty-two joins. Planar subgraphs.

G_4 Urban-rural contagion with assumption of spread within sets of urban and rural communities treated as separate subgraphs. 181 joins. Nonplanar subgraphs.

G_5 Population size with assumption of spread down the hierarchy of population size from largest to smallest centre. Twenty-six joins. Nonplanar graph.

G_6 Population density with assumption of spread through the density hierarchy. Twenty-six joins. Nonplanar graph.

G_7 Journey-to-work contagion with the assumptions that (a) these flows provide a surrogate for spatial interaction between areas and (b) that spread follows high interaction links. Fifty-eight joins. Nonplanar graph.

   To discriminate between the seven graphs, each of the 222-weekly maps was translated into binary 'outbreak/no outbreak' terms. Vertices on each of the 1554 graphs were coded black (B = outbreak) or white (W = no outbreak) and the Moran BW statistics under nonfree sampling (see Cliff and Ord, 1973, pages 4-7) were computed to measure the degree of contagion present in the graph. The greater the degree of correspondence between any graph and the transmission path followed by the diffusion wave, the larger (negative) should become the standard z score for BW joins. In practice there may be practical problems of common links as discussed in Haggett (1976, page 145).

   As a result of testing the spread patterns on the seven different graphs, the following preliminary observations may be made (Haggett, 1976). (1) The spatial diffusion process could be readily separated into inter-epidemic and epidemic periods, with different spread processes dominant in each. (2) During the long periods between main epidemics there is a lower level of contagion on the graphs based upon spatial structure (G_1 through G_4 and G_7), so that any diffusion currents recorded are extremely weak and poorly structured. Infections persist in the larger population clusters (G_5, G_6) and move slowly through the low-density rural areas in a sporadic manner. (3) During the shorter epidemic periods, the general level of contagion is about one-and-a-half times higher, so that spatial processes are more distinctive and easier to monitor (see figure 1.9). (3a) The advance phase of the epidemic, which starts about week 8 in figure 1.9, is marked by a rapid increase both in intensity and spread. Population size becomes a less important determinant; wave and town-country effects increase sharply (G_2, G_4). (3b) At the peak of the epidemic wave, local contagion and regional effects become important, setting up strong areal contrasts between compact clusters of infected/non-infected areas (G_1, G_3). (3c) During the subsequent retreat phase, the falloff in intensity is not associated with a corresponding reduction in geographical area, so that the epidemic appears to decay spatially *in situ*. Contagion in G_1, G_2, and G_3 falls steadily during the retreat phase: one exception is the spatial interaction model, G_7, which shows somewhat higher values for a greater length of time after the peak, which suggests that longer-range contacts between population centres may come later in the history of the epidemic wave.

This suggested history of the diffusion process of an epidemic must remain speculative for a number of reasons. First, it is based on an analysis of a single epidemic wave. Second, it depends on comparisons of the relative magnitudes of z scores for the seven individual models. Direct comparison is made difficult by the different topological structure of the seven graphs. More highly connected graphs lower the power of the join count statistic (see Cliff and Ord, 1973, chapter 7).

A third objection is that the seven models, since they contain varying numbers of common links, contain overlapping and redundant information. Graphs can be tested and redrawn to gain higher degrees of independence, but the finite number of links available limit this process.

All the previous objections are technical. A fourth and more debatable question mark relates to the whole problem of aggregating levels of diffusion behaviour. Most epidemiological work has been confined to diffusion phenomena within very small communities (for example, families, schools, local communities) where individuals may be recognised. Generalisation at a level where a group of several thousand forms the 'individual' *may* be justified by the needs of forecasting public health hazards but leaves other questions of understanding unanswered. Nevertheless, the method has enabled us to show more clearly than with earlier approaches the importance of population size in the early stages of an epidemic, the

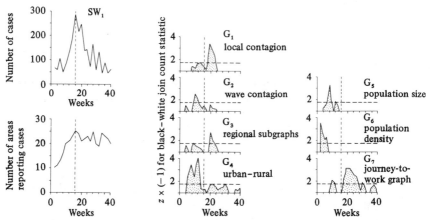

Figure 1.9. Level of autocorrelation among Cornish GRO areas that reported measles cases during wave SW_1. Definitions of the seven graphs, G_1 to G_7, are given in the text. Vertical pecked line indicates epidemic peak in terms of number of reported cases. Horizontal pecked line indicates $\alpha = 0.05$ significance level (one tail) in a test of positive spatial autocorrelation. The contrast between the prepeak phase (population size dominates) and postpeak phase (spatial structure dominates) is very evident.

strengthening of wave and contagion effects during the main epidemic phase, and the *in situ* decay of contagion during the retreat phase of an epidemic.

1.3.4 Autocorrelation measures

Additional insights into the evolutionary history of a diffusion process may be gained from an examination of the structure of the autocorrelation function of the variable of interest (here the reported number of measles cases per head of population) as it varies over space and through time. The theory involved is described in Box and Jenkins (1970, chapter 6), Cliff et al (1975, chapter 8), Bennett (1975a), and Martin and Oeppen (1975). The end product is a *correlogram* plot of some measure of autocorrelation on the ordinate of a Cartesian coordinate system, and the time or space lag on the abscissa.

The correlograms for time series of measles epidemics have an extremely characteristic form in other than very small settlements. Figure 1.10 gives the plots for four Icelandic communities. The autocorrelation functions decay rapidly over the first few time lags and display new peaks, damped in amplitude, at harmonics of the spacing between recurrent epidemics (where these are present). The graph for Reykjavik is a particularly good example. The autocorrelation functions are typical of what Bartlett (1957) has called quasi-stationary recurrent epidemics. The partial autocorrelation functions, which give the autocorrelation between variate values that are k temporal lags apart, with the effect of all other temporal lags held constant, are effectively zero, apart from the cyclical elements after only a few lags. As discussed in Box and Jenkins (1970) and Martin and Oeppen (1975), an autoregressive process displays autocorrelations that tail off approximately exponentially in time (and space), whereas the partials show a cutoff after the temporal (and spatial) lags indicative of the order of the process. We are therefore led to the conclusion that measles epidemics identify as temporally (positive) autoregressive with some cyclical components. If we operate in the frequency, rather than the data, domain, we reach the same conclusion as the power spectrum for Falmouth UD in Southwest England shows (figure 1.11).

If we look at epidemic *spread* (that is, the space–time process, as opposed to the purely temporal recurrence intervals) and use space–time autocorrelation analysis in the manner of Martin and Oeppen (1975), our conclusions are not changed. Figure 1.12 shows the joint space–time autocorrelation surface for the Cornish measles data for the whole time period, with epidemic (weeks 1–50 and 186–205) and interepidemic (weeks 51–185) phases considered separately. Results both for raw data and data first differenced in time [2] are presented. For the raw data, weeks 1–222, the autocorrelations generally decay steadily over time and

---

[2] That is, $\nabla Y_{it} = Y_{it} - Y_{i, t-1}$, where Y_{it} is the variate value in region i at time t, and ∇ is the difference operator.

space, whereas the partials cut off roughly at lag 1 in space and lag 2 or 3 in time[3]. With the differenced data, the autocorrelations cut off after lag 1 in time; we cannot say much about the behaviour of the partials.

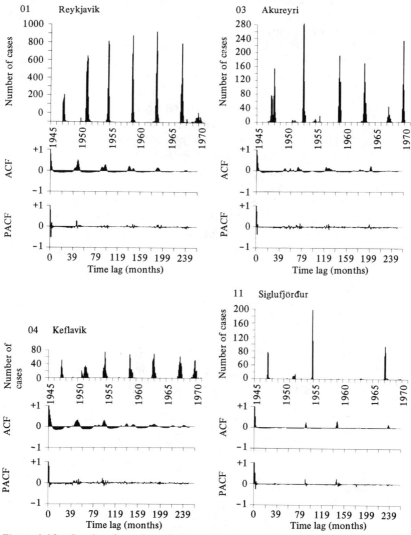

Figure 1.10. Graphs of number of reported cases and their associated autocorrelation functions (ACF) and partial autocorrelation functions (PACF) for four Icelandic communities (Reykjavik, Akureyri, Keflavik, Siglufjörður).

[3] In computing the partials under the Martin and Oeppen procedure, the order of entry of time effects was their natural serial order (that is, first $t-1$, then $t-2$, and so on). In space, arbitrary entry of spatial lags was allowed since there is no natural serial ordering as in the time domain.

The evidence apparently points to a positive autoregressive scheme of low order [AR(2)] in time for the raw data, with only weak confirmation of a contagious component of spread [AR(1)] among adjacent communities. This identification of a positive autoregressive process is no surprise in a series with a large number of zero entries. See section 1.4.1 for a further discussion of this problem.

If the second spread component postulated for measles, namely a hierarchical one between central places, is present this will be detected in autocorrelation analysis by repeaking of purely spatial correlograms and space-time correlograms at spatial lags that correspond with the separation between central places in the study area. Figure 1.13 gives the average spatial correlogram for the whole Southwest for the epidemic and interepidemic phases, and also the percentage of positive and negative autocorrelations recorded at each spatial lag for the time periods indicated. In the epidemic periods, the increase in positive autocorrelation at lags 9 and 12 for SW_1 and at lags 4, 12, and 13 for SW_2 corresponds roughly to the separation on the graph of the Southwest between urban areas with a population in excess of 50 000 shown in figure 1.3, namely, Greater Bristol, Plymouth, the towns of West Cornwall, Gloucester–Cheltenham, and the towns of the Southeast Dorset/South Wiltshire area.

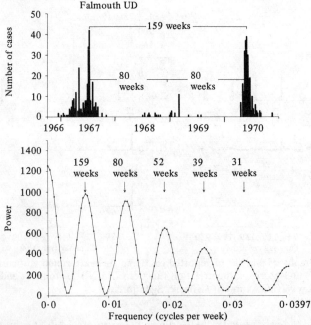

Figure 1.11. Number of reported cases and corresponding power spectrum for a Cornish GRO area, Falmouth UD.

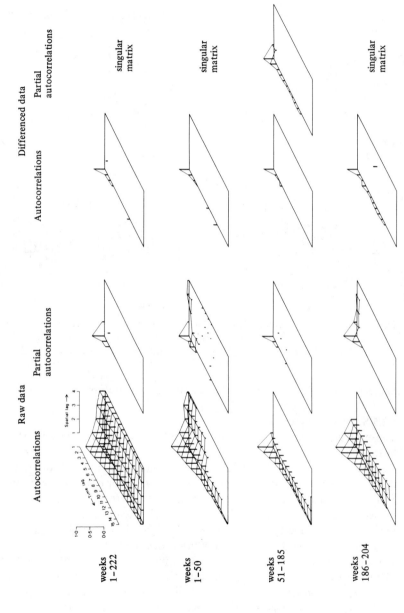

Figure 1.12. Space–time correlograms for the whole of Southwest England over a 222-week period. Note the stronger spatial structure in the epidemic phases, SW$_1$ (weeks 1–50) and SW$_2$ (weeks 186–204) than in the interepidemic phase (weeks 51–185).

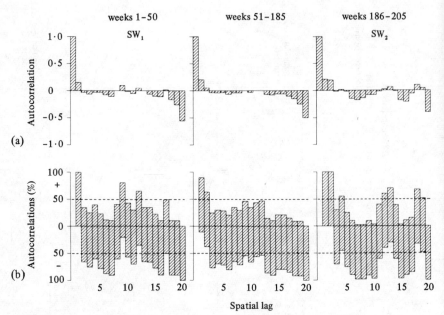

Figure 1.13. Spatial correlograms for Southwest England for epidemic and inter-epidemic phases. (a) Average autocorrelation recorded at each spatial lag. (b) Percentage of positive and negative autocorrelations recorded at each spatial lag. The increase in positive autocorrelation at lags 9 and 12 in SW_1, and lags 4, 12, and 13 in SW_2, compared with the interepidemic period, corresponds roughly to the spacing between towns of population in excess of 50000 on a graph of the Southwest.

1.3.5 Summary of process

The empirical work described in this section has shown the importance both of population size and spatial effects in the diffusion process. Generally, larger places will tend to have outbreaks first, all other things being equal, simply because they have larger susceptible populations. Spatially, contagion, or wave spread, from the larger central places seems to dominate. The evidence for hierarchical spread through the central-place system seems at best tenuous, although clearly some longer-distance leaps ahead of the diffusion waves must occur (cf Mollison, 1972). Therefore on the present evidence it seems best to think of a contagious diffusion process outwards from source areas. Quite how we account for the introduction of the disease into smaller source areas in a large nonisolated urban system, from urban centres in which measles is endemic, without postulating some central-place effect, is not clear; we simply have not been able to find very strong evidence for spatial hierarchical effects in the data sets we have examined.

1.4 Modelling the diffusion process

1.4.1 Autoregressive and moving-average models

As we have seen in section 1.3.4, identification of the generating process for measles data generally leads to the conclusion that a low-order space–time autoregressive (STAR) model is appropriate, if we are interested in short-term forecasting, when working with the raw data [or a moving-average (MA) scheme when working with first differenced data]. One of the simplest models we can use if we are not concerned with the cyclical recurrences, is the exponential smoothing model, which is the forecast equation for a first-order integrated moving-average (IMA) process. We use the term 'integrated' like Box and Jenkins (1970) to indicate the possibility of repeated application of a difference operator to the data to eliminate trend, seasonal components, and so on. Cliff et al (1975, section 10.2.2) have proposed the following space–time exponential smoothing model.

Let Y_{it} be the number or rate of reported measles cases in region i at time t and let $\overline{Y}_{i-s}$ denote the arithmetic mean of the observations which are sth spatial nearest neighbours of i. We can define the spatial (that is, purely cross-sectional) weighted moving average for the ith region as

$$\check{Y}_i = \sum_{s=0}^{l} (1-\phi)^s \overline{Y}_{i-s} \bigg/ \sum_{s=0}^{l} (1-\phi)^s , \qquad i = 1, 2, ..., n , \qquad (1.3)$$

where

l is the maximum spatial lag,

ϕ is a weighting constant, $0 \leqslant \phi \leqslant 1$, and

$\check{\ }$ is used to denote averaging over space.

Note that in this equation we are reducing the spatial information to a one-dimensional series (cf a time series) by computing $\overline{Y}_{i-s}$. These spatial averages are computed for each region in each time period, and the forecasting equation for region i is then obtained by smoothing the spatial averages for that region through time according to the model

$$\tilde{Y}_{it} = \sum_{k=0}^{m} [(1-\gamma)^k] \check{Y}_{i, t-k} \bigg/ \sum_{k=0}^{m} (1-\gamma)^k . \qquad (1.4)$$

Here

$\sim$ is used to denote the temporal smoothing of the spatial averages $\check{Y}_{it}$,

m is the maximum temporal lag, and

γ is a weighting constant, $0 \leqslant \gamma \leqslant 1$.

For large m equation (1.4) reduces to

$$\tilde{Y}_{it} = \gamma \check{Y}_{it} + (1-\gamma) \tilde{Y}_{i, t-1} , \qquad (1.5)$$

which is more convenient to update than equation (1.4) as new observations are made. The attractions of the model are its ease of updating and its dependence on the two intuitively appealing weighting constants ϕ and γ.

These parameters are commonly evaluated by direct search. The one-time-period-ahead forecast for the ith area is simply the latest available value of $\check{Y}_{it}$.

The model has been applied to areas both in Southwest England and Iceland. For the former, we look at the results for Truro RD and for the latter, Reykjavik [see figures 1.14 and 1.15(a)]. In the case of Truro RD, $\phi = 0 \cdot 55$ and $\gamma = 0 \cdot 65$ were the best values for the weighting constants in terms of goodness-of-fit at the calibration stage, implying dependence upon fairly 'near' time and space lags. This is consistent with the autocorrelation results reported in section 1.3.4. In the case of Reykjavik, $\phi = 0 \cdot 85$ and $\gamma = 0 \cdot 95$, which indicates the even more dramatic importance of 'near' space lags. The temporal smoothing coefficient for Reykjavik is much greater than that for Truro RD because for the former we are dealing with monthly data and for the latter with weekly data. From a forecasting point of view the most obvious feature of figures 1.14 and 1.15(a) is the slavish 'one step in arrears' nature of the forecasts; we can produce extremely good forecasts of last week! This arises because of the emphasis given in the model to recent observations.

As we might expect, 'blind' autoregressive models fare no better in terms of their forecasting performance than do exponential smoothing models. One of the simplest space–time autoregressive models is of the form

$$Y_{i,\,t+1} = \beta_0 + \beta_{1i} Y_{it} + \beta_{2i} \sum_j w_{ij} Y_{jt} + \epsilon_{i,\,t+1}, \qquad \begin{aligned} i &= 1, 2, ..., n, \\ t &= 1, 2, ..., T. \end{aligned} \qquad (1.6)$$

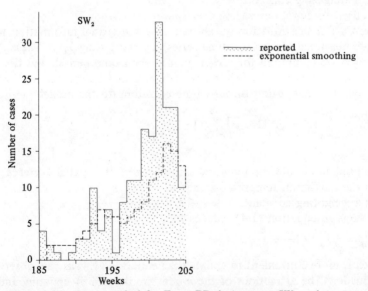

Figure 1.14. Cases notified for Truro RD during wave SW_2 and one-step-ahead forecasts using space–time exponential smoothing model.

Here β_0 to β_2 are parameters to be estimated. β_0 is the constant (if fitted), β_1 is the coefficient of the purely temporal autoregressive term, and β_2 is the coefficient of the space–time covariances; the $\{w_{ij}\}$ are prespecified structural weights which take on values greater than zero if region j is believed to influence region i and zero otherwise. The parameters of the model are thus fixed through time but may be separately estimated for each region. The $\{\epsilon_{i,\,t+1}\}$ are regression residuals of the usual kind. Fitting only the purely temporal autoregressive term in equation (1.6) to the data for Reykjavik and then using the model with the estimated value, $\hat{\beta}_1$, to produce one-step-ahead forecasts yields the results shown in figures 1.15(a), 1.15(b); the consistent lagging one time period behind is again the dominant feature.

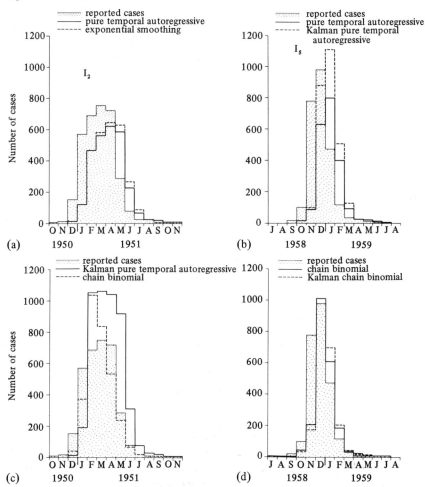

Figure 1.15. Cases notified for Reykjavik during waves I_2 and I_5 and one-step-ahead forecasts from models indicated.

These difficulties with autoregressive/moving-average models arise for two main reasons. First, the model identification procedure is so swamped by the strings of zero cases reported in the interepidemic phases that, as noted, a simple autoregressive structure is almost invariably reported for raw data by identification methods based on the Box–Jenkins and the Martin–Oeppen approaches (see section 1.3.4). The crucial assumption of the Box–Jenkins methods of a continuous series of observations is being broken. We are learning little about the spread process involved, and we are really only doing elaborate curve fitting. Second, the models used are fixed in the parameters, which means that their real strength should lie in forecasting recurrent measles epidemics which, as discussed in section 1.3.4, form a quasi-stationary cyclical waveform. It is a severe test of the models to use them to forecast within-epidemic behaviour, which is highly nonstationary—compare, for example, the build-up and fade-out phases in the number of cases reported during an epidemic in figure 1.15. These comments suggest that it may be worth exploring two main alternative approaches, namely (1) Kalman filtering to try to overcome the non-stationarity problem by an adaptive parameter model, and (2) overtly process-based models which incorporate information on levels of infective and susceptible individuals and contact patterns between them. We turn first to Kalman filtering.

1.4.2 Kalman filter models

Basic review papers which examine in detail the structure of Kalman filter models are those of Harrison and Stevens (1976) and Martin (1978). We specify first the general dynamic linear model (GDLM) as follows.

Suppose we have a system of parameters, θ_t, evolving through time as in the *system equation*

$$\theta_t = G\theta_{t-1} + v_t , \tag{1.7}$$

where $v_t \sim N(0, V_t)$, and that we observe y_t according to the *observation equation*

$$y_t = X_t\theta_t + u_t , \tag{1.8}$$

where $u_t \sim N(0, U_t)$. In equations (1.7) and (1.8) we make the following definitions:

t is the time index, $t = 0, 1, 2, ...,$

y_t is the $(n \times 1)$ vector of observations (the dependent variable) made at time t,

θ_t is the $(k \times 1)$ vector of process parameters at time t,

X_t is the $(n \times k)$ matrix of independent variables known at time t,

G is the $(k \times k)$ known matrix of structural weights defining the movement of the parameters from one moment to the next,

v_t is a $(k \times 1)$ random normal vector, and

u_t is an $(n \times 1)$ random normal vector; v_t, u_t are taken to have zero means and variances known at time t, given by $V_t = E(v_t v_t^T)$ and $U_t = E(u_t u_t^T)$, where T denotes the transpose operator.

Let $H_t = (y_1, y_2, ..., y_t, X_1, X_2, ..., X_t)$ be the history of the process up to and including time t. We now state a very powerful result known as the *Kalman filter* algorithm.

Suppose that at time $t = 0$ the prior distribution of θ_0 is $N(m_0, C_0)$. Then the posterior distribution of θ_t given H_t is also normally distributed; that is, $(\theta_t | H_t) \sim N(m_t, C_t)$, where

m_t, $t = 0, 1, ...$, is a $(k \times 1)$ column vector, and

C_t, $t = 0, 1, ...$, is a $(k \times k)$ positive definite matrix calculated from the following recursion. Let

$$\left.\begin{array}{ll} \hat{y}_t = X_t G m_{t-1}, & e_t = y_t - \hat{y}_t, \\ R_t = GC_{t-1}G^T + V_t, & \hat{\Sigma}_t = X_t R_t X_t^T + U_t, \\ A_t = R_t X_t^T (\hat{\Sigma}_t)^{-1}. & \end{array}\right\} \quad (1.9)$$

Then

$$m_t = G m_{t-1} + A_t e_t, \quad \text{and} \quad C_t = R_t - A_t \hat{\Sigma}_t A_t^T. \quad (1.10)$$

For one-step-ahead forecasting we proceed as follows:

$$(\theta_{t-1} | H_{t-1}) \sim N(m_{t-1}, C_{t-1}),$$

and

$$\theta_t = G\theta_{t-1} + v_t,$$

where v_t is independent of $(\theta_{t-1} | H_{t-1})$. Therefore from equations (1.9)

$$(\theta_t | H_{t-1}) \sim N(G m_{t-1}, GC_{t-1}G^T + V_t)$$
$$= N(G m_{t-1}, R_t).$$

Now, as given in equation (1.8), $y_t = X_t \theta_t + u_t$, where u_t is independent of $(\theta_t | H_{t-1})$. Therefore from equations (1.9)

$$(y_t | H_{t-1}) \sim N(X_t G m_{t-1}, X_t R_t X_t^T + U_t)$$
$$= N(\hat{y}_t, \hat{\Sigma}_t).$$

Hence the quantities in equations (1.9) have the following meaning:

$\hat{y}_t$ is the one-step-ahead forecast of y_t given H_{t-1},

e_t is the one-step-ahead forecast error given H_{t-1},

R_t is the variance of θ_t given H_{t-1},

$\hat{\Sigma}_t$ is the variance of the one-step-ahead forecast given H_{t-1}.

Cliff and Ord (1971) considered adaptive parameter modelling of the sort implied by equations (1.7) and (1.8), but not within the general Kalman format. Geographical applications of the Kalman filter approach to forecast levels of population and economic activity in the Northwest of England are given in Bennett (1975a; 1975b; 1975c; 1975d; 1975e).

Harrison and Stevens (1976) provide details of the way in which the models described in section 1.4.1 can be written in a Kalman format. As an illustration we take the simple temporal autoregressive model defined for a single region. On adapting equation (1.6) we have Y_{t+1}, the number of reported measles cases at time $t+1$, as

$$Y_{t+1} = \beta_1 Y_t + \epsilon_{t+1}, \qquad t = 1, 2, ..., T, \tag{1.11}$$

where $\epsilon_{t+1} \sim N(0, Y_t \sigma^2)$. If we rewrite equation (1.11) in the form of the GDLM, we have

$$\left.\begin{array}{ll} Y_t = \beta_{1t} Y_{t-1} + u_t, & u_t \sim N(0, Y_{t-1}\sigma^2), \\ \beta_{1t} = \beta_{1, t-1} + v_t, & v_t = 0. \end{array}\right\} \tag{1.12}$$

In the notation of the GDLM,

$$\theta_t \equiv \beta_{1t}, \qquad X_t \equiv Y_{t-1}, \qquad G \equiv 1, \qquad U_t \equiv Y_{t-1}\sigma^2, \qquad V_t \equiv 0,$$

and equations (1.9) and (1.10) become

$$\hat{y}_t = Y_{t-1}m_{t-1}, \qquad e_t = Y_t - \hat{y}_t,$$
$$R_t = C_{t-1}, \qquad \hat{\Sigma}_t = Y_{t-1}^2 R_t + Y_{t-1}\sigma^2,$$
$$A_t = R_t Y_{t-1}(\hat{\Sigma}_t)^{-1}, \qquad m_t = m_{t-1} + A_t e_t, \quad \text{and} \quad C_t = R_t - A_t^2 \hat{\Sigma}_t,$$

where m_t and C_t are now scalars. In this model we are using the Kalman filter rapidly to update the parameters as fresh data become available, rather than postulating a process for the movement of the parameters through time. The model parameters are allowed to adapt through time therefore in a simple-minded fashion. It is an improvement over the usual autoregressive model (1.6) where parameter estimates, once made, are commonly used for prediction as if they are known exactly, and they are not usually subject to frequent revision.

To use the algorithm, initial estimates for m_0, C_0, and σ^2 have to be provided for the region(s) of interest. The model was fitted to the data for Reykjavik. Months 1 to 36 of the series, which include the first major outbreak, I_1, were treated as a calibration period. The formation of estimates for m_0, C_0, and σ^2 is discussed in Ball (1977, page 45) and the values obtained from the calibration period were $\hat{m}_0 = 1 \cdot 00$, $\hat{C}_0 = 0 \cdot 051$, and $\hat{\sigma}^2 = 21 \cdot 88$. The model was then used to make one-step-ahead forecasts for I_2, and the results are shown in figure 1.15(c). Again, despite the adaptive parameter form, the forecast number of cases follows the actual figures but one month in arrears. The effect becomes more marked with time. In addition the magnitude of the epidemic is over-estimated. Ball and Cliff (1979) show that, because of the form of the model used, the Kalman parameter estimates, $\hat{m}_t$, become less sensitive with time and closely resemble in shape the reported number of cases.

Modification of equations (1.12) to allow β_1 to vary stochastically through time [that is, with $v_t \sim N(0, V_t)$] resulted in reduction of the lag effect but led to an even grosser overestimate of the magnitude of the epidemic. The problems are once more primarily due to the lack of a continuous series of observations, with the result that the Kalman filter estimates of the parameters switch from near zero to near one between interepidemic and epidemic phases, that is, the filter acts almost like a dummy variable.

It is evident that reasonable forecasts for epidemic spread are extremely difficult to obtain within the traditional time-series framework, and we therefore now consider explicit process-based spread models as an alternative. Indeed, we wonder whether our movement in this direction is mirroring the same debate which has taken place in econometrics since 1970 as to whether Box–Jenkins methods or economic time-series modelling are 'better' for forecasting economic processes. Thus Prothero and Wallis (1976, page 468) report the following:

> "Granger and Newbold (1975) commented that 'so far the sparring partner [the time-series model] is consistently out-pointing the potential champion [the econometric model]."
> "Christ (1975) reports comparisons in which ARIMA forecasts 'are uniformly the poorest', so perhaps the potential champion is reaching match fitness".

1.4.3 Process-based models

(a) *The Hamer–Soper Model.* We consider first a single-region model. At any time t, we assume that the total population in the region can be divided into three classes: namely the population at risk or *susceptible population* of size $P^S(t)$, the *infected population* of size $P^I(t)$, and the *removed population* of size $P^R(t)$. Where the sizes of the P^S, P^I, and P^R populations can be referred to unambiguously without the (t) notation, it will be dropped for simplicity. The removed population is taken to be composed of people who have had the disease, but who can no longer pass on the disease to others because of recovery, isolation on the appearance of overt symptoms of the disease, or death. Four types of transition are allowed.

(1) A susceptible being infected by contact with an infective.

(2) An infective being removed. We assume that infection confers life-long immunity to further attack after recovery, which is reasonable in the case of measles.

(3) A susceptible 'birth'. This can either come about through a child growing up into the critical age range (that is, reaching about six months of age), or through a susceptible entering the population by migration into the region from outside.

(4) An infective entering the P^I population by migration into the region from outside. For simplicity we assume that there is no migration out of the region.

Suppose that transition i occurs at the rate r_i ($i = 1, 2, 3, 4$); that is, in a small time interval $(t, t + \delta t)$ the probability of transition i occurring is $r_i \delta t + o(\delta t)$, where $o(\delta t)$ means a term of smaller order (that is, considerably smaller) than δt. All events are assumed to be independent and to depend only on the present state of the population. The probability density of the time between any pair of successive transitions is

$$r \exp(-rt), \tag{1.13}$$

where

$$r = \sum_{i=1}^{4} r_i, \tag{1.14}$$

and the probability that the next transition is of type i is

$$\frac{r_i}{r}, \qquad i = 1, 2, 3, 4. \tag{1.15}$$

We assume, in the transition types 1–4, that the infection rate is proportional to the product $P^S P^I$, that the removal rate is proportional to P^I, and that the birth and immigration rates are constant. We can thus prepare the table of transitions given in table 1.5. Note that, in contrast to the ARIMA approach, the structure of the process is explicitly incorporated in the model.

The model was first put forward in its deterministic form by Hamer (1906) and was studied extensively by Soper (1929). Haggett (1972; 1975; 1978) has considered the utility of the model in a regional setting. The main drawback of this deterministic model is that it leads to damping of successive epidemic waves which is not observed in practice—see figure 1.16. Soper believed erroneously that this damping could be eliminated by allowing for the fact that individuals exposed to measles undergo a latent period before becoming infectious, as shown in figure 1.1, instead of assuming, as in the deterministic model described, that they transmit infectious material for the entire time between initial exposure and removal. However, even with this modification and, in addition, modifications to allow for seasonal variations in the infection rate and for spatial factors, damping still persists although it is reduced.

We therefore consider the more realistic stochastic formulation of the model. Even this relatively simple model is surprisingly intractable analytically, and except in special cases such as $\nu = \eta = 0$ it is best

Table 1.5. Transition types and rates.

Type of transition	Rate
$P^S \to P^S - 1, P^I \to P^I + 1, P^R \to P^R$	$\lambda P^I P^S$
$P^S \to P^S, P^I \to P^I - 1, P^R \to P^R + 1$	μP^I
$P^S \to P^S + 1, P^I \to P^I, P^R \to P^R$	ν
$P^S \to P^S, P^I \to P^I + 1, P^R \to P^R$	η

studied using Monte Carlo techniques. It is, however, possible to see intuitively how the model operates. An infective is isolated after an average period of $1/\mu$ days, and while he is infectious he causes new infections at the rate of λP^S per day. If we ignore the changes in P^S during this period, one infective infects an average number of $\lambda P^S/\mu$ ($= \kappa$, say) susceptibles before he is removed. From the theory of the simple birth and death process, we would expect that, when $\kappa \leqslant 1$, a small epidemic would die out. However, when $\kappa > 1$, a small epidemic will spark off a major outbreak, although, of course, as the epidemic spreads P^S will fall and $\lambda P^S/\mu$ can become less than unity. Thus the general pattern will be that the susceptible population will build up (transition type 3) to around the critical population size $P^S = \mu/\lambda$, when an epidemic will spread until the susceptible population falls sufficiently for the epidemic phase to pass [cf Kendall's (1957) Pandemic Threshold Theorem]. The cycle will then repeat itself. In large communities where measles is endemic, the period between epidemic peaks is of approximate length $\mu/\lambda\nu$, the mean time for the birth of μ/λ susceptibles. In smaller communities, where there is fade out, the period is longer because once the critical susceptible population size is reached there is a delay until the disease is introduced again into the region from outside (see Bartlett, 1956).

To adapt the model to handle regional interactions, the transitions given in table 1.5 must be defined for each region. We also wish to allow for the division of the incubation period into latent and infectious components by revising table 1.5 to define the set of transitions shown in table 1.6. Each region in an n-region system therefore has its own susceptible (P^S), latent (P^L), infectious (P^I), and removed (P^R) populations; that is, we have

$$\{(P_i^S, P_i^L, P_i^I, P_i^R), \; i = 1, 2, ..., n\}.$$

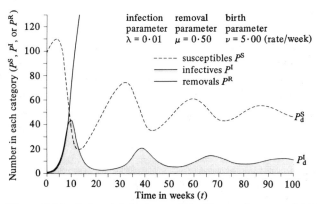

Figure 1.16. Deterministic form of the Hamer–Soper model for an artificial data set showing the curves for the P^S, P^I, and P^R populations. Note the characteristic damping with time of the P^S and P^I curves to the levels P_d^S and P_d^I respectively.

Of the parameters, ζ and μ depend only on the nature of the disease and so can be taken to be regionally invariant. Interregional interaction is entered into the model by defining the probability rate of infection for region i, λ_i, as

$$\lambda_i = \sum_{j=1}^{n} \lambda_{ij} P_i^S P_j^I , \qquad i = 1, 2, ..., n . \tag{1.16}$$

This sum includes the term $\lambda_{ii} P_i^S P_i^I$, as in the single-region model, and covers the ith intraregional interaction. The remaining terms handle interactions between region i and other regions j. Again for simplicity we assume that there is no migration between regions, although there is no difficulty in adding migration terms to the model. At the aggregate level, distance decay ideas imply that the probability of contact between a person in P^I and a person in P^S will fall off with increasing distance between them. A plausible first step in modelling the $\{\lambda_{ij}\}$ is to use the gravity formulation

$$\lambda_{ij} \propto \frac{1}{d_{ij}^2} , \qquad i \neq j , \tag{1.17}$$

where d_{ij} is the distance between the (demographic) centroids of regions i and j.

The diagonal terms $\{\lambda_{ii}\}$, the within-region probability interaction rates, represent the proportion of the total population of region i with which a susceptible in region i makes contact. λ_{ii} can be taken to be inversely proportional to the total population of region i by the rough argument that any given susceptible might be expected to have approximately the same size of acquaintance and kinship circle (that is, contact group at home, school, etc), and to have the same level of risk of infection as a result, whatever the population size of the region.

Murray and Cliff (1977) have applied the regionally based model to the Bristol area data described in section 1.2.2. Calibration details are given in the reference. Some of the results appear in figure 1.17. The model is able to accommodate the broad differences in the phase and amplitude characteristics of the epidemic waves for Kingswood and Keynsham. Although the model does not forecast day-to-day variability in reported cases with precision, the outstanding feature is the fact that forecasts of recurrences either coincide with, or are just in advance of, actual recurrences.

Table 1.6. Revised table of transition types and rates.

Type of transition	Rate
$P^S \rightarrow P^S - 1, P^L \rightarrow P^L + 1, P^I \rightarrow P^I, P^R \rightarrow P^R$	$\lambda P^I P^S$
$P^S \rightarrow P^S, P^L \rightarrow P^L - 1, P^I \rightarrow P^I + 1, P^R \rightarrow P^R$	ζP^L
$P^S \rightarrow P^S, P^L \rightarrow P^L, P^I \rightarrow P^I - 1, P^R \rightarrow P^R + 1$	μP^I
$P^S \rightarrow P^S + 1, P^L \rightarrow P^L, P^I \rightarrow P^I, P^R \rightarrow P^R$	ν
$P^S \rightarrow P^S, P^L \rightarrow P^L, P^I \rightarrow P^I + 1, P^R \rightarrow P^R$	η

This stands in contrast to the 'one week behind' characteristic of the time-series models. From a control point of view it might be argued that it is more important to have good advance warning of the approach of an epidemic than it is to know the detailed behaviour once the wave has arrived.

(b) *Chain binomial models.* We suppose that it is possible to record the state of the system at times $t = 0, 1, 2, \ldots$. Ideally the time interval between recording points should correspond in length to the serial interval of the disease, which ensures that we do not witness multiple epidemic cycles in a single time period provided that each infected individual is isolated after the appearance of symptoms. The total population at the beginning of time period t, which we denote by $P(t)$, will contain $P^S(t)$ susceptibles, $P^I(t)$ infectives who can transmit the disease to susceptibles, and $P^R(t)$ removals. Removals are infectives who have ceased to pass on the disease to susceptibles because of, say, recovery or isolation. During the tth time period, $P(t)$ may be modified by the addition of $P^A(t)$ arrivals (births and/or immigrants) and the loss of $P^D(t)$ departures (deaths and/or emigrants). Finally, let $P^{\text{new}}(t)$ denote the number of new cases which occur in time period t.

The following accounting identities may then be written down for each region (Cliff and Ord, 1978):

$$P(t+1) = P(t) + P^A(t) - P^D(t),$$

and

$$P(t) = P^S(t) + P^I(t) + P^R(t) .$$

$$(1.18)$$

If $\alpha_t P^A(t)$ denotes the number of individuals among the new arrivals who are infectives, and $\delta_t P^D(t)$ denotes the number of infectives among the departures, then

$$P^I(t+1) = P^I(t) + P^{\text{new}}(t) - [P^R(t+1) - P^R(t)] + \alpha_t P^A(t) - \delta_t P^D(t) . \quad (1.19)$$

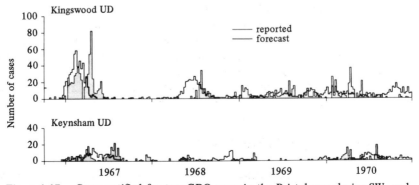

Figure 1.17. Cases notified for two GRO areas in the Bristol area during SW$_1$ and SW$_2$ and one-step-ahead forecasts from a stochastic version of the Hamer–Soper model.

In practice we may wish to drop the last two terms as being negligible. However, the arrivals term is vital if we wish to consider diseases reimported into an area after local fade-out; these arrivals are the individuals who trigger a recurrence of the disease when the number of susceptibles is high.

In common with Bailey (1975) we say that there has been *adequate contact* by time t between the jth susceptible, S_j (a member of population P^S), and the infectives $\{I_i\}$ (of population P^I), if S_j contracts the disease during period t. The contact may have occurred one or more time periods earlier, depending upon the length of the serial interval and its relationship to the length of the data recording interval.

We now define the probability that S_j is identified as an infective during time period t as $\Theta_j(t)$. One way of specifying $\Theta_j(t)$ might be to assume the following:

(a) the serial interval is relatively constant in length and equal to the interval between observations, so that S_j is identified as being infected during period t if he has sustained adequate contact during period $t-1$;

(b) the disease is transmitted only by adequate contact between a susceptible and an infective;

(c) adequate contact with a single infective is sufficient to transmit the disease;

(d) contacts between S_j and different infectives are independent events;

(e) there is homogeneous mixing between the susceptibles and infectives *within* a region at some standard rate of contacts per unit time and (possibly) mixing between regions at a different (lower) rate.

Then

$$\Theta_j(t) = 1 - \text{prob (no contact with any infective)}$$

$$= 1 - \prod_{i=1}^{P^I(t)} \{1 - \Theta_{ij}(t)\}, \tag{1.20}$$

where $\Theta_{ij}(t) = \text{prob (contact between } I_i \text{ and } S_j)$. Clearly in equation (1.20), contacts between S_j and the different I_i are regarded as independent events [assumption (d)], and we need to consider only the infectives in circulation at time t [assumption (a)]. The form of equation (1.20) is then given by assumption (c), with assumption (b) guaranteeing that the infection cannot arise from other sources. Finally, assumption (e) will allow simplification of equation (1.20) in particular cases.

It follows from equation (1.20) that the probability generating function (pgf) for the number of new cases, $P^{\text{new}}(t+1)$, during period $t+1$ is given by

$$g(z) = \prod_{j=1}^{P^S(t)} [1 - \Theta_j(t) + z\Theta_j(t)]. \tag{1.21}$$

Equation (1.21) represents a more general version of the Reed–Frost chain binomial model. The assumptions underlying the model will vary in plausibility from one application to the next, although assumption (a) is likely to be the most difficult to meet. Ways of relaxing assumption (a) are considered in Cliff and Ord (1978).

Fitting model (1.21) to the Reykjavik data [see Ball (1977) for calibration details] yields the one-step-ahead forecasts shown in figures 1.15(c) and 1.15(d). Although the forecasts have the 'one-lag-behind' effect' in the initial stages, by the epidemic peaks, they are in phase with the observed series and decay in much better harmony with the observed data than do the time-series models of figures 1.15(a) and 1.15(b). Further discussion of the chain binomial model appears in Cliff and Ord (1978).

The stochastic Hamer–Soper and chain binomial models seem to be better than the time-series models in avoiding the 'one-step-behind' effect. However, the parameters of the process-based models we have considered are fixed through time and so it is natural to ask whether we can combine the Kalman filter method with them. This has been attempted for the chain binomial model by Ball (1977) and Ball and Cliff (1979), where mathematical details are given. The one-step-ahead forecasts are given in figure 1.15(d), and do not give any real improvement over the straight-forward chain binomial projections.

1.4.4 Model comparisons

The quite clear implication of the work described in this section is that process-based models are likely to be a better bet for forecasting than blind ARIMA models. This is principally because information on the size of the susceptible population is incorporated in the former, and it guarantees that as the susceptible population is exhausted the forecast epidemic will ultimately fade out in the manner observed in reality. There is no such guarantee in the case of the ARIMA models. Indeed, if we continued the forecasts of the kind of model shown in figure 1.14 further into the 'future', we should find that the trend of upwards growth would be perpetuated as forecasts themselves were used as the basis for further forecasts. The time-series models considered here include no damping out feature, with the result that they are basically 'trend predictors'. The following remark probably applies equally well to diffusion modelling as to econometric modelling:

"Overall the econometric approach, which makes the interrelationships between variables quite explicit, provides a better-fitting representation than the univariate time series approach, in which certain interrelationships are implicit, on our interpretation, but the general interrelatedness is ignored in estimation. The extent to which taking this into account would improve the relative performance of the time series approach remains to be investigated. In the meantime we conclude that to take account of the specific interdependence of economic variables, whether or not these variables can be individually represented as ARIMA processes, provides a more fruitful approach to model-building" (Prothero and Wallis, 1976, page 485).

1.5 Conclusion

In this chapter we have tried to identify the space–time regularities in measles epidemics from data for the Southwest of England and Iceland. These data have then been analysed with the use of Box–Jenkins and explicitly process-based models. There is strong evidence for a spatially contagious diffusion process for measles at the geographical scale studied. Although we suspect there must also be a hierarchical spread through the urban size system, we have been unable to find unequivocal evidence for this in the data sets analysed.

From a forecasting point of view, it is fairly clear that although the Box–Jenkins methods may be perfectly good with continuous data series, the long strings of zeros in the interepidemic phases of measles outbreaks means that they are unlikely to be helpful models for epidemic spread. The dependence upon the recent history of the process built into the models means the epidemics will already have arrived before the change from zero cases is detected and allowed for; hence we notice the 'one-time-period-in-arrears' effect. We have argued that knowledge of when an epidemic is about to fire is more important from the viewpoint of vaccination control programmes than is a knowledge of the detailed history of an epidemic when it is in progress. This implies that the process-based Hamer–Soper and chain binomial models, which avoid the lag effect, are likely to be a more useful class of models upon which to concentrate in future research.

If we study figures 1.15(a), 1.15(c), 1.15(d), and figure 1.17 again, it is interesting to note the way in which forecasts improve as we alter the models as shown below.

Model	Figure	Comment
Pure temporal autoregressive	1.15(a)	one-step-in-arrears effect
Chain binomial	1.15(c), 1.15(d)	forecast based purely on time series; but process base means that although forecasts start in arrears, model 'locks on' because of self-regulating relationship between P^S and P^I population sizes
Stochastic Hamer–Soper	1.17	follows epidemic cycles satisfactorily because of P^S and P^I population size relationship. In addition, because spatial information is added through λ_{ij} parameter, the epidemic in Kingswood and Keynsham is fired at the correct moment (no lag effect) because of signal from Bristol lead region where disease is endemic

Although the number of measles cases has dropped dramatically since the mid-1960s, when major vaccination programmes were introduced, the study conducted is of more than academic interest. Vaccination has had the effect of shifting the age distribution of cases upwards and there is weaker evidence for an absolute increase in the age-specific incidence in junior and senior high-school children in the United States. Although the rate of serious complications amongst measles cases is low (for example, about 1 per 1000 for measles encephalitis) this too shows a small but positive increase with age. The need to identify the spatial risk zones to contain measles outbreaks remains, and perhaps becomes more important with a partly vaccinated population since more higher-age susceptibles (with greater risk of complications) will be present in the population. It is hoped that the spatial forecasting models discussed in this chapter may be able to make a small contribution towards this goal.

Acknowledgements. The authors wish to acknowledge that the research reported here is part of a programme supported by funds from the Social Science Research Council. Field work in Iceland is being supported by the Nuffield Foundation (Social Sciences Small Grants Scheme). We are particularly grateful to Mr Roy Versey of the Department of Geography in the University of Cambridge for cartographic work and supporting research.

References
Bailey N T J, 1957 *The Mathematical Theory of Epidemics* (Charles Griffin, London)
Bailey N T J, 1975 *The Mathematical Theory of Infectious Diseases* (Charles Griffin, London)
Ball F, 1977 *Measles Outbreaks in Iceland* unpublished project, Diploma in Statistics, University of Cambridge, Cambridge, England
Ball F, Cliff A D, 1979 "The use of Kalman filter methods in diffusion modelling" (forthcoming)
Bartlett M S, 1956 "Deterministic and stochastic models for recurrent epidemics" *Proceedings of the Third Berkeley Symposium on Mathematical Statistics and Probability* **4** 81-100
Bartlett M S, 1957 "Measles periodicity and community size" *Journal of the Royal Statistical Society A* **120** 48-70
Bartlett M S, 1960 *Stochastic Population Models in Ecology and Epidemiology* (Methuen, London)
Benjamin B, 1968 *Health and Vital Statistics* (Allen and Unwin, London)
Bennett R J, 1975a "The representation and identification of spacio-temporal systems: an example of population diffusion in North-West England" *Institute of British Geographers, Publications* **66** 73-94
Bennett R J, 1975b "Dynamic systems modelling of the North-West region: 1. Spatio-temporal representation and identification" *Environment and Planning A* **7** 525-538
Bennett R J, 1975c "Dynamic systems modelling of the North-West region: 2. Estimation of the spatio-temporal policy model" *Environment and Planning A* **7** 539-566
Bennett R J, 1975d "Dynamic systems modelling of the North-West region: 3. Adaptive parameter policy model" *Environment and Planning A* **7** 617-636
Bennett R J, 1975e "Dynamic systems modelling of the North-West region: 4. Adaptive spatio-temporal forecasts" *Environment and Planning A* **7** 887-898
Black F L, 1966 "Measles endemicity in insular populations: critical community size and its evolutionary implications" *Journal of Theoretical Biology* **11** 207-211
Box G E P, Jenkins G M, 1970 *Time Series Analysis, Forecasting and Control* (Holden-Day, San Francisco)
Christ C F, 1975 "Judging the performance of econometric models of the U.S. economy" *International Economic Review* **16** 54-74
Cliff A D, Haggett P, Ord J K, Bassett K, Davies R B, 1975 *Elements of Spatial Structure: A Quantitative Approach* (Cambridge University Press, London)

Cliff A D, Ord J K, 1971 "A regression approach to univariate spatial forecasting" in *Regional Forecasting* Eds M Chisholm, P Haggett, A E Frey (Butterworth, London) pp 47-70

Cliff A D, Ord J K, 1973 *Spatial Autocorrelation* (Pion, London)

Cliff A D, Ord J K, 1978 "Forecasting the progress of an epidemic" in *Dynamic Modelling for Space-Time Systems* Eds R L Martin, N J Thrift, R J Bennett (Pion, London) pp 191-204

Granger C W J, Newbold P, 1975 "Economic forecasting: the atheist's viewpoint" in *Modelling the Economy* Ed. G A Renton (Heinemann, London) pp 131-147

Haggett P, 1972 "Contagious processes in a planar graph: an epidemiological application" in *Medical Geography* Ed. N D McGlashan (Methuen, London) pp 307-324

Haggett P, 1975 "Simple epidemics in human populations: some geographical aspects of the Hamer-Soper diffusion models" in *Processes in Physical and Human Geography: Bristol Essays* Eds R F Peel, M Chisholm, P Haggett (Heinemann, London) pp 373-391

Haggett P, 1976 "Hybridizing alternative models of an epidemic diffusion process" *Economic Geography* **52** 136-146

Haggett P, 1978 "Regional and local components in elementary space-time models of contagious processes" in *Timing Space and Spacing Time in Socio-Economic Systems* Eds T Carlstein, D N Parkes, N J Thrift (Edward Arnold, London) pp 19-34

Hamer W H, 1906 "Epidemic diseases in England" *Lancet* **1** 733-739

Harrison P J, Stevens C F, 1976 "Bayesian forecasting" *Journal of the Royal Statistical Society B* **38** 205-247

Kendall D G, 1957 "La propogation d'une épidémie au d'un bruit dans une population limitée" *Publications of the Institute of Statistics, University of Paris* **6** 307-311

Keyfitz N, Flieger W, 1968 *World Population: An Analysis of Vital Data* (Chicago University Press, Chicago)

Martin R L, 1978 "Kalman filter modelling of time-varying processes in urban and regional analysis" in *Dynamic Modelling for Space-Time Systems* Eds R L Martin, N J Thrift, R J Bennett (Pion, London) pp 104-126

Martin R L, Oeppen J E, 1975 "The identification of regional forecasting models using space-time correlation functions" *Institute of British Geographers, Publications* **66** 95-118

Mollison D, 1972 "Possible velocities for a simple epidemic" *Advances in Applied Probability* **4** 233-257

Murray G D, Cliff A D, 1977 "A stochastic model for measles epidemics in a multi-region setting" *Institute of British Geographers, Publications, New Series* **2** 158-174

Preusser H, 1976 *The Landscape of Iceland: Types and Regions* (W Junk, The Hague)

Prothero D L, Wallis K F, 1976 "Modelling macroeconomic time series" *Journal of the Royal Statistical Society A* **139** 468-500

Scott P, 1971 *The Adaptation of Knox's Technique for Use on Aggregated Data and Its Application to Assess the Utility of Graphs in Epidemiology* unpublished term paper, Department of Geography, University of Bristol, England

Soper H E, 1929 "Interpretation of periodicity in disease prevalence" *Journal of the Royal Statistical Society A* **92** 34-73

Stocks P, 1949 "Sickness and the population of England and Wales, 1944-47" *Studies in Popular Medical Subjects* **2**

Stocks P, Karn M N, 1928 "A study of the epidemiology of measles" *Annals of Eugenics* **3** 361-398

Thorarinsson S, 1961 "Population changes in Iceland" *Geographical Review* **51** 519-533

Regional dynamics in British unemployment, and the impact of structural change

L W Hepple

2.1 Introduction

This chapter presents an exploratory analysis of time variations in regional responses to national unemployment fluctuations in Great Britain. A number of studies have estimated regional unemployment responses, but there is considerable evidence for structural changes in the British economy and these changes may have led to time variation in the regional responses. After outlining the nature of regional response models and reviewing the evidence for structural change, recursive least-squares techniques are used as the basis for general tests of parameter variation. Time-varying parameter models are then outlined and applied to the regional responses. The final section outlines some possible directions for further work.

2.2 Regional unemployment responses

Regional response models examine the ways in which different regions react to national economic fluctuations. The recent literature on regional unemployment responses springs largely from two papers by Thirlwall (1966) and Brechling (1967). The basic feature of their models was to regress the unemployment level of each region against the national unemployment rate

$$E_t^u = \beta_1 + \beta_2 E_t^{u\text{\tiny\it{n}}} + \epsilon_t , \qquad t = 1, 2, ..., T , \qquad (2.1)$$

where
E_t^u is unemployment in a specific region at time t,
$E_t^{u\text{\tiny\it{n}}}$ is unemployment in Great Britain as a whole at time t, and
ϵ_t is a stochastic disturbance term.
β_2 is a measure of regional cyclical sensitivity. If $\beta_2 > 1$ then the region is more sensitive to cyclical fluctuations in unemployment than Great Britain as a whole, whereas if $\beta_2 < 1$ the region is less sensitive than the nation. Brechling (1967) further argued that β_1 could be interpreted as a measure of relative 'structural unemployment', though this interpretation has limitations. Brechling's formulation also allowed for possible lead–lag effects, seasonal effects (through dummy variables), and for trends in structural unemployment by including linear and quadratic time-trend variables.

   The Thirlwall and Brechling studies, estimated for the standard regions of Great Britain by making use of data for the 1950s and the first half of the 1960s, revealed the cyclical sensitivity of the high unemployment regions such as Scotland, Wales, North and Northwest England, and the low regional responses of the South and the Midlands. The studies confirmed that the traditionally depressed regions not only suffered from long-term high unemployment but also bore the brunt of economic recessions during the postwar sequence of 'stop-go' in the British economy.

The approach initiated by Thirlwall and Brechling has since been applied and extended in a large number of studies of the magnitude and timing of regional responses, both in Britain and elsewhere. Examples are the studies of subregional responses in Northwest England (Campbell, 1975) and Northeast England (Hepple, 1975), and abroad the studies of King and his associates in the USA (for example, King et al, 1969) and of Van Duijn (1975) in the Netherlands.

2.3 Structural change in the British labour market

There is considerable evidence of structural change in the British labour market after the severe deflation by the Labour government in July 1966. In only six months, unemployment rose from 260000 to 600000, and since then levels of unemployment have been higher than would be expected in terms of aggregate output or measures of the demand for labour (the registered vacancy figures). A number of different explanations of this shift have been put forward and each can produce some supporting evidence (Bowers et al, 1972). On the demand side a 'shakeout' of previously hoarded labour has been suggested (Knight and Wilson, 1974), and on the supply side explanations include an increased willingness to accept unemployment, and to search longer before accepting new employment, owing to the Redundancy Payments Act of 1965 and the introduction of earnings-related benefits in September 1966, and an increase in the supply of young labour as the effects of the postwar birth boom became felt in the labour market (Foster, 1974). None of the suggested causes has won acceptance as the sole cause, and Bowers et al (1970, page 60) comment: "given the plethora of explanations, the similarity of their expected consequences, and the virtual simultaneity of their appearance, an eclectic position is the only one at present tenable".

Whatever the balance of causal influences, this shift will be mirrored (to varying degrees) in shifts in regional and industrial unemployment responses. Bowers et al (1970) have examined this by estimating regional unemployment response models using data for 1960 to 1969 and introducing dummy shift and slope effects in the last quarter of 1966. They found upward changes in regional unemployment rates against the national rate in Wales and the Southwest, lesser changes in the Southeast and the Midlands, and downward changes in Scotland and the Northwest and to a lesser extent in the North.

The economic discussion of a discrete shift in 1966–1967 has assumed stable labour market and regional relationships before and after the shift. However, a number of writers have argued that since 1967 (and particularly during the 1970s) there have been fundamental structural changes in the

British economy (Bacon and Eltis, 1976), and that these have further altered regional labour market relationships. Chisholm in particular has argued that "in the field of regional policies a major *secular* change has occurred/is occurring in the relevant economic conditions" (Chisholm, 1976, page 201). The structural changes are numerous: the impact of changing technology and productivity, exchange devaluation and changing export competitiveness, the rise in oil prices (and consequent rise in demand for coal) the shift of employment out of manufacturing, and the very success of earlier regional policy in altering regional employment structures.

The regional unemployment aspect of this argument is that as national unemployment has risen past 600000, then 1 million, and now 1·5 million, the structural changes have led to a tendency for regional unemployment rates to converge towards the national average (Chisholm, 1976; Keeble, 1977). The traditionally depressed regions of Scotland, Wales, and North and Northwest England are bearing less of the brunt of the economic recession, and more of the unemployment is in newly emerging pockets (associated with structural problems of particular industries) in the Midlands and the South. The consequence for spatial policies is that they should be less concerned with broad interregional disparities, and more concerned with specific problems of industrial growth (Chisholm, 1976) and the urban system (Keeble, 1977).

In the face of these arguments (and the further question as to whether regional relationships were in fact stable *before* 1966) the present chapter sets out to examine structural change in regional unemployment responses in Great Britain, not in terms of a discrete shift at a known or unknown time point, but through general tests for departure from constant relationships, and it explores the nature of the regional parameter changes and evolution.

2.4 Model specification and data
The analysis in this chapter is based on data for the period January 1950 to September 1977. The model employs quarterly data. Monthly data would allow a finer analysis of changes in the time-series relationships, but for an exploratory study this is offset by the need for a more complex model specification. In a monthly model, seasonal variation is more difficult to estimate and explicit recognition of regional–national leads and lags is needed since these are mainly of the order of zero to three months.

The specification of the regional response model is in terms of a linear arithmetic response, with dummy variables to account for seasonal effects. For constant coefficients the model is

$$E_t^u = \beta_1 + \beta_2 E_{t2}^{un} + \beta_3 E_{t3}^{un} + \beta_4 E_{t4}^{un} + \beta_5 E_{t5}^{un} + \epsilon_t , \tag{2.2}$$

where
E_t^u is the level of unemployment in a particular region at time t, and
E_{t2}^{un} is the level of unemployment in Great Britain at time t.
E_{t3}^{un}, E_{t4}^{un}, and E_{t5}^{un} are the seasonal dummy variables, which take a value of

one in the relevant quarters and zero elsewhere. β_1 is therefore a coefficient of 'structural unemployment' centred on the winter quarter of the year, and β_2 is a coefficient of regional cyclical response.

To construct consistent regional time series it was necessary to aggregate some of the present standard regions of Great Britain. The Southeast and East Anglia were grouped to form a larger Southeast region, and the East Midlands was grouped with Yorkshire and Humberside to form an East Midlands–Yorks region. The other regional divisions are Scotland, Wales, Northwest England, North England, Southwest England, and the West Midlands. For each region the monthly number of total unemployed (mainly derived from the *Department of Employment Gazette*, though some West and East Midlands data for 1963–1965 were derived from other Department of Employment sources) was converted to the unemployment rate, by making use of the June survey of total employed and unemployed as the denominator for each year. The monthly series were then aggregated into quarterly averages centred on February, May, August, and November. Throughout the chapter the Roman numerals I, II, III, and IV denote the quarters of the year. The final series for each region thus runs from 1950:I to 1977:III.

For recent years a small number of monthly unemployment figures are unobtainable because of industrial action in the Department of Employment. The months are December 1974, December 1975, and November and December 1976. In some cases official estimates were available, and these were then used; in other cases values were interpolated in line with the trend in adjacent months. In addition, in February 1972 a power strike led to exceptionally high unemployment figures in all regions for that one month: the figures for this month were adjusted in line with adjacent months to prevent this one observation having a drastic influence on parameter estimates.

2.5 Recursive least-squares techniques

The diagnostic tests for structural changes in time-series regression suggested by Brown et al (1975) are based on the properties of residuals generated by the techniques of sequential or recursive ordinary least-squares (ROLS) estimation outlined by Plackett (1950).

Consider the general linear model based on T time periods

$$y = X\beta + e, \tag{2.3}$$

where

y is the $(T \times 1)$ vector of observations on the dependent variable;

X is the $(T \times K)$ matrix of observations on the K independent variables;

β is the $(K \times 1)$ vector of regression coefficients; and

e is the $(T \times 1)$ vector of disturbances, assumed to have mean zero and variance σ^2.

Treating each row (time period) of equation (2.3) separately gives the form

$$Y_t = x_t^T \beta_t + \epsilon_t , \qquad t = 1, 2, ..., T , \tag{2.4}$$

where

Y_t and ϵ_t are now scalars,

x_t^T is a row vector of K observations on the independent variables at time t, and

β_t is the $(K \times 1)$ vector of regression coefficients at time t.

If the regression relationship is constant over time, then

$$\beta_1 = \beta_2 = ... = \beta_T = \beta . \tag{2.5}$$

ROLS estimation consists of estimating the T-period relationship of equation (2.3) by sequentially updating the estimates a row at a time for $t = K+1, K+2, ..., T$. The initial estimate must be based on the first K observations to produce a full-rank solution. By denoting $\hat{\beta}_{t-1}$ as the vector of regression coefficient estimates at time $t-1$ and P_{t-1} as the matrix $(X^T X)^{-1}$ based on $t-1$ observations, the updating is generated by

$$\hat{\beta}_t = \hat{\beta}_{t-1} + K_t(Y_t - x_t^T \hat{\beta}_{t-1}) , \tag{2.6}$$

where

$$K_t = \frac{P_{t-1} x_t}{d} , \tag{2.7}$$

$$d = 1 + x_t^T P_{t-1} x_t , \tag{2.8}$$

$$P_t = P_{t-1} - K_t x_t^T P_{t-1} , \qquad t = K+1, K+2, ..., T . \tag{2.9}$$

The updating of $\hat{\beta}_{t-1}$ is based on a correction that uses the one-step prediction error between $\hat{Y}_t$ and actual Y_t. K_t, the gain vector, controls the weight of new information and tends to a null vector as t increases. The ROLS procedure also generates a set of recursive residuals

$$\tilde{\epsilon}_t = \frac{Y_t - x_t^T \hat{\beta}_{t-1}}{d^{1/2}} , \qquad t = K+1, ..., T . \tag{2.10}$$

The $(T-K)$ recursive residuals have the desirable property that under the assumption of constant regression coefficients, they are independent and uncorrelated, with mean zero and variance σ^2, the same as the true disturbances ϵ in equation (2.3). These uncorrelated residuals provide the basis for a number of test statistics. It should be noted that the ROLS estimation is best implemented, not by direct use of the algebra of equations (2.6) to (2.9), but by using a factorised form that avoids explicit use of P_t. Gentleman (1973) has given an algorithm for this, and since it is also one of the most efficient numerical methods for estimating the full model [equation (2.3)], the recursive residuals emerge as an almost cost-free by-product of an efficient regression algorithm. During the updating,

t statistics, sums of squared errors, and R^2 values can also be easily derived (Riddell, 1975).

Brown et al (1975) derive two general tests for nonconstant regression coefficients based on the recursive residuals. The 'cusum' test uses the cumulative sum series of recursive residuals

$$V_t = \sum_{m=K+1}^{t} \frac{\tilde{\epsilon}_m}{s} , \qquad t = K+1, ..., T , \tag{2.11}$$

where s is the estimated standard deviation of the disturbances. Under the null hypothesis of stable regression coefficients, it can be shown that V_t is approximately normally distributed with mean zero and variance $t - K$. If two-sided 95% significance levels are defined for $t = K+1, ..., T$, the null hypothesis is rejected if the V_t series crosses these significance lines. The second test employs the 'cusum of squares', the cumulative sum of squares series of the recursive residuals

$$S_t = \sum_{m=K+1}^{t} \frac{\tilde{\epsilon}_m^2}{s^2} , \qquad t = K+1, ..., T . \tag{2.12}$$

The distribution of $1 - S_t$ can be shown to follow a beta distribution under the null hypothesis (Garbade, 1977) and significance lines defined.

These two tests are not parametric tests of stability but general tests for departure from constancy, and it is useful to employ both tests since they may have power against different alternative time-varying structures. There is some evidence that the cusum of squares test is poorer than the cusum test at detecting incrementally changing coefficients as against discrete shifts (Garbade, 1977). ,

Recursive residuals have been used by Harvey and Phillips as the basis for a number of other tests in regression analysis: serial correlation (Phillips and Harvey, 1974), heteroscedasticity (Harvey and Phillips, 1974), and functional misspecification (Harvey and Collier, 1977). For the present analysis two of these provide useful adjuncts to the cusum and cusum of squares tests for parameter variation. The heteroscedasticity test, based on the ratio of the sum of squares of the last $\frac{1}{3}T$ residuals to that of the first $\frac{1}{3}T$, is a useful complement to the cusum of squares test. Harvey and Collier's ψ test for functional misspecification is an exact test closely related to the cusum test, and for data ordered with respect to time provides a test for structural change. The statistic

$$\psi = \left[(T-K-1)^{-1} \sum_{t=K+1}^{T} (\tilde{\epsilon}_t - \bar{\tilde{\epsilon}}_t)^2 \right]^{-\frac{1}{2}} (T-K)^{-\frac{1}{2}} \sum_{t=K+1}^{T} \tilde{\epsilon}_t , \tag{2.13}$$

where $\bar{\tilde{\epsilon}}$ is the arithmetic mean of the recursive residuals, follows a t distribution under the null hypothesis.

2.6 Regional applications

The regional response model given in equation (2.2) was estimated by ROLS techniques for each of the eight regional divisions of Great Britain, and the recursive residual tests were applied. The tests found evidence of structural change in *all* eight regions. The cusum test found significant parameter change at the 5% level in every region, though the degree of departure varied markedly. In Wales the cusum series becomes significant in 1952:I and remains significant until the end of the sample, whereas in the Southeast only one cusum value, that for 1966:II, is beyond the 5% line. The cusum of squares test detects significant changes in all regions except the West Midlands and the Northwest. The heteroscedasticity test is significant in all except the West Midlands, Northwest, and Wales, whereas Harvey and Collier's ψ test finds significant misspecification in the West Midlands, Northwest, Wales, and the Southwest.

The pattern of test results can give some indication of the type of parameter variation. A tendency to incrementally changing coefficients is suggested in Northwest England and the West Midlands by the insignificance of the cusum of squares series combined with a highly significant ψ test and cusum series. The very weak cusum significance, combined with a highly significant cusum of squares series and heteroscedasticity test, suggests the Southeast may be a case of shifting residual variance rather than parameter shifts. Graphical presentation of the cusum and cusum of squares series can also provide useful indications of when changes occurred. For example, figure 2.1 shows the cusum plot for North England and reveals the change in relationship after 1958.

Insight into the dynamics of the regional parameter shifts can best be obtained by plotting the $\hat{\beta}_t$ estimates against time for each parameter in each region. Here we are mainly interested in shifts in the cyclical sensitivity of regions to national fluctuations. However, it could be

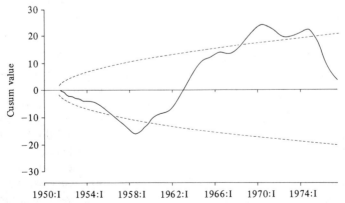

Figure 2.1. Cusum plot for ROLS model of North England. The dashed lines are the 5% confidence lines for the null hypothesis.

misleading to focus solely on the $\hat{\beta}_{t2}$ coefficients. For example, if there is an exogenous increase in structural unemployment in a region, entirely unrelated to national fluctuations, the increase in the $\hat{\beta}_{t1}$ coefficient will inevitably also mean a change in the $\hat{\beta}_{t2}$ coefficient for some time. It is therefore useful to combine changes in $\hat{\beta}_{t1}$ and $\hat{\beta}_{t2}$ by calculating the *expected* level of regional unemployment for a specified level of national unemployment, that is, to calculate $\hat{E}_t^u = \hat{\beta}_{t1} + \hat{\beta}_{t2} E^{uN}$, where E^{uN} is national unemployment fixed at some level. Here a 3% national level is used. By ignoring the $\hat{\beta}_t$ coefficients for the seasonal dummies, the $\hat{E}_t^u$ series is standardised on the winter quarter, so eliminating seasonal variation. The $\hat{E}_t^u$ series reveals whether regional unemployment is converging to or diverging from the national average over time.

Interpretation of the plots of $\hat{\beta}_t$ and $\hat{E}_t^u$ faces some difficulties. The $\hat{\beta}_t$, $t = K+1, ..., T$, are not drawn from a standard distribution. The early $\hat{\beta}_t$ are based on a small number of observations, and the $\hat{\beta}_t$ will be erratic even under the assumption of constant β_t until sample size increases and the relationship stabilises [Young (1969) provides excellent demonstrations of this]. Later in the $\hat{\beta}_t$ series there is the opposite problem: the K_t or gain matrix becomes more and more a null matrix, so that new information has less and less effect on the $\hat{\beta}_t$. In other words, new information has a greater impact early in the time series and is damped later on, so that the $\hat{\beta}_t$ series may lead us to overinterpret early shifts in β_t and not detect genuine later shifts.

Given these qualifications, the $\hat{\beta}_t$ nevertheless provide very valuable insights into the evolution of regional responses. Only a small selection of the graphical plots can be given here, but the results of the others can be summarised. It is useful to divide the analysis between the regions of higher unemployment (Scotland, Wales, North and Northwest England), where regional unemployment has generally been above the national average and cyclical sensitivity greater than unity, and the regions of lower unemployment (Southeast, Southwest, West Midlands, East Midlands–Yorks).

2.6.1 Regions of higher unemployment

The most dramatic pattern is that for Northwest England. The cyclical response coefficient $\hat{\beta}_{t2}$ shows a progressive and stepped fall from a peak of 3·26 in 1953–1954 to a low of 1·16 in 1971, with a rise since to 1·29 (figure 2.2). The major steps in the fall occur between 1954:III and 1955:III and with the onset of the national recessions in 1958, 1962–1963, and 1966:IV. The $\hat{\beta}_{t1}$ coefficient rises as a mirror image, but the $\hat{E}_t^u$ series for a constant national rate of 3% parallels the cyclical coefficient, and shows a fall from 6·1% in 1953–1954 to 3·4% in the late 1960s. The very high values during the early 1950s are due to the sudden collapse of the Lancashire textile industry in the 1951–1952 slump (Miles, 1968).

The ROLS updating for Scotland and North England reveals rather different parameter evolution. In Scotland the cyclical response is close to unity for much of the 1950s, though structural unemployment is high, but with the 1958 recession $\hat{\beta}_{t2}$ rose rapidly to 1·5 and $\hat{\beta}_{t1}$ fell from 1·7 to 0·8, with expected regional unemployment $\hat{E}_t^u$ rising from 4·6% to 5·3%. The 1962–1963 recession produced a further slight rise. The 1966 recession, however, resulted in a fall of $\hat{\beta}_{t2}$ to 1·27, with $\hat{E}_t^u$ dropping from 5·4% to 5·0%. The 1971–1972 recession produced some upturn in $\hat{\beta}_{t2}$, but this has since fallen back to 1·1, with the expected level of regional unemployment now at 4·8%.

The pattern for North England is similar to Scotland, except for the timing. The major rise in cyclical sensitivity, from 1·0 to 1·7, occurs in 1962–1963 rather than in 1958, and continues to rise slightly during the 1966–1967 recession, peaking at 1·9 in 1970 and falling since to 1·36 in 1977:III. Expected levels parallel this: 3·9% in 1959–1962, 5·2% in 1970, and 4·7% in 1977. Both Scotland and North England therefore show a clear tendency for cyclical sensitivity to increase during the early 1960s, with a fall back towards the levels of the 1950s during the present decade—a tendency more marked in Scotland than North England. However, they show very different reactions to the 1966 deflation. The ROLS estimates for Wales show a less oscillating pattern: after a rise in cyclical sensitivity (and expected levels) in 1954–1956 to a $\hat{\beta}_{t2}$ value of 1·42 in 1961, there has been a slow decline through to 1977, with $\hat{\beta}_{t2}$ falling to 1·1 and $\hat{E}_t^u$ falling marginally from 4·9% to 4·5%.

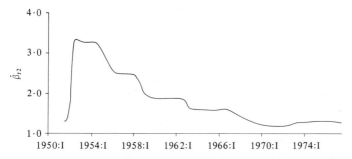

Figure 2.2. Recursive least-squares estimates of cyclical response $\hat{\beta}_{t2}$ for Northwest England.

2.6.2 Regions of lower unemployment

Amongst the regions of lower unemployment, the most dramatic pattern of parameter variation is shown by the West Midlands. There is a reversal of the pattern for Northwest England. The cyclical response of the West Midlands has risen in a series of steps from 0·47 in 1957:II to a peak of 1·35 in 1972:II, with subsequent drop to 1·2. The upturn steps occur during the recessions of 1958, 1962–1963, 1966–1967, and 1971–1972 (figure 2.3).

This is paralleled by the trend in expected unemployment: 1·34% in 1956, 2·7% in 1972, and 2·6% in 1977. The ROLS updating thus indicates that the West Midlands is still a region of lower unemployment, but with a strong rise in cyclical sensitivity.

Southwest England shows a similar evolution: cyclical sensitivity rises from 0·48 in 1956–1957 to 0·97 in 1971 (with a slight fall between 1961 and 1966), and after a fall in 1972–1974, a rise to 1·11 in 1977. The expected level of unemployment follows this, growing from 2·3% in 1956–1957 to 3·0% in 1967, and 3·2% in 1977. The East Midlands–Yorks region has a less clear pattern, but there is a definite rise both in $\hat{\beta}_{t2}$ and $\hat{E}_t^u$ after 1966: $\hat{\beta}_{t2}$ rises from 1·0 before 1966 to 1·22 in 1969–1974 and $\hat{E}_t^u$ rises from 2·4% to 2·7%. As might be expected from the recursive residual tests, the Southeast region is the hardest to interpret: $\hat{\beta}_{t2}$ rises during the expansion period 1955–1958 from 0·5 to 0·62, but falls back to 0·53, before rising to 0·59 after 1966, again falling after 1971, but rising with the 1975 recession to 0·71 in 1977. $\hat{E}_t^u$ follows a similar pattern but the variation is only between 1·95% and 2·1%.

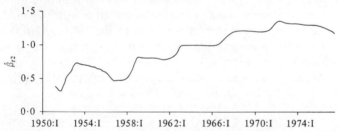

Figure 2.3. Recursive least-squares estimates of cyclical response $\hat{\beta}_{t2}$ for the West Midlands.

2.6.3 General comments

It is important not to oversimplify the specific details of regional parameter evolution when interpreting the ROLS results. However, there are a number of general trends in the regional responses that can be described with some confidence. The first is the tendency for both cyclical sensitivity and expected unemployment levels in the regions of higher unemployment to fall towards the national norm, and for those of the regions of lower unemployment to move upwards towards the norm, so that regional responses are becoming more similar. This is in line with Chisholm's arguments and with the results of Bowers et al (1970). However, these changes are not all focussed on a discrete shift in 1966–1967, though a number of relations do change at that time. Parameter shifts and trends are present both before and after 1966. The rising regional sensitivity of the West Midlands starts well before 1966, and the fall in cyclical sensitivity in Northwest England is largely over by 1966. Moreover, the tendency towards convergence is really a comparison of the 1970s with the period

1958–1966, because regional sensitivity in Scotland and the North rose in the late 1950s and early 1960s.

It has been suggested by Black and Slattery (1975) that apparent shifts in regional unemployment responses after 1966 may in fact be simply movements along *nonlinear* response functions, which appear as parameter shifts in linear response models. To examine this a number of experiments were run, by adding quadratic terms or using logarithmic forms in equation (2.2) and estimating them by ROLS. The results were generally unsatisfactory, showing greater parameter stability in some regions for particular time periods but greater instability in others, though this is a topic worth further investigation.

The ROLS estimates also show some marked differences in the regional responses to individual economic recessions, but before pursuing the interpretation of detailed parameter variations, the limitations of the ROLS estimates must be recalled. The 'damping' effect of the ROLS updating may lead to neglect or underestimation of some changes, and to overcome this it is valuable to examine the regional unemployment responses with the use of techniques and statistical models that explicitly allow for time variation in the parameters. These time-varying parameter methods are outlined and applied in the following sections.

2.7 Time-varying parameter models

Time-varying parameter models and estimation techniques have been extensively developed in the engineering literature on optimal control problems (Astrom, 1970; Jazwinski, 1970; Sage and Melsa, 1971). In particular they have been employed for tracking and orbital determination in the American lunar and interplanetary space programmes (for example, Bierman and Thornton, 1977). More recently these methods have been introduced into the economics literature (Athans, 1974; Cooley et al, 1977; Garbade, 1977), and small-sample properties of a number of methods have been investigated by Bennett (1976). Here I shall consider a small selection of methods that grow directly out of the basic ROLS procedure of section 2.5, and I shall not examine other highly promising methods such as those of Cooley and Prescott (1976).

Two different approaches to time-varying parameter estimation can be identified. In the first, an *explicit* model of time variation and its structure is estimated. In the second, no explicit model of the time variation is constructed, but the estimation is structured to allow the parameter vector to be updated by new information, whilst old information is either discarded or weighted to have less impact. In this second category, two methods are considered here: exponential weighting and moving-window regression.

2.7.1 Exponential weighting

Exponential weighting allows the parameter estimates to be kept open to new information (and so track parameter evolution) by 'aging' the earlier data so that recent information has most impact (Young, 1969). The weight given to past values, the 'memory' of the model, is a monotonic decreasing function of time, with the rate of decay determined by an exponential parameter α. This parameter is incorporated into the ROLS estimation, to give the updating equations

$$\hat{\beta}_t = \hat{\beta}_{t-1} + K_t(Y_t - x_t^T\hat{\beta}_{t-1}) \, , \tag{2.14}$$

where

$$K_t = \frac{\alpha}{(1-\alpha)^2} \frac{P_{t-1}x_t}{d} \, , \tag{2.15}$$

$$d = 1 + \frac{\alpha}{1-\alpha} x_t^T P_{t-1} x_t \, , \tag{2.16}$$

$$P_t = \frac{1}{1-\alpha} P_{t-1} - K_t x_t^T P_{t-1} \, , \qquad t = K+1, ..., T \, . \tag{2.17}$$

When $\alpha = 0$ this reduces to the constant parameter ROLS model; as α approaches $1 \cdot 0$ the weight given to new information at the expense of old becomes greater. If α is set too high, the parameter estimates will not only track genuine parameter variation but will also detect spurious variation from the residual noise of the system: each hiccup in the disturbances will be transmitted to the $\hat{\beta}_t$.

2.7.2 Moving-window regression
In moving-window regression, a rectangular weighting function or window is used instead of an exponential function. All observations within the window are *equally* weighted, but outside the window, observations carry zero weight. This amounts to fitting the regression to a short segment of M successive observations, and then moving this segment along the time series for $M, M+1, ..., T$, adding one new observation and discarding the oldest observation at each stage in the updating. M must be equal to or greater than K, and large enough to prevent the $\hat{\beta}_t$ being dominated by residual noise: if M is set too small, the $\hat{\beta}_t$ will track the disturbance fluctuations as well as genuine parameter variation. The moving-window techniques can be implemented by using the ROLS equations and then allowing for the dropping of an observation (Brown et al, 1975); however, a more efficient recursive computational technique can be derived and is given by Belsley (1973).

2.7.3 Kalman filter models
The exponential weighting and moving-window regression techniques are rather blunt instruments for estimating time-varying parameters. A more structured and elegant approach (with associated disadvantages as well as advantages) is provided by the family of models based on the Kalman filter algorithm (Sage and Melsa, 1971). These involve the explicit formulation of a model of the parameter variation. Here the Kalman

filter approach is outlined in terms of statistical estimation (Young, 1974), rather than in the state-space form usually employed in the control engineering literature.

When the β_t in a model of the form of equation (2.4) are assumed to be time-varying, a very wide range of types of parameter evolution can be represented by the general autoregressive moving-average structure

$$\beta_t = \Phi\beta_{t-1} + \Gamma u_{t-1}, \qquad u_t \sim N(0, Q), \tag{2.18}$$

where Φ and Γ are $(K \times K)$ matrices of autoregressive and moving-average parameters, and u_t is a random disturbance vector with mean zero and covariance matrix Q. When the elements of these matrices are known *a priori*, the Kalman filter allows the recursive estimation of the β_t in a similar fashion to the ROLS procedure. In most economic and social science applications the dynamics of parameter evolution are not known *a priori* in such a structured form, and it is useful to consider simplified formulations. The most basic model is to assume that Q is diagonal and that Φ and Γ are identity matrices. Parameter evolution is then governed by the equation

$$\beta_t = \beta_{t-1} + u_{t-1}. \tag{2.19}$$

This is the random-walk model. The parameters undergo random fluctuations u_{t-1} between time periods, with the diagonal elements of Q determining the variance of these fluctuations. This form of the Kalman filter leads to a recursive estimator only marginally different from the ROLS algorithm. The updating formulae are

$$\hat{\beta}_t = \hat{\beta}_{t-1} + K_t(Y_t - x_t^T\hat{\beta}_{t-1}), \tag{2.20}$$

where

$$K_t = \frac{P_{t|t-1}x_t}{d} \tag{2.21}$$

$$d = 1 + x_t^T P_{t|t-1} x_t, \tag{2.22}$$

$$P_{t|t-1} = P_{t-1} + Q, \tag{2.23}$$

$$P_t = P_{t|t-1} - K_t x_t^T P_{t|t-1}. \tag{2.24}$$

The only change from the ROLS estimator is the inclusion of equation (2.23) and its use in place of P_{t-1} in the other equations. As with the original ROLS procedure, efficient computer implementation of the updating avoids the formation and updating of the P_t matrix, but uses a covariance factorisation method rather than explicitly following equations (2.20) to (2.24) (Bierman and Thornton, 1977).

The diagonal elements of Q, which specify the level of noise driving the parameter variations, are the only information that has to be supplied to the recursive estimator, and Bennett (1976) has found this model one of the most robust and reliable in tracking parameter variations. Unlike the exponential-weighting and moving-window techniques, the random-walk specification can be used in models where some of the parameters are allowed to evolve and others are assumed to be constant, that is, by specifying some diagonal elements of Q to be zero. The need to specify the individual elements of Q is the drawback of the explicit model: if the elements are set too small, then $\hat{\beta}_t$ will smooth over genuine parameter variations, and if too large, $\hat{\beta}_t$ will track fluctuations in the ϵ_t disturbances. Methods such as the Sage–Husa filter exist that recursively estimate the Q elements, but Bennett (1976) found them slow to converge and they are not employed here. This question is discussed further in section 2.9.

2.8 Regional applications

The three time-varying parameter methods—exponential weighting, moving-window regression, and the random-walk Kalman filter—were used to estimate each of the regional responses. The estimators were deliberately structured to be relatively open to new information, so that they would detect any parameter changes, even at some risk of also tracking disturbance fluctuations. The exponential parameter α was set to $0 \cdot 3$. For the moving-window regressions a sample length of eighteen quarters was employed. This length was chosen, in the light of Briscoe and Roberts's study of structural change in UK employment functions (Briscoe and Roberts, 1977), because four-and-a-half years is the average length of the postwar business cycle in the UK (O'Dea, 1975). For the Kalman filter estimator, the diagonal elements of Q were set on the basis of the parameter variability and residual variance evidenced by the other methods.

An encouraging result was the broad similarity in regional parameter evolution for the three methods. In addition the $\hat{\beta}_t$ plots were not very sensitive to moderate changes in the controls or Q values. In particular, in the Kalman model, setting the Q values for the seasonal dummies to zero, that is, assuming constant seasonal effects, had little effect on the evolution of the other $\hat{\beta}_t$ elements. On comparing the three methods, for the control levels chosen, the exponential model gave the most rapid (or erratic) tracking and the Kalman filter (for comparable levels of residual variance) gave the smoothest parameter evolution.

The parameter evolution shown by these methods is, as expected, more erratic than that given by the ROLS estimates, and qualifies some of the ROLS conclusions. In particular the time-varying parameter models show the tendency for regional responses to vary quite markedly over the business cycle and between different cycles, so that where there would be a unidirectional trend in the ROLS estimates, the time-varying $\hat{\beta}_t$ estimates show oscillations around a rising or falling trend.

If we first look at the regions of higher unemployment, the Kalman filter results for Northwest England (figure 2.4) indicate that much of the fall in cyclical sensitivity (and expected level of unemployment) occurred

earlier than the ROLS estimates suggest, with the Kalman $\hat{\beta}_{t2}$ falling to
1·8 by 1956:IV and the expected level to 4·0%. The cyclical sensitivity
continues to fall in 1962–1963 and 1966–1967, and from 1967 to 1971
the Northwest is very close to the national norm both in cyclical sensitivity
and expected level. Since 1971 some rise has taken place, with $\hat{E}_t^u$ rising
to 3·8%.

For Scotland, the time-varying methods show the large variations in $\hat{\beta}_{t2}$
in response to different business cycles: a very marked rise in 1958, a
strong fall in 1966–1967, and an equally strong rise in 1971–1972. The
plot of $\hat{E}_t^u$ for the moving-window regression (figure 2.5) demonstrates,
however, that behind these oscillations the expected level is distinctly
lower (around 5·0%) during the 1970s than during the first half of the
1960s. North England also reveals oscillations, but again confirms the
overall fall from the expected levels of the early 1960s. The time-varying
results for Wales also show a more oscillating plot than the ROLS estimates,
but reflect the general downward trend both in cyclical sensitivity and
expected level, with $\hat{E}_t^u$ falling from around 5·3% in 1961 to 4·2% in 1977.

The time-varying parameter estimates for the regions of lower
unemployment again qualify, but do not overturn, the results of the ROLS
estimates. The Kalman filter results for the West Midlands (figure 2.6)
again show the stepped rise in cyclical sensitivity until the early 1970s,
though there is a tendency to fall back after the rises, and the exponential

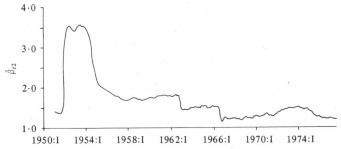

Figure 2.4. Kalman filter estimates of time-varying cyclical response $\hat{\beta}_{t2}$ for Northwest
England.

Figure 2.5. Moving-window regression estimates of $\hat{E}_t^u$, the expected level of regional
unemployment for a national level of 3%, for Scotland.

estimates give a very erratic $\hat{\beta}_{t2}$ pattern. The results do, though, confirm
the upward trend in expected unemployment towards the national level.

   The time-varying methods confirm the ROLS pattern for East Midlands–
Yorks and emphasise the stronger cyclical response after 1966:IV, with
the expected level of unemployment rising from $2 \cdot 4\%$ to $2 \cdot 8\%$. The
results for Southwest England, however, qualify the ROLS interpretation.
They show that after the swift rise in cyclical sensitivity and expected
level in 1956–1957, the ROLS estimates underestimate the *fall* both in $\hat{\beta}_{t2}$
and $\hat{E}_t^u$ from 1958 to 1966, though there is an upward trend since then.
The time-varying results for Southeast England are again difficult to give a
general interpretation to, though they follow the same ROLS pattern and
emphasise the rise in $\hat{\beta}_{t2}$ and $\hat{E}_t^u$ during 1966:IV to 1971.

   The general conclusions from application of the time-varying parameter
models do not conflict with the ROLS results of section 2.6. However,
the ROLS estimates do tend to exaggerate the smoothness of the trends at
the expense of the oscillating regional responses to individual business
cycles and the state of the cycle. The overall tendency towards regional
convergence is more marginal than the ROLS estimates suggest. The
differential regional responses to specific economic downturns are worth
commenting on. With the onset of each economic recession, in some
regions cyclical sensitivity rises, and (as a necessary corollary of the
construction of the national unemployment series) in other regions it falls.
However, the regional pattern of these oscillations is difficult to predict
from cycle to cycle, and it would be interesting economic history to trace
the impact of each recession in terms of industrial sectoral reactions and the
regional expression of these reactions. During the 1958 and 1962–1963
recessions it was in Scotland and North England (and to a lesser extent
the West Midlands) that sensitivity increased. The 1966 deflation led to
increased regional responses in the South and Midlands, but Scotland again
responded strongly to the 1971–1972 recession while areas like the
Southwest and Southeast responded less. The most recent rise in national
unemployment, since 1975, has seen the cyclical sensitivity of the Southwest
and Southeast again rising.

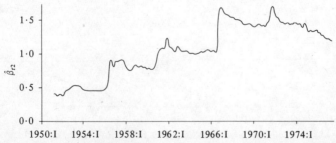

Figure 2.6. Kalman filter estimates of time-varying cyclical response $\hat{\beta}_{t2}$ for the West
Midlands.

2.9 Conclusions

This exploratory study of structural change in regional unemployment responses in Great Britain since 1950 has produced some clear evidence of time variation in the parameters. Overall, these confirm the tendency to convergence in regional unemployment levels and cyclical responses. However, this convergence is only a tendency, and it would be wrong to use the evidence to argue that regional disparities in unemployment and cyclical sensitivity are no longer important or relevant to regional policy. In detail the results show the complexity of the parameter changes, both in terms of timing and in terms of very different regional responses to specific business cycles.

The present analysis has been an exploratory one, and given the evidence of regional parameter variation that has been obtained, there are some useful directions further work might take. It is important to remember that the regional unemployment responses calibrated in this paper are descriptive rather than causal. They cannot in themselves reveal the mechanisms of unemployment change or the reasons for parameter variations. The national evidence discussed in section 2.3 suggests that the search for specific causes may be difficult, if not impossible, but it would be useful to disaggregate the present analysis—into subregional units, sex and age structure, and, if possible, industrial sectors. This is not just a general plea for more detailed work, but is because parameter evolution may well be due to aggregation and composition effects: the changing intraregional makeup of unemployment leading to a changing overall regional response to national fluctuations. At least some of these changing production and demand effects could be elucidated by more disaggregated recursive analysis.

There are also some useful technical extensions that may be worth pursuing. First, in selecting values for the window functions in the exponential and moving-window methods and diagonal Q values for the Kalman filter, one does run the risk of inadvertently tracking disturbance noise as well as real parameter variation, and it would be useful to be able to estimate the Q value *within* the model. The most satisfactory technique is that introduced by Garbade (1977), who suggests embedding the Kalman filter within a nonlinear optimisation to find the maximum-likelihood Q values. Garbade's algorithm works well but is computationally expensive (since the recursive Kalman filter must be evaluated for each trial set of Q values). Such a method would have been unjustified at the exploratory stage of analysis but, with the evidence of interesting parameter variation, it may prove useful in further studies. Second, the present analysis takes no account of serial correlation in the disturbances. There is evidence of serial correlation and this could lead to the parameters tracking oscillations in the disturbances, and it would be useful to incorporate serial correlation into the estimation.

The Kalman filter and other methods discussed in this chapter are one-sided models in that in forming estimates $\hat{\beta}_t$ they use information up to and including time t, but not any information from after time t. This is of course the only possibility for on-line analysis or for forecasting, but in studies such as the present one, a retrospective analysis over the last twenty-eight years, it may be useful to employ all the information before and after time t in forming $\hat{\beta}_t$ through the 'smoothing' algorithms closely related to (but more complex than) the Kalman filter. Useful procedures are given by Garbade (1977) and Cooley et al (1977).

References

Astrom K J, 1970 *Introduction to Stochastic Control Theory* (Academic Press, New York)

Athans M, 1974 "The importance of Kalman filtering methods for economic systems" *Annals of Economic and Social Measurement* 3 49-64

Bacon R W, Eltis W A, 1976 *Britain's Economic Problem: Too Few Producers* (Macmillan, London)

Belsley D A, 1973 "On the determination of systematic parameter variation in the linear regression model" *Annals of Economic and Social Measurement* 2 487-494

Bennett R J, 1976 "Non-stationary parameter estimation for small sample situations: a comparison of methods" *International Journal of System Science* 7 257-275

Bierman G J, Thornton C L, 1977 "Numerical comparison of Kalman filter algorithms: orbital determination case study" *Automatica* 13 23-35

Black W, Slattery D G, 1975 "Regional and national variations in employment and unemployment—Northern Ireland: a case study" *Scottish Journal of Political Economy* 22 195-205

Bowers J K, Cheshire P C, Webb A E, 1970 "The change in the relationship between unemployment and earnings increases" *National Institute Economic Review* 54 44-63

Bowers J K, Cheshire P C, Webb A E, Weeden R, 1972 "Some aspects of unemployment and the labour market 1966-1971" *National Institute Economic Review* 62 75-88

Brechling F P R, 1967 "Trends and cycles in British regional unemployment" *Oxford Economic Papers* 19 1-21

Briscoe G, Roberts C J, 1977 "Structural breaks in employment functions" *The Manchester School of Economic and Social Studies* 45 1-15

Brown R L, Durbin J, Evans J M, 1975 "Techniques for testing the constancy of regression relationships over time" with discussion *Journal of the Royal Statistical Society* 37B 149-192

Campbell M, 1975 "A spatial and typological disaggregation of unemployment as a guide to regional policy—a case study of North-West England 1959-1972" *Regional Studies* 9 157-168

Chisholm M, 1976 "Regional policies in an era of slow population growth and higher unemployment" *Regional Studies* 10 201-213

Cooley T F, Prescott E C, 1976 "Estimation in the presence of stochastic parameter variation" *Econometrica* 44 167-184

Cooley T F, Rosenberg R, Wall K D, 1977 "A note on optimal smoothing for time-varying coefficient problems" *Annals of Economic and Social Measurement* 6 453-456

Foster J I, 1974 "The relationship between unemployment and vacancies in Great Britain (1958-1972): some further evidence" in *Inflation and Labour Markets* Eds D Laidler, D L Purdy (University Press, Manchester) pp 164-196

Garbade K, 1977 "Two methods for examining the stability of regression coefficients" *Journal of the American Statistical Association* **72** 54-63

Gentleman W M, 1973 "Least squares computations by Givens transformations without square roots" *Journal of the Institute of Mathematics and its Applications* **12** 329-336

Harvey A C, Collier P, 1977 "Testing for functional misspecification in regression analysis" *Journal of Econometrics* **6** 103-119

Harvey A C, Phillips G D A, 1974 "A comparison of the power of some tests for heteroscedasticity in the general linear model" *Journal of Econometrics* **2** 307-316

Hepple L W, 1975 "Spectral techniques and the study of interregional economic cycles" in *Processes in Physical and Human Geography: Bristol Essays* Eds R Peel, M Chisholm, P Haggett (Heinemann, London) pp 392-408

Jazwinski A H, 1970 *Stochastic Processes and Filtering Theory* (Academic Press, New York)

Keeble D E, 1977 "Spatial policy in Britain: regional or urban?" *Area* **9** 3-8

King L J, Casetti E, Jeffrey D, 1969 "Economic impulses in a regional system of cities" *Regional Studies* **3** 213-218

Knight K G, Wilson R A, 1974 "Labour hoarding, employment and unemployment in British manufacturing industry" *Applied Economics* **6** 303-310

Miles C, 1968 *Lancashire Textiles: A Case Study of Industrial Change* (Cambridge University Press, London)

O'Dea D J, 1975 *Cyclical Indicators for the Post-War British Economy* (Cambridge University Press, London)

Phillips G D A, Harvey A C, 1974 "A simple test for serial correlation in regression analysis" *Journal of the American Statistical Association* **69** 935-939

Plackett R L, 1950 "Some theorems in least-squares" *Biometrika* **37** 149-157

Riddell W C, 1975 "Recursive estimation algorithms for economic research" *Annals of Economic and Social Measurement* **4** 397-406

Sage A, Melsa J, 1971 *Estimation Theory with Applications to Communication and Control* (McGraw-Hill, New York)

Thirlwall A P, 1966 "Regional unemployment as a cyclical phenomenon" *Scottish Journal of Political Economy* **13** 205-219

Van Duijn J J, 1975 "The cyclical sensitivity to unemployment of Dutch provinces, 1950-1972" *Regional Science and Urban Economics* **5** 107-132

Young P C, 1969 "Applying parameter estimation to dynamic systems" *Control Engineering* **16**(10) 119-125, **16**(11) 118-124

Young P C, 1974 "Recursive approaches to time series analysis" *Bulletin of the Institute of Mathematics and its Applications* **10** 209-224

3

Subregional Phillips curves, inflationary expectations, and the intermarket relative wage structure: substance and methodology

R L Martin

"The Phillips curve has been an empirical finding in search of a theory, like Pirandello characters in search of an author"

J Tobin (1972, page 9)

"... the authors of Phillips curves would do well to label them conspicuously 'Unstable. Apply with extreme care'".

A R Rees and M T Hamilton (1967, page 70)

3.1 Background: Methodenstreit over the Phillips curve

Few areas of economic enquiry have aroused more theoretical, statistical, and econometric debate than the so-called Phillips curve, which in its conventional form (Phillips, 1958; Lipsey, 1960) posits an inverse relationship or trade-off between money wage inflation and the excess demand for labour as proxied by the unemployment rate. Despite an enormous volume of theoretical and empirical analysis, the content and validity of the Phillips curve still remain contentious issues[1]. The central dispute has concerned the role of excess demand in the process of wage determination, although the focus of the controversy has shifted several times. At the level of the national aggregate economy, studies of different countries and time periods have not always confirmed this inverse relationship, but rather have often yielded ambiguous and conflicting results. By comparison very few studies have examined the wage inflation–unemployment relation in regional and urban labour markets, but such evidence as is available is similarly inconclusive. Thus while some analyses of urban markets in the US have found support for local Phillips curves (Albrecht, 1966; 1970; Jackson, 1972; Kaun and Spiro, 1970; Muth, 1968), others have failed to obtain significant estimates of the hypothesised relationship (King and Forster, 1973; Marcus and Reed, 1974; Smith and Patton, 1971; Smith and Smith, 1972; Weissbrod, 1974). Equally variable and indecisive results have typified studies of regional markets in the UK (Cowling and Metcalf, 1967; Hart and MacKay, 1977; MacKay and Hart, 1975; Metcalf, 1971; Thirlwall, 1969; 1970; Webb, 1974).

[1] Useful surveys are contained in Bergstrom et al (1978), Gordon (1977), Laidler and Parkin (1975), Rothschild (1971), and Trevithick and Mulvey (1975). A critical evaluation of Phillips curve theory with particular reference to local markets is given in Martin (1979).

All of these analyses show that the slopes of local Phillips curves are neither convex nor identical as assumed in aggregate models (for example, Lipsey, 1960; Archibald, 1969); quite often the curves are not significantly different from zero, in some cases they are positively sloped, and frequently differ according to the period over which they are estimated (for example, Albrecht, 1970). There would thus appear to be many wage–unemployment relations in the economy, a feature which may have implications for the observed instability of the aggregate Phillips curve, for:

"If the observed responses to unemployment in the different markets differ greatly, not much faith can be put in the aggregate relationship. It is simply something which has been observed, and there is little reason to believe that the same relationship will be observed again ... . If the statistical analysis reveals such differences, then the original doubt about the stability of an aggregate Phillips curve was not unwarranted" (Albrecht, 1966, page 290).

In fact it was one manifestation of this instability that ushered in the major change of focus in the Phillips curve debate. In the middle to late 1960s the Phillips relation appeared to 'break down' and to lose its explanatory power. Since then, wage inflation and unemployment have tended either to move independently or, more recently, to move upward together, that is, to be positively related. The most controversial explanation of these events has come from the theoretical discussion of the neoclassical monetarists, especially Friedman (1966; 1968; 1975) and Phelps (1965; 1967; 1968; 1970). This school maintains that there is no stable law of disequilibrium dynamics of the sort assumed in the naive Phillips curve. It is argued that it is the *expected real* wage that is bid up or down by competitive pressures, depending on the existence of excess demand or supply of labour. According to this theory workers and firms both have rational expectations, which implies that in the long run price (or wage) inflation is perfectly anticipated, and that the wage adjustment mechanism responds fully to such perfectly anticipated inflation (that is, there is no 'money illusion')[2]. The implications of this 'expectations-augmented' Phillips curve is that the only situation in which inflation is fully anticipated, and the real wage is constant, is when unemployment is

---

[2] Friedman's model refers to the expected rate of change in prices whereas Phelps's argument is in terms of the expected rate of change in wages. Although the two formulations differ, they nevertheless reach the same conclusions. Absence of 'money illusion' in the economic system should be interpreted as an assumption that markets display inflation neutrality. This derives from the proposition that individuals and firms behave rationally and account fully for anticipated inflation because their demand and supply schedules are 'free of money illusion', that is, they depend on relative (expected) prices and real (expected) income.

at its 'natural' rate, or that rate which

> "would be ground out by the Walrasian system of general equilibrium
> equations, provided there is embedded in them the actual structural
> characteristics of the labour and commodity markets, including market
> imperfections, stochastic variability in demands and supplies, the cost of
> gathering information about job vacancies and labour availabilities, the
> cost of mobility, and so on" (Friedman, 1968, page 8).

Under this model, increases in labour demand are considered as a shock
to the equilibrium situation. The tendency is for firms initially to increase
money wages and to adjust as the uncertainty concerning wage levels and
rates of change dissolves through observation of market reaction. At the
same time, once a steady rate of money wage increase is obtained, firms
quickly attain their desired vacancy levels. Employees, recognising that all
wages are increasing at a constant rate, in step with price increases, will
adjust their job-search behaviour, and unemployment will increase until
the equilibrium or 'natural' level is reached. As long as no other shocks
occur, the equilibrium rate of wage change will closely approximate
productivity growth plus the expected and observed growth in prices, and
will be unrelated to the unemployment rate. Further, with the assumption
of a direct relationship between wage and price inflation, all inflation in
this steady state would be expected and constant. A clear distinction is
thus made in this approach between the short run and the long run.
Should unemployment fall and remain below the 'natural' rate, persistent
underestimation of the equilibrium value of the real wage results, which in
turn causes constant revision of expectations on the part both of employees
and employers, and a steadily accelerating rate of inflation as the Phillips
curve shifts upwards (hence the term 'accelerationist theory'). Conversely,
should unemployment exceed and remain above the 'natural' rate, ever-
decreasing rates of change will ensue. Thus any trade-off between wage
changes and unemployment is a transient or short-run phenomenon and
highly unstable at that, for eventually as expectations become fully
adjusted to actual magnitudes, the Phillips curve disappears and the locus
of steady-state equilibrium points is vertical at the 'natural' rate of
unemployment[3]. In the long run, therefore, the level of unemployment
will fluctuate randomly about that rate which results from frictional
unemployment in equilibrium in a normally functioning labour market.

[3] The essential feature of this neoclassical theory is, then, the removal of the postulate
of complete information from an otherwise intact Walrasian system. In the strict
Walrasian economy, complete information enables the system to satisfy two conditions
for equilibrium: first, that there is no nonprice rationing—the market clearance
condition; second, that there exist no unanticipated ('surprise') possibilities in wage
and price information. In the weaker variant of Friedman, Phelps, et al, expectations
are correctly adaptive; thus equilibrium in this supposedly 'non-Walrasian' system
denotes a state in which wages and prices on average are found—over space and over
time—to be what they were expected to be.

Although this model is based upon rather restrictive assumptions, and fails to account for certain observed characteristics of labour market adjustment (see Corina, 1972; Martin, 1979), the proposition that in the long run the economic system is 'inflation neutral' raises some interesting questions concerning the temporal behaviour of the intermarket earnings structure within the spatial economy. For it is a well-known empirical fact within both Britain and the USA that despite persistent differences in unemployment, regional earnings differentials have remained virtually constant over the postwar period (Brechling, 1973; MacKay and Hart, 1975; Martin, 1979; Reynolds and Taft, 1956; Thirlwall, 1970). Stability of the regional relative earnings hierarchy over time implies, of course, that regional wages tend to increase at roughly the same rate. Numerous explanations of this apparent interdependence of regional wage changes have been suggested, which range from hypotheses that rely on the proposition that labour markets are noticeably less efficient in high unemployment areas because of structural and frictional imbalances, through neoclassical arguments that emphasise the role of expected wages and potential regional migration, to 'institutional' accounts in which 'earnings spread', 'spillover effects', and multilocation bargaining are central concepts. As yet, however, relatively little is known about the role of price expectations in the regional and subregional wage adjustment process. Now the claim of the accelerationist model that in a situation of no money illusion and perfect anticipations, changes in prices would be offset by corresponding changes in money wages may have immediate implications for the mechanisms that hold the interregional and interurban earnings structure together. For example, if we assume that each local market in the space economy is faced by similar price changes, and further that price changes are aggregated over a similar vector of commodity weights, then given the 'no money illusion–perfect expectations' condition, regional earnings changes may be kept in line by similar reactions to expected changes in the national cost of living index. If this view is taken, the existing structure of interregional differentials would then be interpreted as reflecting nonpecuniary elements, any regional differences in the cost of living, the cost of moving and other barriers to geographical mobility, industrial structure, and so on.

The current lack of detailed theoretical and empirical work on the impact of inflationary expectations on wages in local markets is doubly regrettable since analysis of local market reactions may help to account for the widely varying results obtained with the 'augmented' Phillips curve model for the aggregate economy. The accelerationist theory suggests that the coefficient of an included price expectations variable should be unity. It is probably fair to say that the majority of macrostudies have yielded a coefficient of less than unity and one which, moreover, varies markedly with the time period concerned. If this apparent failure of the accelerationist model is in fact accepted it suggests perhaps that there are submarkets in the economy that do not adjust completely, or that partial adjustment is typical of all submarkets, and hence that the expectations-augmented Phillips curve as such is underspecified as a model of wage inflation in a multimarket economy. However, a number of difficult econometric problems complicate the testing of the accelerationist theory. Empirical work in this area involves simultaneous testing of three hypotheses: that unemployment affects wage bargains in a specific way, that expectations are formed in a specific way, and that they interact with unemployment in the manner predicted. It is evident that in order to test the 'no money illusion' (price coefficient of unity) hypothesis, one has to assume as a maintained hypothesis that the anticipated rate of inflation will be equal to the actual rate on the average. Analytically one can restrict the discussion to the 'no money illusion' hypothesis because the formation of correct expectations is part of the very definition of the long-run steady-state equilibrium. Empirically, however, the test of the coefficient of the price variable is clearly dependent on the implicit assumption that inflationary expectations are correctly measured and that they approach the actual inflation rate in the long run.

The most common specification of the expectations formation mechanism is the adaptive expectations model, wherein the expected rate of inflation is a geometrically declining distributed lag function of all past actual rates of inflation. It can be shown that the adaptive expectations model leads to minimum mean-square error predictions which are 'optimal' for certain types of nonstationary stochastic processes (cf Muth, 1960; Theil and Wage, 1964). Muth also notes (page 299) that under these conditions "the best forecast for the time period immediately ahead is the best forecast for any future time period", an important result for the inclusion of price expectations in labour market analysis, since the problems of wage bargains being struck at discrete intervals are automatically dealt with in such a framework. However, the adaptive expectations model has also received considerable criticism. For example, Sargent (1971) has forcefully asserted that distributed lag schemes which constrain the weights to sum to one, as does the adaptive expectations model, impose a considerable downward bias on the price coefficient and hence do not provide a valid test of the accelerationist hypothesis. In particular, the validity of the assumption of nonstationarity of the inflation rate time series is questioned and it is shown that when this is approximated by a covariance-stationary process, the weights must sum to less than unity. Although this point is related to empirical tests of the theory, the implication is that only a sustained constant inflation would be sooner or later fully anticipated. In all other cases, there would be a tendency to underextrapolate, with the elasticity of expectations (with respect to past price changes) being less than unity. A sum of weights that is less than unity would be equivalent to a maintained hypothesis where the anticipated rate of inflation does not tend to be equal to actual inflation in the long run.

Such difficulties have led some writers to relate 'rational expectations' in the sense of Muth (1961)[4] and the accelerationist theory. But since a vertical long-run Phillips curve only requires that expected price inflation equals actual inflation in the long term, and that the price inflation coefficient is unity, this theory is consistent with *any* hypothesis of the formation of expectations (including the adaptive model) which leads to the equality between actual and expected inflation in full equilibrium. The assumption of 'rational expectations' can be a *sufficient* condition for the generation of a vertical Phillips curve under certain restrictions, but it is not a *necessary* condition for the natural rate theory to hold. The basic point is that specific hypotheses about expectations are related not so much to the accelerationist model itself, but to its econometric testing. Even if the expectations generating mechanism is fairly stable, we are unlikely to model it properly, so that the synthesised expectational variable will incorporate errors which may bias econometric results.   -
Moreover, if the expectations mechanism is in fact unstable, as some evidence seems to suggest, this will further undermine the adequacy of existing tests of the relevant hypotheses.

It is against this background of theoretical and methodological debate that the present chapter seeks to examine a simple expectations-type Phillips curve model for a system of local labour markets, with particular attention paid to the intermarket relative wage structure. The model is developed in the next section. The econometrics of price expectations is taken up in section 3.3, where questions of 'structural stability' are also considered. This discussion suggests the need to test for parametric instability in the estimated model: section 3.4 outlines one possible approach that uses tests based upon recursive residuals. The empirical part of this study, which relates to a set of urban markets in the USA, is contained in section 3.5.

3.2 A Phillips-type model for subregional markets

Although the evidence on the stability of the interindustry wages and earnings structure is extensive (for example, Cullen, 1956; Haddy and Tolles, 1957; Papola and Bharadwaj, 1970; Turner, 1957), considerable diversity of opinion exists regarding the role of wage relativities in the inflation process. One line of enquiry is concerned with the influence of macroeconomic variables, such as unemployment and inflation, on the relative wage structure (Wachter, 1970a; 1970b). Other studies (for example, Eckstein and Wilson, 1962; Throop, 1968) have been directed

---

[4] In this version, it is assumed that rational economic agents believe that their expectations are derived from the reduced form of the model of the economy that the theorist himself believes to be supported by the evidence. In other words, additional economic variables must be included: the exogenously determined variables of the relevant structural system. This form of the rational expectations hypothesis is probably far too strong.

to the effect of relative wages on the general wage and price level, whereas
still others have sought to examine relative wage movements within multi-
sector models of inflation (for instance, Akerlof, 1969; Okun, 1973;
Ross and Wachter, 1973). At a more basic level, the problem of explaining
the behaviour of relative wages is a "classic question in economics ... the
extent to which ... organised power has exerted its will over market forces"
(Kerr, 1957, page 173). The implied distinction between 'market' and
'nonmarket' influences on such a view has been the source of continuing
controversy, particularly as to their respective impact. As Reder remarks:

> "There are two general approaches to the theory of wage structure.
> One is the market theory or the competitive hypothesis, the other is
> what we might roughly term institutional .... For short-run purposes,
> an economic wage theory will probably never prove satisfactory. In a
> given year, wage rates are determined by the whole range of forces
> economists call institutional .... However, for longer periods ... or for
> making inter-area comparisons there may be more to the competitive
> hypothesis ... (Reder, 1958, pages 84–85).

Under the competitive or market approach, the relative wage structure
influences wage movements via the supply of labour, which for a given
micromarket is assumed to have the form

$$E_i = \phi_i(w_1, w_2, ..., w_i, ..., w_n, p, z) , \qquad i \in N , \qquad (3.1)$$

where
E_i is the supply of labour in the ith market,
w_k is the money wage in the kth market,
$N \equiv \{1, 2, ..., n\}$ is the set of markets,
p is the general (economywide) price level, and
z is a portmanteau vector reflecting differences in nonpecuniary
   advantages in different labour markets.
If we partition equation (3.1) and suppress the vector z into the functional
form, we have

$$E_i = \zeta \left(\frac{w_i}{p}, \frac{w_i}{w_1}, ..., 1, ..., \frac{w_i}{w_n} \right) , \qquad i \in N . \qquad (3.2)$$

Equation (3.2) is in fact a general statement of the neoclassical labour
supply function, recognising as it does the complex web of interdependence
which exists between labour groups in distinct markets. The supply of
labour in market i depends not only on the real wage that is offered in
that market, but also on the pattern of perceived (or expected) differentials
between the ith market and the $(n-1)$ remaining markets. Clearly the
supply equation (3.2) specifies that wage comparability as measured by
relative wages is an economic criterion. Further, since the supply function
(3.2) is couched entirely in terms of *real* wages and wage *differentials*, it
holds for wage changes as well as for wage levels. Thus the competitive

model predicts that wages will move in like fashion in different markets as a result of intermarket supply shifts manifested by migration in response to changes in the relative net advantages of various areas—"Potential mobility is the ultimate sanction for the interrelation of wage rates ..." (Hicks, 1932, page 79). This interpretation of the relative wage structure in the wage change process, then, fuses labour mobility and wage adjustment into a single problem: the behaviour of the intermarket wage structure is the major determinant of labour movement amongst occupations, firms, and regions, and in turn the wage structure is shaped by actual or potential mobility of labour[5]. Such a formulation forms the basis of Brechling's (1973) model of the 'dynamic market interdependence' of regional wage changes. [Although, by using expected rather than actual wage changes as between regions, his model is a variant of Phelps's and others (1970) analysis in which firms are assumed to adjust wages in relation to expected wage changes elsewhere in an attempt to achieve and maintain their desired labour recruitment levels.]

Now this 'market' interpretation seems to be conceptually inadequate and incongruous with the facts. After all, the substance of this approach is that there exist supply and demand schedules (and equilibria) which can range freely through time. Thus, given appropriate shifts in these functions, an analysis of changes in relative wages based upon a comparison of two positions of equilibrium may describe positive, negative, or even zero association between these variables. Accordingly, almost any development of the intermarket wage structure may be interpreted, after the event, as being consistent with the model. Secondly, the competitiveness of labour supply appears to be of minor empirical relevance: not only is the elasticity of interarea (urban and regional) labour mobility with respect to wage differentials insufficient to account for the behaviour of relative wages, but also, to the extent that labour is organised into unions and other bargaining groups, conditions of bilateral monopoly may be thought to prevail. Thirdly, the competitive approach implies that the crucial regions in the propagation of wage inflation in the space economy will be those that are the wage leaders, since it is only changes in such regions which cause a widening of wage differentials that will result in migration, which in turn will tend to restore relative wages in the areas of low wages (Brechling, 1973). But although there are some areas and regions which experience both low unemployment and high wages, this relationship is generally weak, and virtually absent in the case of the USA. Finally, the explanation of the behaviour of the wage structure in terms of a spectrum of long-run supply prices coexists rather uncomfortably with the observation that long-run conditions in the product market are directly associated with the wage level. Here the importance of the impact effects of group bargaining—given widely differing elasticities of product demand—cannot be ignored.

[5] Or, to quote Reynolds (1951, page 207): "One moves around in a closed circle of predictable relationships and results".

This leads us to the 'nonmarket' or institutional approach in which relative wage positions and behaviour are interpreted in terms of historically and economically arbitrary institutional and 'power' forces. This view has a long history in labour economics, and in fact formed part of Keynes's departure from neoclassical orthodoxy, as is obvious from his statement that labour does not exist separately from groups of individuals who are all trying to defend and improve their relative income position:

> "... the struggle about money wages precisely affects the distribution of the aggregate real wage between different labour groups, and not its average amount per unit of employment ... the effect of combination on the part of a group of workers is to protect their *relative* wage. The *general* level of real wages depends on the other forces of the economic system" (Keynes, 1936; page 14).

Keynes stresses that the operation of emulation and comparison reactions, particularly as these are expressed through collective bargaining and other centralised wage contracts, puts great store by the preservation of a stable pattern of wage differentials. And he argues that many such bargains and contracts are likely to take place on an industrywide or multiplant basis, which results in pronounced interconnection between different labour markets. The terminology used to describe this process has varied considerably—wage leadership, demonstration effects, spillover, key bargains, wage pattern adjustment, equitable or coercive comparisons, and wage diffusion. Common to all these terms is the notion that wage relationships between markets (occupations, firms, regions) can be derived from political influence theory, which maintains that traditional economic variables are, at the very least, supplemented by social and institutional variables if a complete explanation of wage movements is to be obtained. Thus Maher (1961) has argued that 'comparability' has been a fundamental criterion for wage changes within basic manufacturing industries in the USA, such comparisons being the complex result of: the input–output nexus of the industries; their similar technological and economic constitution; their geographical concentration and proximity; the concentration of workers into multilocational unions; and the similarity of the unions involved and their competition for representation of similar workers which causes a common type of response to bargaining stimuli. Maher's study bears some resemblance to the earlier analyses of Ross (1948) and Dunlop (1957) with their respective concepts of *orbits of coercive comparison* and *wage contours.*

On considering the UK labour market, Crossley (1966) believes that collective bargaining and convention operate to moderate short-run market behaviour and that "their tendency is to maintain wages from year to year in closer correspondence with the long-run equilibria than would result under the competitive hypothesis taken by itself" (Crossley, 1966, page 227).

There is no necessary impairment of market efficiency involved since, in striving towards a satisfactory relationship, unions and other labour groups may seek to reconcile their 'orbits of coercive comparison' with the employers' competitive constraints by searching for and following, tacitly or otherwise, a mutually tolerable wage pattern.

It is an established fact of the industrial relations experience of most Western economies that 'nonmarket' comparability effects constitute a large proportion of wage claims. If we adopt Hicks's (1974) terminology, we could say that local labour markets are characterised by 'fix-price' rather than 'flex-price' behaviour. That is, individual labour markets respond to pressures which are largely external to their own particular supply and demand conditions, this being most pronounced in periods of slack demand. In conditions of more buoyant demand, different groups of workers, and hence different local markets, may attempt, with varying success, to become autonomous units. As Lerner and Marquand (1963, page 11) put it:

"Increases in earnings in industrial groups which are growing fast lead to pressures for increases in groups which are growing more slowly; increases in one works of a company create pressures for increases on grounds of 'comparability' or to restore 'fair relativities' in the region where the works are, in works of the company elsewhere, and hence in other regions. A relatively low level of demand for labour is not sufficient to prevent increases in earnings on grounds of comparability with earnings in other regions with a high level of demand for labour. Consequently, in most regions there is little direct connection between the level of earnings and the demand for labour. Only where excess demand for labour is particularly strong—perhaps above a certain threshold level—does it exert a direct upward pressure on earnings. Where the excess demand is less great, the effect of institutional factors upon earnings is predominant".

The Friedman–Phelps Phillips model and hypotheses based on such nonmarket 'spillover effects' need not be thought of as representing entirely conflicting views; but it is clear that the latter cannot be tested adequately at the aggregate (macro) level because of the emphasis on the role of industry—and regional—specific factors such as the 'gap' between local earnings and aggregate (economywide) earnings. In a multimarket formulation, inclusion of a relative wage factor makes at least as much sense as exclusive reliance on the market variables of excess demand and price expectations. With these considerations in mind, we hypothesise that the wage adjustment function in a given subregional labour market can be written in terms of the extended Phillips curve:

$$\frac{\dot{w}_i}{w_i} = f_i \left[e_i^{xs}, \left(\frac{\dot{p}}{p}\right)^e, \left(\frac{w_i}{w}\right)_{t-k} \right], \qquad i \in N , \tag{3.3}$$

where

$\dot{w}_i/w_i$ is the proportionate rate of change of the average wage in the ith
         subregional market,

$(\dot{p}/p)^e$ is the expected rate of change in the aggregate (economywide)
         price level,

e_i^{XS} is the local relative excess demand for labour defined as

$$e_i^{XS} = g_i \frac{(e_i - E_i)}{E_i} \; ; \qquad g_i(0) = 0 \, , \qquad g_i' > 0 \, , \tag{3.4}$$

in which e_i and E_i are the quantities of labour demanded and supplied,
and $N \equiv \{1, 2, ..., n\}$ is the set of local markets. The variable $(w_i/w)_{t-k}$
measures the ratio of wages in the ith local market as a proportion of the
average level of wages in the system; it is intended to capture the relative
standing of the workers in a particular market *vis-à-vis* the 'average'
worker, and thus represents the intermarket 'comparison' feedback on
the wage determination process of an individual market. Since wage
decisions and negotiations occur infrequently, relative wage adjustments
involve a lot of catching up and leapfrogging; the length of the time lag k
is an institutional datum reflecting the discrete nature of this wage change
process.

In accordance with Phillips (1958) it is generally assumed that excess
demand in the labour market should as a first approximation be represented
by the unemployment rate e^u; more specifically it is assumed that there is a
stable inverse transformation such that $e_i^{XS} = \theta_i(e_i^u)$. With the assumption
that the function in equation (3.3) and this excess demand–unemployment
mapping are linear, we have

$$\frac{\dot{w}_i}{w_i} = a_{i0} - a_{i1}e_i^u + a_{i2}\left(\frac{\dot{p}}{p}\right)^e - a_{i3}\left(\frac{w_i}{w}\right)_{t-k'} , \qquad i \in N \, . \tag{3.5}$$

If in a given market the average wage is below that elsewhere, this may
provide grounds for demanding increases in that market. The converse
may also apply, which leads us to expect a negative coefficient for the
relative wage variable in equation (3.5).

Although this model, apart from the inclusion of a term measuring the
relative wage position of a given market, is very similar to the standard
'natural rate' (expectations-augmented) Phillips curve, it does contain some
interesting implications about the long-run interregional (or interurban)
wage structure in the space economy. Long-run equilibrium in the
Friedman model requires *two* conditions: (1) that expectations are
correct in the long run, $(\dot{p}/p)_t^e \rightarrow (\dot{p}/p)_t$ as $t \rightarrow \infty$; and (2) that the wage
adjustment mechanism responds fully to anticipated inflation, $a_{i2} = 1$ in
equation (3.5). In the multimarket space economy a third condition of
long-run equilibrium is that of stability of the intermarket wage structure,

that is,

$$\left(\frac{w_i}{w}\right)_{t-k} = \left(\frac{w_i}{w}\right)_t, \qquad \text{for all } t . \tag{3.6}$$

If we further adopt a simple cost-plus relationship for price changes, such that the rate of increase in prices is equal to the economywide rate of wage increase, $\dot{w}/w$, less the trend rate of growth of productivity, $\dot{q}/q$:

$$\left(\frac{\dot{p}}{p}\right)_t = \left(\frac{\dot{w}}{w}\right)_t - \left(\frac{\dot{q}}{q}\right)_t, \tag{3.7}$$

then substitution of equations (3.6) and (3.7) into equation (3.5), given perfect expectations, yields

$$\frac{\dot{w}_i}{w_i} = a_{i0} - a_{i1} e_i^u + a_{i2}\left(\frac{\dot{w}}{w} - \frac{\dot{q}}{q}\right) - a_{i3}\frac{w_i}{w}, \qquad i \in N . \tag{3.8}$$

Also, in order for equation (3.6) to hold, $\dot{w}_i/w_i$ must be equal to $\dot{w}/w$ in equation (3.8) for all $i = 1, ..., n$. I turn now to the implications of the accelerationist hypothesis as applied to this model.

If wage bargains are made in real terms so that $a_{i2} = 1$ $(i = 1, ..., n)$, then from equation (3.8)

$$0 = a_{i0} - a_{i1} e_i^u - \frac{\dot{q}}{q} - a_{i3}\frac{w_i}{w}, \qquad i \in N , \tag{3.9}$$

from which the equilibrium or natural rate of unemployment for a given market is determined as

$$e_i^{u*} = \frac{a_{i0} - a_{i3}(w_i/w) - (\dot{q}/q)}{a_{i1}}, \qquad i \in N , \tag{3.10}$$

and its equilibrium relative wage position is

$$\left(\frac{w_i}{w}\right)^* = \frac{a_{i0} - a_{i1} e_i^u - (\dot{q}/q)}{a_{i3}}, \qquad i \in N . \tag{3.11}$$

Thus if the accelerationist hypothesis is correct and wages adjust fully to perfectly anticipated inflation in each and every market, the model described here implies that the wage structure in the space economy is independent of the rate of inflation. This conclusion, which would come as no surprise to advocates of the Friedman model, is a multimarket analogue of the natural rate of unemployment thesis.

But suppose instead that the strict version of the natural rate theory does not hold, that is, $0 \leqslant a_{i2} < 1$, $i = 1, ..., n$ (with $a_{i2} = 0$ corresponding to the simple or naive Phillips curve). Long-run equilibrium of the system still requires that the rate of change of wages in each market is the same, so that the equilibrium relative wage position of a given subregional market is then given as

$$\left(\frac{w_i}{w}\right)^* = \frac{a_{i0} - a_{i1}e_i^u - a_{i2}(\dot{q}/q) - (1-a_{i2})(\dot{w}/w)}{a_{i3}} , \tag{3.12}$$

from which it is seen that if the response to expected inflation is not complete, the equilibrium wage structure in the space economy depends upon the rate of inflation.

Further implications of this simple model can be derived if we assume that the unemployment rate in a given market can be expressed as a simple linear function of the economywide unemployment rate, that is,

$$e_i^u = \mu_i e^u ; \qquad \mu_i > 0 , \qquad i \in N . \tag{3.13}$$

Equation (3.13) says that an increase in e^u has the effect of increasing unemployment in each market and that the magnitude of these effects is independent of the value of e^u [6]. Therefore, by combining equations (3.13) and (3.12), we have

$$\left(\frac{w_i}{w}\right)^* = \frac{a_{i0} - c_i e^u - a_{i2}(\dot{q}/q) - (1-a_{i2})(\dot{w}/w)}{a_{i3}} , \tag{3.14}$$

where $c_i = a_{i1}\mu_i$. Now it is definitionally true that aggregating across all markets,

$$\sum_{i \in N} \frac{L_i}{L} \left(\frac{w_i}{w}\right)^* = 1 , \tag{3.15}$$

where L_i/L is the labour force weight (assumed fixed) for the ith market; hence

$$\sum_{i \in N} \frac{L_i}{L} d\left(\frac{w_i}{w}\right)^* = 0 . \tag{3.16}$$

Thus if we hold productivity change constant we have

$$\sum_{i \in N} \frac{L_i}{L} d\left(\frac{w_i}{w}\right)^* = \sum_{i \in N} \frac{L_i}{L} \left[-c_i d e^u - (1-a_{i2}) d\left(\frac{\dot{w}}{w}\right) \right] = 0 , \tag{3.17}$$

which yields the equilibrium aggregate trade-off between wage inflation and unemployment of

$$\frac{d(\dot{w}/w)}{de^u}\bigg|_{\Delta(w_i/w)^* = 0, \, i = 1, ..., n} = \sum_{i \in N} \frac{L_i}{L} c_i \bigg/ \sum_{i \in N} \frac{L_i}{L}(1-a_{i2}) = \frac{\bar{c}}{1-\bar{a}_2} . \tag{3.18}$$

---

[6] In other words, equation (3.13) implicitly assumes that the spatial distribution of relative unemployment is constant, an assumption which seems to be supported by the observation that the ranking of regions by unemployment tends to remain remarkably stable over long periods of time (see, for example, Brechling, 1967).

This equation indicates that, under certain assumptions, the aggregate trade-off between wage inflation and unemployment can be regarded as an 'average' trade-off. A more interesting implication of this formulation, however, is that it indicates the *distortion* in the relative wage structure generated by inflation. From equation (3.14), if all else is held constant,

$$\frac{d(w_i/w)^*}{d(\dot{w}/w)} = \frac{1}{a_{i3}}\left[-c_i\frac{de^u}{d(\dot{w}/w)} - (1-a_{i2})\right]$$

$$= \frac{c_i}{a_{i3}}\left[\frac{(1-\bar{a}_2)}{\bar{c}} - \frac{(1-a_{i2})}{c_i}\right] , \tag{3.19}$$

which gives the direction of response of the relative wage position of the ith market to the change in the rate of expected inflation. But from equation (3.17),

$$\sum_{i \in N} \frac{L_i}{L} \frac{d(w_i/w)^*}{d(\dot{w}/w)} = 0 , \tag{3.20}$$

which implies that a_{i2} must be negatively correlated with c_i across individual markets, since if the response to expected inflation, a_{i2}, and the response to unemployment (excess demand), c_i, were positively related, then the right-hand side of equation (3.19) would be positive when aggregated across all markets, which would violate equation (3.20). In other words, if the responses of individual subregional markets differ, they must differ systematically.

I have argued that wage inflation in regional and subregional (urban) markets is influenced both by local market conditions and external relative factors. Thus a measure of the relative wage structure was added to an expectations-type Phillips curve to reflect the idea that under a given set of overall demand conditions, groups of workers in various markets will attempt to obtain a larger wage if they believe they are relatively underpaid. This simple model contains several implications for wage behaviour in a multimarket context. Given that there exists a geographical (regional and urban) wage structure, steady-state equilibrium requires that the rate of wage increase be equal across all markets. It was shown that if wage bargains are made in real terms then the intermarket wage structure is independent of the rate of inflation. If, however, adjustment to expected inflation is incomplete in any of the markets, inflation results in distortion of the wage structure. This in turn implied that if responses differ across markets they should differ systematically; more specifically, some markets should be more highly responsive to price inflation and less responsive to unemployment than other markets. These latter markets should, therefore, be more strongly influenced by the relative wage factor.

3.3 Price expectations and structural instability

One major difficulty which arises in giving empirical content to the model of the previous section is that of modelling the formation of price expectations. In most studies, price expectations are assumed to be some linear combination of past actual price movements only, and hence involve the use of distributed lag models. Of these, the adaptive expectations mechanism has received the most attention; according to this hypothesis, expectations are revised in proportion to the forecast error associated with the expectation of inflation of the last period:

$$\left(\frac{\dot{p}}{p}\right)^e_t = \left(\frac{\dot{p}}{p}\right)^e_{t-1} + \lambda\left[\left(\frac{\dot{p}}{p}\right)_{t-1} - \left(\frac{\dot{p}}{p}\right)^e_{t-1}\right]$$

$$= \lambda\left(\frac{\dot{p}}{p}\right)_{t-1} + (1-\lambda)\left(\frac{\dot{p}}{p}\right)^e_{t-1}, \qquad 0 \leqslant \lambda \leqslant 1, \tag{3.21}$$

which after repeated iterations yields the geometrically declining distributed lag function

$$\left(\frac{\dot{p}}{p}\right)^e_t = \lambda \sum_{k=0}^{\infty} (1-\lambda)^{k-1}\left(\frac{\dot{p}}{p}\right)_{t-k}. \tag{3.22}$$

Muth (1960) shows that equation (3.22) can be interpreted as that expected value of the present rate of inflation which minimises the forecast error $E[(\dot{p}/p)_t - (\dot{p}/p)^e_t]^2$ if the underlying price inflation variable follows a random walk, in which case it can be represented as the sum of a 'permanent' or 'normal' component and a 'transitory' component, namely

$$\left(\frac{\dot{p}}{p}\right)_t = \left(\frac{\dot{p}}{p}\right)^N_t + \eta_t, \qquad \eta_t \sim \text{NID}(0, \sigma^2_\eta), \tag{3.23}$$

where the normal component is assumed to follow a moving-average process

$$\left(\frac{\dot{p}}{p}\right)^N_t = \left(\frac{\dot{p}}{p}\right)^N_{t-1} + \xi_t, \qquad \xi_t \sim \text{NID}(0, \sigma^2_\xi). \tag{3.24}$$

Although numerous studies have found some support for the adaptive expectations hypothesis (for example, Turnovsky, 1970; Turnovsky and Wachter, 1972; Lahiri, 1976), these same analyses have revealed particular limitations of this model. Firstly, when the inflation rate is accelerating, the model is certain to produce inaccurate forecasts. To see this rewrite equation (3.21) as

$$\left(\frac{\dot{p}}{p}\right)^e_t - \left(\frac{\dot{p}}{p}\right)^e_{t-1} = \lambda\left[\left(\frac{\dot{p}}{p}\right)_{t-1} - \left(\frac{\dot{p}}{p}\right)^e_{t-1}\right]. \tag{3.25}$$

If $(\dot{p}/p)_t$ is trended then $(\dot{p}/p)^e_t$ can only be so if $(\dot{p}/p)^e_t$ is consistently less than $(\dot{p}/p)_t$. Thus expectations are certain to be inaccurate when $(\dot{p}/p)_t$ is accelerating. Moreover, $(\dot{p}/p)^e_t$ is a convex combination of $(\dot{p}/p)^e_{t-1}$ and $(\dot{p}/p)_{t-1}$, and so the maximum value it can obtain in these circumstances is that of $(\dot{p}/p)_t$.

Secondly, the empirical results obtained by using the adaptive scheme seem to contradict the explicit assumption that the coefficient of adaptive adjustment, λ, is time-invariant; in fact the evidence suggests that λ increases in times of accelerating inflation and general uncertainty. That this is likely to occur can be seen by noting that λ depends in a known way on the variances of the permanent and transitory components of price inflation[7]

$$\lambda = 1 + \frac{1}{2}\frac{\sigma_\xi^2}{\sigma_\eta^2} - \frac{\sigma_\xi}{\sigma_\eta}\left(1 + \frac{1}{4}\frac{\sigma_\xi^2}{\sigma_\eta^2}\right)^{\frac{1}{2}}. \tag{3.26}$$

If the variances σ_ξ^2 and σ_η^2 are time-varying, then λ will also exhibit temporal instability. For example, under conditions of increasing rates of change in prices, the variance of $(\dot{p}/p)_t^N$ will be large relative to the variance of η_t, so that λ will increase towards unity; in other words, economic agents will become more sensitive to the most recent data. Conversely, when inflation rates are steady, the variance of η_t may exceed that of $(\dot{p}/p)_t^N$, with the result that λ will tend towards zero, and equal weights would be given to all past rates of inflation.

Thirdly, econometric estimation of the model in equation (3.5) with use of the adaptive expectations scheme involves certain statistical difficulties. Adding an error term, ϵ_t, to the model in equation (3.5), substituting equation (3.22) to eliminate the unobservable expectations variable, and applying a Koyck transformation, leads to the reduced form estimation equation

$$\left(\frac{\dot{w}_i}{w_i}\right)_t = a_{i0}\lambda - a_{i1}e_{it}^u + a_{i1}(1-\lambda)e_{i\,t-1}^u + a_{i2}\lambda\left(\frac{\dot{p}}{p}\right)_{t-1} - a_{i3}\left(\frac{w_i}{w}\right)_{t-k}$$

$$+ a_{i3}(1-\lambda)\left(\frac{w_i}{w}\right)_{t-k-1} - (1-\lambda)\left(\frac{\dot{w}_i}{w_i}\right)_{t-1} + \epsilon_{it} - (1-\lambda)\epsilon_{i\,t-1}$$

$$= b_{i0} - b_{i1}e_{it}^u + b_{i2}e_{i\,t-1}^u + b_{i3}\left(\frac{\dot{p}}{p}\right)_{t-1} - b_{i4}\left(\frac{w_i}{w}\right)_{t-k} + b_{i5}\left(\frac{w_i}{w}\right)_{t-k-1}$$

$$- b_{i6}\left(\frac{\dot{w}_i}{w_i}\right)_{t-1} + v_{it}. \tag{3.27}$$

Not only is equation (3.27) overidentified, but also the estimate of the price expectations coefficient, a_{i2}, is obtained only indirectly, as

$$\hat{a}_{i2} = \frac{\hat{b}_{i3}}{1 - \hat{b}_{i6}}, \tag{3.28}$$

so that it is difficult to perform statistical tests on a_{i2}. Further, the error terms of the reduced form equation are not uncorrelated; this, coupled with the presence of a lagged endogenous variable, will result in biased and inconsistent estimates if ordinary least squares (OLS) are applied directly.

[7] It is not necessary to assume that ξ_t and η_t are uncorrelated. If $E(\xi_t\eta_t) = \sigma_{\xi\eta}$ and $E(\xi_t\eta_s) = 0$, $t \neq s$, it is only necessary to replace $\sigma_\xi^2/\sigma_\eta^2$ in equation (3.26) by $\sigma_\xi^2/(\sigma_\eta^2 + \sigma_{\xi\eta})$ (Muth, 1960, page 304).

Only in the unlikely case where the residuals $\hat{\epsilon}_{it}$ follow a first-order autoregressive structure with parameter exactly equal to $(1-\lambda)$ would the problem of inconsistency disappear.

In the light of these limitations, an alternative approach to the modelling of inflationary expectations will be adopted—the so-called 'convergence' or 'return-to-normality' scheme (Brinner, 1977; Kane and Malkiel, 1976), in which price expectations are assumed to be equal to the sum of the 'normal' rate of inflation and a positive fraction, λ, of the deviation of actual inflation from the 'normal' rate [this scheme is similar to the 'convergent parameter' random coefficients regression model developed by Rosenberg (1973b)]:

$$\left(\frac{\dot{p}}{p}\right)_t^e = \left(\frac{\dot{p}}{p}\right)_{t-1}^N + \lambda\left[\left(\frac{\dot{p}}{p}\right)_{t-1} - \left(\frac{\dot{p}}{p}\right)_{t-1}^N\right], \qquad 0 \leqslant \lambda \leqslant 1, \qquad (3.29)$$

or

$$\left(\frac{\dot{p}}{p}\right)_t^e = \lambda\left(\frac{\dot{p}}{p}\right)_{t-1} + (1-\lambda)\left(\frac{\dot{p}}{p}\right)_{t-1}^N, \qquad (3.30)$$

where 'normal' inflation is updated by contrasting the most recent prior conception of normal inflation with lagged values of actual inflation:

$$\left(\frac{\dot{p}}{p}\right)_t^N = \left(\frac{\dot{p}}{p}\right)_{t-1}^N + \sum_k \pi_k\left[\left(\frac{\dot{p}}{p}\right)_{t-k-1} - \left(\frac{\dot{p}}{p}\right)_{t-1}^N\right]$$

$$= \sum_k \pi_k\left(\frac{\dot{p}}{p}\right)_{t-k-1}, \qquad \text{if } \sum_k \pi_k = 1. \qquad (3.31)$$

Substituting equation (3.31) into (3.29) gives the basic form to be included in the estimated wage inflation equation:

$$\left(\frac{\dot{p}}{p}\right)_t^e = \sum_k \pi_k\left(\frac{\dot{p}}{p}\right)_{t-k-2} + \lambda\left[\left(\frac{\dot{p}}{p}\right)_{t-1} - \sum_k \pi_k\left(\frac{\dot{p}}{p}\right)_{t-k-2}\right], \qquad (3.32)$$

and hence

$$\left(\frac{\dot{w}_i}{w}\right)_t = a_{i0} - a_{i1}e_i^u + a_{i2}\sum_k \pi_k\left(\frac{\dot{p}}{p}\right)_{t-k-2} + a_{i2}\lambda\left[\left(\frac{\dot{p}}{p}\right)_{t-1} - \sum_k \pi_k\left(\frac{\dot{p}}{p}\right)_{t-k-2}\right]$$

$$- a_{i3}\left(\frac{w_i}{w}\right)_{t-k} + \epsilon_{it}. \qquad (3.33)$$

Testable hypotheses in this model include the assertion that λ is a positive fraction significantly greater than zero, and that the coefficient of the 'normal' rate of inflation, a_{i2}, is insignificantly different from unity (corresponding to the accelerationist hypothesis). However, in a similar way to the adaptive model, we might expect the convergence parameter, λ, to vary with changes in the time path of inflation and with general uncertainty. Further, this source of unstable behaviour should be distinguished from that of a variable coefficient of adjustment to expectations, a_{i2}. According to Gordon (1972, page 406) a variable coefficient version of the Phillips curve would help reconcile: the steady increase in the size of the elasticity of wages to past inflation; the partial adjustment observed in most postwar econometric studies of wage inflation for earlier sample periods; the 'natural' rate hypothesis that the rate of inflation will accelerate if unemployment is held below the 'natural' rate; and the relative flatness of the Phillips curve to the right of the 'natural' rate evidenced by the absence of any apparent tendency towards accelerating *deflation* in periods of high unemployment.

Numerous alternative specifications of a variable coefficient of adjustment to expected inflation can be formulated; some obvious examples are:

$$a_{i2} = \alpha_i \left(\frac{\dot{p}}{p}\right)_t^e ; \qquad \alpha_i > 0 , \tag{3.34}$$

$$a_{i2} = \alpha_{i0} + \alpha_{i1} \left(\frac{\dot{p}}{p}\right)_t^e ; \qquad \alpha_{i0} < 1 , \quad \alpha_{i1} > 0 , \tag{3.35}$$

$$a_{i2} = \alpha_{i0} + \alpha_{i1} D^e ; \qquad \alpha_{i0} < 1 , \quad \alpha_{i1} > 0 , \tag{3.36}$$

where D^e is a dummy variable of the form

$$D^e = \begin{cases} 1 , & \text{if } \left(\frac{\dot{p}}{p}\right)_t^e \geqslant \left(\frac{\dot{p}}{p}\right)^* , \\ 0 , & \text{otherwise} , \end{cases} \tag{3.37}$$

and

$$a_{i2} = \alpha_{ij} \tag{3.38}$$

where

$$j = \begin{cases} 1 , & \text{if } \left(\frac{\dot{p}}{p}\right)_t^e \leqslant \left(\frac{\dot{p}}{p}\right)^* , \\ 2 , & \text{if } \left(\frac{\dot{p}}{p}\right)_t^e > \left(\frac{\dot{p}}{p}\right)^* . \end{cases} \tag{3.39}$$

In equations (3.37) and (3.39) the value $(\dot{p}/p)^*$ represents some *a priori* 'threshold' rate of inflation. On applying these specifications to the basic wage model, we have, respectively,

$$\left(\frac{\dot{w}_i}{w_i}\right)_t = a_{i0} - a_{i1} e_{it}^u + \alpha_i \left[\left(\frac{\dot{p}}{p}\right)_t^e\right]^2 - a_{i3} \left(\frac{w_i}{w}\right)_{t-k} , \tag{3.40}$$

$$\left(\frac{\dot{w}_i}{w_i}\right)_t = a_{i0} - a_{i1} e_{it}^u + \alpha_{i0} \left(\frac{\dot{p}}{p}\right)_t^e + \alpha_{i1} \left[\left(\frac{\dot{p}}{p}\right)_t^e\right]^2 - a_{i3} \left(\frac{w_i}{w}\right)_{t-k} , \tag{3.41}$$

$$\left(\frac{\dot{w}_i}{w_i}\right)_t = a_{i0} - a_{i1} e_{it}^u + \alpha_{i0} \left(\frac{\dot{p}}{p}\right)_t^e + \alpha_{i1} D^e \left(\frac{\dot{p}}{p}\right)^e - a_{i3} \left(\frac{w_i}{w}\right)_{t-k} , \tag{3.42}$$

$$\left(\frac{\dot{w}_i}{w_i}\right)_t = a_{i0} - a_{i1} e_{it}^u + \alpha_{ij} \left(\frac{\dot{p}}{p}\right)_t^e - a_{i3} \left(\frac{w_i}{w}\right)_{t-k} . \tag{3.43}$$

Gordon (1972) has used a specification similar to equation (3.34), with squared values of expected inflation, whereas equation (3.35) leads to another nonlinear adjustment equation. The dummy variable specification of equation (3.36) is analogous to the threshold 'inflation severity' variable suggested by Eckstein and Brinner (1972), whereas the asymmetric adjustment function in equations (3.38) and (3.39) resembles the threshold 'switching regime' model of Hamermesh (1970).

The difficulty of these and other possible specifications is that the data cannot easily distinguish between several plausible alternatives. But all such formulations do serve to suggest that the usual assumption that the adjustment to expectations coefficient remains unchanged throughout the sample period is a rather strong one, and probably untenable. Since this issue has important implications for the accelerationist hypothesis, the adjustment coefficient should at least be tested for instability.

In addition to this form of 'structural change', there are several other sources of parametric variation in the wage relation specified in equation (3.5) which may be mentioned here. One feature that is likely to lead to instability in this model is that the equation does not allow for the fact that in any one time period only a fraction of the work force receives an increase, or that different groups receive their increases with different frequencies. Thus the rate of change of wages at time t will be proportional to the change in the wages of workers receiving an adjustment in that period, denoted by $(\dot{w_i}/w_i)^a$:

$$\frac{\dot{w_i}}{w_i} = \kappa_{it}\,\rho_{it}\left(\frac{\dot{w_i}}{w_i}\right)^a , \tag{3.44}$$

where

ρ_{it} is the fraction of the work force in the ith market actually receiving the increase, and

κ_{it} is the relative wage of that fraction, w_i^a/w_i.

Substituting equation (3.44) in the wage equation (3.5) yields

$$\frac{\dot{w_i}}{w_i} = a_{i0}\,\kappa_{it}\rho_{it} - a_{i1}\,\kappa_{it}\rho_{it}\,e_{it}^u + a_{i2}\,\kappa_{it}\rho_{it}\left(\frac{\dot{p}}{p}\right)_t^e - a_{i3}\,\kappa_{it}\rho_{it}\left(\frac{w_i}{w}\right)_{t-k} + \kappa_{it}\rho_{it}\,\epsilon_{it} . \tag{3.45}$$

This suggests that the estimated coefficients of the wage equation are likely to display instability owing to variations in the frequency of the bargaining process or its time pattern which lead to trends or fluctuations in ρ_{it} and κ_{it}; also the residuals may be heteroscedastic.

Parametric variation may also arise because of the effect of differential responses to expected inflation, excess demand, and comparability bargaining among the industries or sectors that comprise the urban or regional aggregate. This 'composition effect' can be seen by assuming that a given local market consists of m sectors or industries, so that the average wage is given as

$$w_i = \sum_{j \in M} \frac{L_{ij}}{L_i} w_{ij} \; ; \qquad \sum_{j \in M} \frac{L_{ij}}{L_i} = 1 \; , \qquad M \equiv \{1, 2, ..., m\} \; , \qquad (3.46)$$

where L_{ij}/L_i is the labour force share accounted for by the jth industry or sector, and w_{ij} its wage. Then we may write

$$\frac{\dot{w}_i}{w_i} = \sum_{j \in M} \frac{w_{ij} L_{ij}}{w_i L_i} \frac{\dot{w}_{ij}}{w_{ij}} + \sum_{j \in M} \frac{w_{ij} L_{ij}}{w_i L_i} \left(\frac{\dot{L}_{ij}}{L_{ij}} - \frac{\dot{L}_i}{L_i} \right) \; . \qquad (3.47)$$

The second term on the right-hand side of equation (3.47) is the composition effect resulting from shifts in the distribution of employment and in relative wages of the various sectors. A more important effect can be seen by assuming that the determinants of the wage behaviour of the jth sector are the explanatory variables of the urban or regional relationship, where, by expressing these exogenous variables by the vector x_i,

$$\frac{\dot{w}_{ij}}{w_{ij}} = f_{ij}(x_i) \; , \qquad (3.48)$$

and then using equations (3.47) and (3.48) to obtain

$$\frac{\dot{w}_i}{w_i} = \sum_{j \in M} \frac{w_{ij} L_{ij}}{w_i L_i} f_{ij}(x_i) + \sum_{j \in M} \frac{w_{ij} L_{ij}}{w_i L_i} \left(\frac{\dot{L}_{ij}}{L_{ij}} - \frac{\dot{L}_i}{L_i} \right) \; . \qquad (3.49)$$

From equation (3.49) it is clear that only if the labour force in each sector grows at the same rate as the total urban (or regional) labour force, and responses across sectors to the determinants of wage change are the same, will there be an exact relationship between the behaviour of the individual industries or sectors and the behaviour of the urban (or regional) aggregate. However, if reaction functions differ from industry to industry, then trends or fluctuations in interindustry relative wages, in relative employment size, relative unemployment, and relative size of labour force will be reflected in shifts in the aggregate urban (or regional) relationship. Since a fixed employment-weighted series is rarely available, some form of composition effect and corresponding parametric variation is, therefore, almost certain to be present.

   A third source of structural instability in the estimated wage equation concerns the appropriateness of the unemployment rate as a proxy for excess demand for labour. Recent years have seen the emergence in several countries of a positively sloped partial Phillips curve. A number of writers, the neoclassical monetarists in particular, argue that this change of relationship is more a once-for-all shift than a continuing movement, and is to be explained by a sharp rise in the 'natural rate' of unemployment (that is, voluntary and frictional unemployment), itself attributed to better redundancy and unemployment benefits, minimum wage legislation, and other social arrangements and policies (Brittan, 1975, pages 51–64; Friedman, 1977, page 15; Gray et al, 1975, pages 42–43). It is argued that allowance for this upward shift, by making use of an appropriately redefined unemployment variable, restores the wage inflation–excess demand relationship. However, other workers have shown (for example, Holmes and Smyth, 1970) that a wide range of values of unemployment may be consistent with a given level of excess demand or supply of labour, thus undermining the basic assumption of a stable transformation between excess demand and unemployment and implying more or less continual variation in the wage change–unemployment relation, rather than the step-like break of the monetarist interpretation.

3.4 Regression in the presence of time-varying parameters

The discussion of the previous section should suffice to indicate that there are ample and plausible arguments to suggest that an estimated Phillips curve relation is likely to exhibit parametric (structural) instability—quite apart from the theoretical short-run predictions of the accelerationist theory. Thus some sort of check of the constancy of the parameters would seem to be desirable. Two general types of parameter variation might be present: first, where parameters change at a limited number of points in time, but remain constant between these points; second, where parameters evolve continually over time. It is relatively easy to test for discrete structural shifts at a prespecified number of points (for example, Chow, 1960), but testing for changes at unknown points is rather more difficult, although some progress has been made in the case of a single change at an unknown point (Quandt, 1960; Farley et al, 1975). Recently Brown et al (1975), (hereafter BDE) proposed more general methods for testing for structural changes in regression relationships by using tests based on recursive residuals, and subsequently Garbade (1977) and Harvey and Phillips (1977) have combined and compared this approach with that of time-varying parameter regression (TVPR), that is, dynamic regression models with stochastic parameter structures. In particular, Harvey and Phillips have assessed the behaviour and usefulness of the BDE tests in the presence of random-walk parameters and have shown that these procedures can also be of use in detecting this form of continuous parameter variation. The following discussion draws upon their results.

The BDE tests are applied to a series of recursive residuals which are obtained from the recursive form of the OLS estimator. Let the regression model be

$$Y_t = x_t^T \beta + \epsilon_t , \qquad t = 1, ..., T , \tag{3.50}$$

where x_t and β are $(k \times 1)$ vectors and $\epsilon_t \sim \text{NID}(0, \sigma^2)$. Under the null hypothesis of stable coefficients in equation (3.50), the OLS estimate of β based on observations up to time t is $\hat{\beta}_t = (X_t^T X_t)^{-1} X_t^T y_t$ with estimation covariance matrix $\sigma^2 (X_t^T X_t)^{-1}$, where $X_t = (x_1, x_2, ..., x_t)^T$ and $y_t = (Y_1, Y_2, ..., Y_t)^T$ for $k \leqslant t \leqslant T$. The estimate of β based on $t-1$

observations may be updated when the tth observation is added by

$$\hat{\beta}_t = \hat{\beta}_{t-1} + \frac{(\mathbf{X}_{t-1}^T \mathbf{X}_{t-1})^{-1} \mathbf{x}_t (Y_t - \mathbf{x}_t^T \hat{\beta}_{t-1})}{1 + \mathbf{x}_t^T (\mathbf{X}_{t-1}^T \mathbf{X}_{t-1})^{-1} \mathbf{x}_t} , \qquad (3.51)$$

and

$$(\mathbf{X}_t^T \mathbf{X}_t)^{-1} = (\mathbf{X}_{t-1}^T \mathbf{X}_{t-1})^{-1} - \frac{(\mathbf{X}_{t-1}^T \mathbf{X}_{t-1})^{-1} \mathbf{x}_t \mathbf{x}_t^T (\mathbf{X}_{t-1}^T \mathbf{X}_{t-1})^{-1}}{1 + \mathbf{x}_t^T (\mathbf{X}_{t-1}^T \mathbf{X}_{t-1})^{-1} \mathbf{x}_t} . \qquad (3.52)$$

The recursive residuals are defined as

$$\tilde{\epsilon}_t = \frac{Y_t - \mathbf{x}_t^T \hat{\beta}_{t-1}}{[(1 + \mathbf{x}_t^T (\mathbf{X}_{t-1}^T \mathbf{X}_{t-1})^{-1} \mathbf{x}_t]^{\frac{1}{2}}} , \qquad t = k+1, ..., T , \qquad (3.53)$$

and, under the null hypothesis, follow a serially independent Gaussian process with mean zero and variance σ^2. The minimum number of observations needed to start the recursion is k. Then provided $\mathbf{X}_k$ is of rank k, the initial estimate of β is

$$\hat{\beta}_k = (\mathbf{X}_k^T \mathbf{X}_k)^{-1} \mathbf{X}_k Y_k = \mathbf{X}_k^{-1} \mathbf{y}_k . \qquad (3.54)$$

This estimate can be used to start the recursion.

The recursive residuals $\{\tilde{\epsilon}_t, \ t = k+1, ..., T\}$ are used to compute a number of tests of the null hypothesis of parametric constancy. In the 'cusum' test the cumulative sum series

$$V_t = \hat{\sigma}^{-1} \sum_{j=k+1}^{T} \tilde{\epsilon}_j , \qquad t = k+1, ..., T , \qquad (3.55)$$

where $\hat{\sigma}^2 = \sum_{j=k+1}^{T} \tilde{\epsilon}_j^2 / (T - k)$, is intended to test for a systematic shift in the regression relationship. Under the null hypothesis of stability, V_t is approximately normally distributed with mean zero and variance $(t - k)$. The null hypothesis is rejected if

$$|V_t| > [a(T-k)^{\frac{1}{2}} + 2a(t-k)(T-k)^{-\frac{1}{2}}] , \qquad \text{for any } t \in [k+1, T] ,$$

where a is chosen to obtain the desired confidence level. BDE suggest $a = 1 \cdot 143$ for a 1% confidence level and $a = 0 \cdot 948$ for a 5% level. Harvey (in the discussion to BDE) has suggested that a more powerful test is obtained by estimating σ^2 as

$$\sum_{j=k+1}^{T} \frac{(\tilde{\epsilon}_j - \bar{\tilde{\epsilon}})^2}{T - k - 1} ,$$

where $\bar{\tilde{\epsilon}}$ is the mean of the recursive residuals. An exact test is obtained with

$$\psi = [(T-k-1)^{-1} \sum_{j=k+1}^{T} (\tilde{\epsilon}_j - \bar{\tilde{\epsilon}})^2]^{-\frac{1}{2}} (T-k)^{-\frac{1}{2}} \sum_{j=k+1}^{T} \tilde{\epsilon}_j , \qquad (3.56)$$

which follows a t distribution under the null hypothesis. Although originally proposed as a test for functional misspecification (Harvey and Collier, 1975), its resemblance to the last 'cusum' statistic, V_t, suggests its applicability in the present context.

In the 'cusum of squares' test, the cumulative sum of squares series

$$S_c(t) = \left(\sum_{j=k+1}^{t} \tilde{\epsilon}_j^2 \right) \bigg/ \sum_{j=k+1}^{T} \tilde{\epsilon}_j^2 , \qquad t = k+1, ..., T , \qquad (3.57)$$

is calculated, and is a monotonically increasing sequence of positive numbers with $S_c(T) = 1$. This test is a useful complement to the 'cusum' test, especially when the departure from constancy of β is haphazard rather than systematic. Under the null hypothesis of stability, $1 - S_c(t)$ has a beta distribution with parameters $\alpha = -1 + \frac{1}{2}(T-k)$ and $\beta = -1 + \frac{1}{2}(t-k)$, and thus $S_c(t)$ has mean value $(T-k)/(t-k)$. BDE suggest constructing a confidence interval for $S_c(t)$ as $[(t-k)/(T-k)] \pm c_0$ where c_0 is chosen from table 1 of Durbin (1969) with $\alpha = 0 \cdot 5$ times the confidence level desired and $k = \frac{1}{2}(T-k) - 1$ in that table. If $|S_c(t) - [(t-k)/(T-k)]| > c_0$ for any $t \in [k+1, T]$ the null hypothesis is rejected.

Based on analogues of the recursion (3.51) to (3.52), a further test of parametric constancy is the so-called 'homogeneity test' (BDE) derived from fitting a moving regression over short time segments, with use of a variant of the ordinary analysis of variance test for nonoverlapping groups. The time segments for a moving regression of length n are $(1, n), (n+1, 2n), ..., ((p-2)n+1, (p-1)n), ((p-1)n+1, T)$, where p is the integral part of T/n, and the variance ratio considered, the 'homogeneity statistic', is

$$\mathrm{HS} = \frac{(T-kp)}{(kp-k)} \frac{S_R(1, T) - [S_R(1, n) + S_R(n+1, 2n) + ... + S_R(pn-n+1, T)]}{[S_R(1, n) + S_R(n+1, 2n) + ... + S_R(pn-n+1, T)]} ,$$
$$(3.58)$$

where $S_R(r, s)$ is the residual sum of squares from the regression on the time segment from time r to time s inclusive. This is equivalent to the usual 'between groups over within groups' ratio of mean squares and under the null hypothesis is distributed as $F(kp-k, T-kp)$.

Returning to the 'cusum' and 'cusum of squares' tests, Harvey and Phillips have argued that the efficacy of these tests depends on their behaviour under specific alternative hypotheses, because the recursive residuals themselves exhibit a very different pattern of behaviour under particular forms of stochastic parameter variation. One of the most useful stochastic structures is the multivariate random walk, which when combined with the regression model in equation (3.50) yields

$$Y_t = x_t^T \beta_t + \epsilon_t ,$$

$$\beta_t = \beta_{t-1} + \nu_t , \qquad t = 1, ..., T , \qquad (3.59)$$

x_t, β_t, and ν_t are $(k \times 1)$ vectors of independent variables, coefficients, and random disturbances respectively, and ϵ_t and ν_t are assumed to be mutually and serially uncorrelated with $\epsilon_t \sim \mathrm{NID}(0, \sigma^2)$ and $\nu_t \sim \mathrm{NID}(0, \sigma^2 Q)$. Although more complex models of the dynamic evolution of the βs are possible (see Rosenberg, 1973a; Sage and Melsa, 1971), we shall confine our attention to the random-walk model in the remainder of this section. Since parameter changes are likely to come from a variety of sources, it is reasonable to assume that some of these may be permanent whereas others may not. The random-walk process is sufficiently general that it will encompass both permanent and transitory parameter movements of many types (Cooley and Prescott, 1973), and is capable of generating a wide class of sample paths [see the simulations by Sarris (1973)].

Now when the regression coefficients follow this form of stochastic process it can be shown that (1) the variances of the recursive residuals will tend to increase over time for most sets of Xs, and (2) any positive first-order serial correlation in the Xs will impart positive serial correlation in the recursive residuals. By contrast, the variances of OLS residuals do not, generally speaking, appear to display a distinctive pattern under this alternative, except insofar as the Xs exhibit marked patterns over time. The OLS residuals will, however, tend to be serially correlated. The fact that the variances of recursive residuals tend to increase with t if random-walk parameters are present, suggests that when using the 'cusum of squares' statistic, a one-sided test should be used against this type of alternative hypothesis. Another test which makes use of this increasing variance property is based on a simple F statistic for heteroscedasticity. Let $S_R(\frac{1}{3})$ and $S_R(\frac{3}{3})$ denote the sums of squares of the first and last third of the recursive residuals sequence respectively. Then

$$H = \frac{S_R(\frac{3}{3})}{S_R(\frac{1}{3})} \tag{3.60}$$

follows an F distribution with $[\frac{1}{3}(T-k), \frac{1}{3}(T-k)]$ degrees of freedom under the null hypothesis. If random-walk parameters are present, H will tend to be large, so a one-sided test is appropriate. The presence of serial correlation in the recursive residuals may be tested by using the 'modified' Von Neuman Ratio (VNR) (Theil, 1971, page 219):

$$\text{VNR} = \frac{(T-k)}{(T-k-1)} \sum_{t=k+2}^{T} (\tilde{\epsilon}_t - \tilde{\epsilon}_{t-1}) \Big/ \sum_{t=k+1}^{T} \tilde{\epsilon}_t^2 , \tag{3.61}$$

where the normal definition of the denominator, $\sum_{t=k+1}^{T} (\tilde{\epsilon}_t - \bar{\tilde{\epsilon}})^2$ has been replaced by $\sum_{t=k+1}^{T} \tilde{\epsilon}_t^2$ because in the presence of random-walk parameters the recursive residuals will tend to have the same sign, so that $\bar{\tilde{\epsilon}}$ may be nonzero. Hence the normal Von Neuman Ratio statistic would involve a loss of power.

The powers of all the recursive residuals tests have been studied by Harvey and Phillips by making use of Monte Carlo methods for a wide range of situations, but with particular attention focussed on random-walk parameters, and their main conclusions can be summarised as follows.

The serial correlation tests, the Durbin–Watson test and the Von Neuman Ratio based on recursive residuals, have a high power against almost all models in which the parameters follow random-walk processes. Since the recursive residuals tend to have steadily increasing variances in such situations, the H test for heteroscedasticity can be fairly powerful as well. For the same reason the 'cusum of squares' test is also effective, particularly if a one-sided test is formulated. The 'cusum' and ψ tests are also quite powerful against random-walk parameters, although they tend to be less effective than the H test and the 'cusum of squares' test when the explanatory variables show a steady increase over time. Both of these sets of tests appear to be reasonably robust to serial correlation in the disturbances of fixed parameter models, provided that the serial correlation is not too pronounced.

When the model is correctly specified, the Von Neuman Ratio and heteroscedasticity H test are independent, so that the use of these two in conjunction may enable one to distinguish between random-walk parameters and serial correlation to a limited extent. A significant value of the VNR statistic together with a nonsignificant H test is likely to indicate fixed parameters with serially correlated disturbances. If both the VNR and H tests are significant, however, this could be an indication either of random-walk parameters or of very strong serial correlation. The chances of distinguishing between the two types of misspecification appear to be more favourable when the X variables are trended, for the H test tends to be more powerful in this situation.

In the empirical analyses that follow, in addition to presenting estimates based both upon standard OLS and the Cochrane–Orcutt transformation, parametric stability will also be checked for by using the tests discussed in this section. Attention will be confined to testing for instability; a more thorough, but perforce more lengthy, study would involve detailed graphical investigation to determine the location of significant structural changes together with, if necessary, explicit modelling of the parameter trajectories.

3.5 Estimates for a system of metropolitan labour markets
The focus of the empirical part of this study is on the behaviour of wage inflation in the manufacturing sector of certain metropolitan labour markets in the industrial region of central northeast USA (figure 3.1). Since consistent wage data for individual industries are not available at this local scale, attention will be confined to manufacturing industry as a whole. The original wage data for each city consist of monthly series of average weekly earnings of production workers[8]. Although this earnings index includes piece rates, bonuses, and overtime payments, it is the best local wage series that is available. This series was converted into straight-time hourly earnings (that is, excluding overtime payments) in the following way. Let w^E denote total average weekly earnings, h_s average standard weekly hours, and h_a average total weekly hours, then an estimate of

[8] Data on average earnings and length of work week in metropolitan areas were supplied by the Bureau of Labor Statistics, US Department of Labor, Washington.

average straight-time hourly earnings can be derived as

$$w_t = \frac{w_t^{\mathrm{E}}}{h_{\mathrm{s}} + 1 \cdot 5(h_{\mathrm{a}} - h_{\mathrm{s}})} \,, \qquad (3.62)$$

where it is assumed that overtime hours are paid at $1 \cdot 5$ times the standard wage and where standard weekly hours for each city were estimated as a weighted series by applying the city's proportions of manufacturing employment in durable and nondurable industry to the national average standard weekly hours series for these two major subdivisions of the manufacturing sector.

The basic unemployment data refer to the monthly total civilian unemployment rate by city, more detailed figures for manufacturing being unavailable [9]. The price level data used is the monthly national average 'consumer price index for urban wage earners and clerical workers' (all items) [10]. All of these monthly series were transformed to quarterly data by forming averages over the appropriate three-month periods.

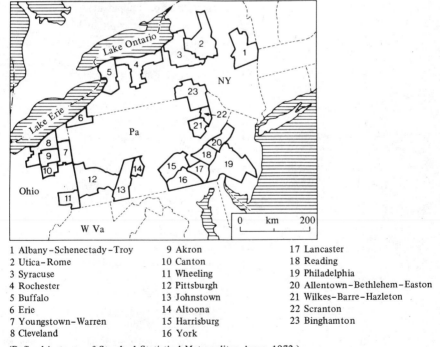

1 Albany–Schenectady–Troy	9 Akron	17 Lancaster
2 Utica–Rome	10 Canton	18 Reading
3 Syracuse	11 Wheeling	19 Philadelphia
4 Rochester	12 Pittsburgh	20 Allentown–Bethlehem–Easton
5 Buffalo	13 Johnstown	21 Wilkes–Barre–Hazleton
6 Erie	14 Altoona	22 Scranton
7 Youngstown–Warren	15 Harrisburg	23 Binghamton
8 Cleveland	16 York	

(Defined in terms of Standard Statistical Metropolitan Areas, 1972.)

Figure 3.1. The urban labour markets selected for analysis.

[9] Monthly unemployment rates for each city were extracted from *Area Trends in Employment and Unemployment*, Bureau of Labor Statistics, US Department of Labor.
[10] Published in *Monthly Labor Review*, Bureau of Labor Statistics, US Department of Labor. This index is an average of fifty-six metropolitan areas.

Table 3.1. Hourly earnings and dominant industry group in manufacturing, by city, 1972.

City	Average hourly earnings ($)	Dominant industry group	Percentage of total city manufacturing employment
Albany – Schenectady – Troy	3·99	Machinery, including electrical and transportation equipment	41·5
Utica – Rome	3·48	Machinery, including transportation equipment	40·2
Syracuse	4·01	Machinery, including transportation equipment	49·1
Rochester	4·38	Instruments and related products	50·6
Buffalo	4·52	Machinery, including electrical and transportation equipment	36·6
Erie	3·80	Machinery, including electrical equipment	41·7
Youngstown – Warren	4·58	Primary and fabricated metal products	60·6
Cleveland	4·31	Machinery, including electrical and transportation equipment	38·3
Akron	4·49	Rubber and plastics products	42·7
Canton	4·25	Primary and fabricated metal products	44·1
Wheeling	3·81	Primary and fabricated metal products	32·4
Pittsburgh	4·31	Primary and fabricated metal products	49·5
Johnstown	4·01	Primary and fabricated metal products	65·2
Altoona	3·19	Textiles, apparel and leather	29·6
Harrisburg	3·38	Machinery, including electrical and transportation equipment	24·4
York	3·21	Machinery, including electrical and transportation equipment	30·1
Lancaster	3·44	Machinery, including electrical and transportation equipment	21·6
Reading	3·44	Machinery, including electrical and transportation equipment	26·7
Philadelphia	3·82	Machinery, including electrical and transportation equipment	25·5
Allentown – Bethlehem – Easton	3·67	Primary and fabricated metal products	25·6
Wilkes – Barre – Hazleton	2·98	Textiles, apparel and leather	39·7
Scranton	3·08	Textiles, apparel and leather	38·0
Binghamton	3·61	Primary and fabricated metal products	49·7

Before discussing the construction of the dependent and independent variables in the wage model, I shall first examine some general features of the interurban wage structure. Table 3.1 gives the average hourly earnings (adjusted for overtime) in manufacturing by city for 1972, together with an indication of the interurban differences in industrial structure in terms of dominant employment category. A distinct urban earnings hierarchy is clearly evident, ranging from $4·58 per hour in Youngstown–Warren to $2·98 per hour in Wilkes–Barre–Hazleton. Although there appears to be some relationship between position in the hierarchy and dominant industry type, this relationship is by no means perfect, and earnings rank seems to be as much related to geographical location of the city (compare table 3.1 with figure 3.1).

The behaviour of the urban earnings hierarchy is summarised in table 3.2, where the markets are ranked according to their position in the earnings structure for selected years from 1961 to 1976. The ranking of cities for each year is compared to the ranking in 1961 by calculating the Spearman rank correlation coefficient (r_s). Also shown are the average annual unemployment rates over the period 1960–1976. In all periods there are some markets which show changes in their relative position. Nevertheless, given the time span over which the comparison is made, and given that the the time span over which the comparison is made, and given that the comparisons are based on average hourly earnings which in the short run are likely to be more volatile than any other wage index (for example, wage rates), the degree of stability which emerges is quite remarkable; all the Spearman rank correlation coefficients are positive and significant at the 1% level. To the extent that any observable instability is present it occurs in the latter part of the period (when inflation was accelerating). A second feature is that the urban areas which are always near the top of the earnings hierarchy show a definite geographical concentration (in the Cleveland–Pittsburgh area). Finally, there appears to be no systematic overall relationship between subregional unemployment and wage leadership (position in the earnings hierarchy).

Having established that a stable interurban relative wage structure exists, I shall turn now to consideration of the proposed wage inflation model. By means of the quarterly series referred to above, the dependent and independent variables were defined as follows:

$$\left(\frac{\dot{w_i}}{w_i}\right)_t = \frac{w_{it} - w_{i\,t-4}}{w_{i\,t-4}} \, , \tag{3.63}$$

$$\left(\frac{\dot{p}}{p}\right)_t = \frac{p_{it} - p_{i\,t-4}}{p_{i\,t-4}} \, , \tag{3.64}$$

$$\bar{e}_{it}^{u} = \frac{1}{4} \sum_{j=0}^{3} e_{i\,t-j}^{u} \, , \tag{3.65}$$

Table 3.2. Ranking of urban labour markets by average hourly earnings in manufacturing, selected years; and average rate of unemployment.

City	Earnings rank							Average unemployment rate (%) 1960–1976
	1961	1963	1966	1969	1972	1974	1976	
Albany–Schenectady–Troy	8=	7	9	9	10	10	11	4·5
Utica–Rome	14=	15	15	16	16	19	16=	6·6
Syracuse	11	8=	8	8	8=	9	9	5·2
Rochester	8=	8=	5	5	4	5=	5	3·9
Buffalo	4	4	3	3	2	2	2	6·8
Erie	12	11	12	13	13	11	12	5·7
Youngstown–Warren	1	2	2	2	1	1	1	7·3
Cleveland	5	5	6=	4	5=	5=	4	7·4
Akron	2	1	1	1	3	4	8	4·3
Canton	6	6	6=	7	7	5=	6	5·9
Wheeling	7	10	11	12	12	12	10	6·9
Pittsburgh	3	3	4	6	5=	3	3	5·8
Johnstown	10	12	10	10	8=	8	7	7·6
Altoona	20	20	20	20	21	21	21	7·3
Harrisburg	18=	19	19	19	19	16	16=	3·6
York	21	21	21	21	20	20	20	5·1
Lancaster	18=	17	17=	17=	17=	18	19	3·2
Reading	17	18	17=	17=	17=	15	15	3·6
Philadelphia	13	13	13	11	11	14	13	5·6
Allentown–Bethlehem–Easton	14=	14	14	15	14	13	14	4·2
Wilkes-Barre–Hazleton	23	23	23	23	23	23	22	4·5 a
Scranton	22	22	22	22	22	22	23	4·2 a
Binghamton	16	16	16	14	15	17	18	4·8
r_S		+0·94	+0·90	+0·88	+0·82	+0·81	+0·82	

a Average refers to the period 1960–1973.   = Denotes equal ranking.

and

$$\left(\frac{w_i}{\overline{w}}\right)_{t-k} = \left(\frac{w_i}{\overline{w}}\right)_{t-4}, \qquad \text{where} \quad \overline{w}_t = n^{-1}\sum_{i=1}^{n} w_{it}, \quad n = 23. \qquad (3.66)$$

The four-quarter rate-of-change variable used in equation (3.63) is, of course, only one of many possible measures of wage inflation. It has been criticised on the grounds that if this specification is used when in fact it is the wrong one, it can introduce serial correlation into the error terms of the model (Black and Kelejian, 1972; Hamermesh, 1970; Rowley and Wilton, 1973). This will bias upwards tests of statistical significance, but will not bias the coefficient estimates unless there are lagged terms of the dependent variable in the equation. A frequently proposed alternative measure is the one-quarter rate of change, $(\dot{w}_i/w_i)_t = (w_{it} - w_{it-1})/w_{it-1}$. However, this variable tends to amplify any 'noise' in the data, and may exhibit marked seasonality. We have chosen to adopt equation (3.63) in the belief that annual adjustment of wages is the norm in the manufacturing sector[11].
The urban relative wage variable in equation (3.66) is taken to be the ratio of the wage of a given market to the mean wage of all markets in the regional system, and the length of the 'spillover' lag, k, was assumed to be coincident with the period between wage changes and was thus set equal to four quarters. One final problem concerns the specification of the weights in the estimate of the 'normal' rate of price inflation to be used in the expectations variable. *Faute de mieux* the weights $\{\pi_k\}$ were imposed *a priori* on the assumption of an Almon first-degree polynomial distributed lag for six quarters with a zero constraint at the seventh quarter; this corresponds to the imposition of linearly declining weights, which sum to unity, as indicated in table 3.3.

The time-series behaviour of wage inflation and unemployment in each metropolitan market is shown in figure 3.2 (in the cases of Scranton and Wilkes–Barre–Hazleton, unemployment data are available only to the end of 1973). All cities show a similar overall pattern of fluctuating but generally untrended wage inflation up to the mid-1960s, and thereafter an accelerating inflation rate characterised by distinct cyclical movements.

Table 3.3. Distributed lag structure of 'normal' inflation rate, $\left(\dfrac{p}{p}\right)_t^N = \sum_k \pi_k \left(\dfrac{p}{p}\right)_{t-k-1}$.

Lag weights								Number of quarters	Mean lag
0	1	2	3	4	5	6	7		
0·0	0·286	0·238	0·190	0·143	0·095	0·048	0·0	6	2·67

[11] For the four-quarter method to be correct, it must also hold that the distribution of annual wage changes be uniform across each quarter. I have already discussed the problem of parametric instability which can arise as a result of violation of this assumption.

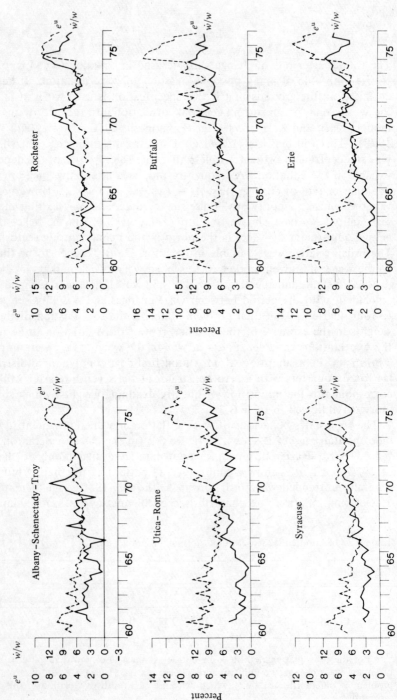

Figure 3.2. Wage inflation and unemployment, by city; quarterly series, 1960–1976. $\dot{w}/w$ is the annual rate of change in average hourly earnings (excluding overtime) for production workers in manufacturing, defined as $(w_t - w_{t-4})/w_{t-4}$, e^u is the total civilian unemployment rate.

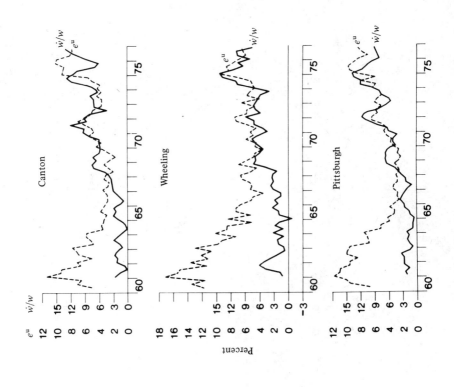

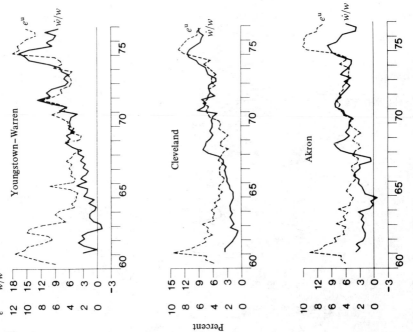

Figure 3.2 (continued)

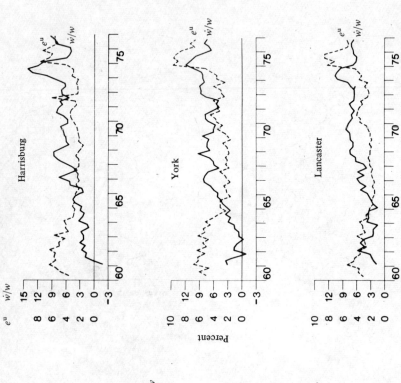

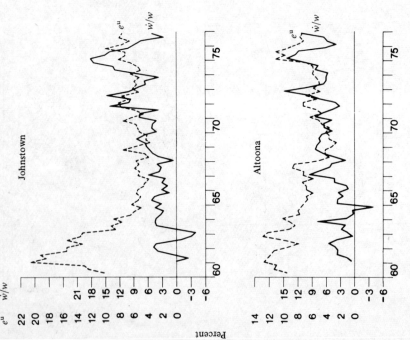

Figure 3.2 (continued)

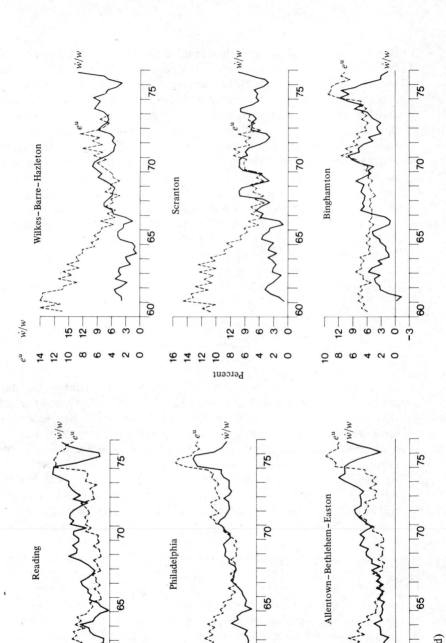

Figure 3.2 (continued)

Beyond these broad features, however, the temporal variability of the twenty-three wage change series shows considerable contrasts. The most irregular time curves tend to be associated both with the smaller centres and those that have a high degree of industrial specialisation. The least variable series are shown by medium-to-large markets and those with a broader spectrum of manufacturing industry. The occasional negative rates of wage inflation in certain of the areas reflect marked intersectoral shifts in employment. The main characteristic feature of the quarterly unemployment series is their decline from high levels at the beginning of the period to minima in the late 1960s followed by steadily rising values throughout the 1970s. All of the series exhibit distinct short-term cyclical fluctuations and seasonal variation, although the amplitude of these fluctuations varies from centre to centre. Finally, no clear inverse relationship between wage inflation and unemployment is apparent from casual inspection of the graphs.

The wage inflation model was first estimated by OLS applied to the equation

$$\left(\frac{\dot{w}_i}{w_i}\right)_t = \beta_{i0} + \beta_{i1}\bar{e}^u_{it} + \beta_{i2}\left(\frac{\dot{p}}{p}\right)^N_{t-1} + \beta_{i3}\left[\left(\frac{\dot{p}}{p}\right)_{t-1} - \left(\frac{\dot{p}}{p}\right)^N_{t-1}\right] + \beta_{i4}\left(\frac{w_i}{\bar{w}}\right)_{t-4} + \epsilon_{it}$$

$$(3.67)$$

The results are given in table 3.4. Virtually all of the Durbin–Watson statistics lie in the inconclusive range $(1 \cdot 75 > DW > 0 \cdot 8)$ and suggest that some degree of serial correlation may be present in the regression error terms. Thus the regressions were also estimated by using the Cochrane–Orcutt technique to give fixed initial value maximum-likelihood estimates for a first-order autoregressive error specification, AR(1). As can be seen from table 3.5, all of the regressions have a nonzero positive coefficient for this error process, and although the Durbin–Watson statistic (calculated with use of the adjusted disturbance terms) has increased, in most cases it is still below the upper bound required for insignificance $(DW_U = 2 \cdot 2)$.

If we look at the complete set of regressions, our results do not give much support for a Phillips-type relationship between wage inflation and unemployment. In the OLS regressions of table 3.4, all but three of the equations have the correct negative sign for the coefficient of the unemployment variable, but the relationship is significantly different from zero in only eleven cases. When the model is reestimated with allowances for serial correlation, the number of markets in which there is a significant inverse relationship between wage change and unemployment drops to five. In most of the urban markets, therefore, the rate of wage inflation appears to be independent of the local excess demand for labour as proxied by the unemployment rate. However, it is not possible to put complete confidence in these results because the total civilian unemployment rate may not be the most appropriate surrogate for excess demand for labour in manufacturing, and because of the possible temporal instability of the coefficient.

The evidence for the convergent expectations model is mixed, both with respect to the 'normal inflation' hypothesis and the 'partial convergence' hypothesis. The OLS estimates show a significant, positively-signed coefficient of normal inflation in all cases, but in only ten of the markets is the coefficient insignificantly different from unity, the value required under the 'natural rate' (accelerationist) hypothesis, although in most of these latter equations wage inflation seems to be independent of unemployment as predicted for the case of no 'money illusion'. These results, which are closely replicated in the Cochrane–Orcutt regressions, suggest that the rate of wage inflation does adjust to past rates of price inflation but that complete adjustment as postulated by the 'natural rate' model does not hold for all submarkets; rather partial adjustment appears to be the typical response at the subregional scale. The indifferent results for the partial convergence hypothesis may reflect the possible (even probable) temporal instability of the adjustment coefficient (β_{i4}), particularly in periods of rapid inflation or general uncertainty.

Finally, the results give strong support for the relative wage comparison hypothesis. The coefficient for this variable has the predicted sign and is significantly different from zero in every market in the OLS regressions. Although the influence of this variable is reduced in the Cochrane–Orcutt estimates, the coefficient remains significant in most equations and is of the correct sign in all cases. One interesting feature is that, with the exception of Youngstown–Warren (the market at the top of the earnings hierarchy—see table 3.1), the role of this factor in the wage inflation process tends to be greater in those markets characterised by partial adjustment to inflationary expectations. It was argued in section 3.2 that distortion of the relative wage structure would occur under conditions of partial adjustment to price inflation expectations. Our empirical analyses suggest that the 'catch-up' effect of relative wage comparisons is an important mechanism by which workers rectify such distortions and is an integral part of the bargaining process. Of course, our findings relate to the use of a general comparison variable—the ratio of the average wage of an individual market to the mean wage of the regional system of markets— and a more comprehensive analysis might seek to determine whether the comparison effect was more localised geographically, that is whether wage adjustments of this sort operate in terms of individual interurban spillovers and pattern bargains. The difficulty with this approach, however, is that comparisons may be made between different pairs or groups at different times; if so, it would be misleading to choose any one reference market or group of markets once-for-all. Given a lack of information on such intermarket interactions, it seems preferable to include a more aggregate wage as the basis for comparison, one which allows for a broader type of spillover effect.

Table 3.4. Basic OLS estimates (period of fit: 1961/I–1976/IV). The estimated standard error of each coefficient is given in parentheses.

City	b_{i0}	b_{i1}	b_{i2}	b_{i3}	b_{i4}	$\bar{R}^2$	DW	Standard error	Number of observations
Albany–Schenectady–Troy	3·24** (0·97)	−0·14 (0·23)	0·67** (0·08)	0·07 (0·20)	−0·34** (0·13)	0·54	1·59	1·71	64
Utica–Rome	2·77** (0·74)	−0·11 (0·12)	0·74** (0·05)	0·33** (0·14)	−0·38** (0·10)	0·76	1·35	1·19	64
Syracuse	2·30** (0·61)	0·03 (0·14)	0·71** (0·06)	0·51** (0·20)	−0·62** (0·14)	0·73	1·29	1·20	64
Rochester	4·42** (0·67)	−0·62** (0·20)	0·91** a (0·06)	0·15 (0·86)	−0·29** (0·12)	0·81	1·31	1·14	64
Buffalo	2·02** (0·57)	0·01 (0·08)	0·93** a (0·06)	0·47** (0·14)	−0·37** (0·19)	0·80	1·44	1·30	64
Erie	2·93** (0·59)	−0·16* (0·08)	0·78** (0·05)	0·41** (0·15)	−0·47** (0·15)	0·80	1·39	1·25	64
Youngstown–Warren	2·39** (0·89)	−0·12 (0·13)	0·91** a (0·09)	0·52** (0·22)	−0·71** (0·09)	0·79	1·68	1·86	64
Cleveland	2·77** (0·53)	−0·30* (0·17)	0·94** a (0·05)	0·41** (0·13)	−0·45** (0·16)	0·83	1·46	1·24	64
Akron	3·61** (0·69)	−0·39** (0·16)	0·75** (0·07)	0·23 (0·18)	−0·64** (0·14)	0·63	1·68	1·41	64
Canton	2·87* (0·74)	−0·14 (0·12)	0·91** a (0·08)	0·27 (0·19)	−0·79** (0·14)	0·58	1·68	1·59	64
Wheeling	1·42 (1·06)	0·03 (0·09)	0·90** a (0·10)	0·59** (0·16)	−0·62** (0·28)	0·79	1·79	1·86	64
Pittsburgh	3·17** (0·67)	−0·16* (0·08)	0·76** (0·07)	0·49** (0·14)	−0·79** (0·10)	0·85	1·55	1·38	64

Johnstown	1·99* (1·25)	−0·02 (0·11)	0·89**[a] (0·14)	0·31** (0·09)	−0·40** (0·10)	0·64	1·71	2·78	64
Altoona	4·19** (1·53)	−0·21 (0·17)	0·74** (0·12)	0·26 (0·20)	−0·68** (0·21)	0·71	1·85	2·14	64
Harrisburg	4·01** (0·73)	−0·36** (0·18)	0·69** (0·06)	0·33* (0·17)	−0·59** (0·11)	0·75	1·29	1·43	64
York	4·70** (0·51)	−0·69** (0·13)	0·79** (0·05)	0·18 (0·14)	−0·42** (0·10)	0·80	1·39	1·18	64
Lancaster	3·35** (0·53)	−0·18 (0·17)	0·66** (0·07)	0·45** (0·16)	−0·39** (0·11)	0·64	1·19	1·28	64
Reading	3·29** (0·61)	−0·36** (0·18)	0·70** (0·07)	0·24 (0·21)	−0·68** (0·17)	0·66	1·25	1·70	64
Philadelphia	2·90** (0·56)	−0·17* (0·15)	0·93**[a] (0·05)	0·40** (0·14)	−0·30** (0·09)	0·81	1·42	1·99	64
Allentown–Bethlehem–Easton	3·17** (0·63)	−0·25* (0·14)	0·74** (0·06)	0·27 (0·22)	−0·69** (0·19)	0·70	1·25	1·50	64
Wilkes-Barre–Hazleton	2·36* (1·23)	−0·04 (1·14)	0·95**[a] (0·17)	0·31 (0·27)	−0·61** (0·13)	0·65	1·29	1·35	52[b]
Scranton	3·08** (1·48)	−0·11 (0·13)	0·93**[a] (0·21)	0·35 (0·30)	−0·51** (0·12)	0·60	1·50	1·54	52[b]
Binghamton	4·37** (1·13)	−0·47** (0·26)	0·76** (0·08)	0·54** (0·21)	−0·63** (0·11)	0·67	1·49	1·55	64

* Significant at 90% level.　** Significant at 95% level.

[a] Insignificantly different from unity at 95% level.

[b] Period of fit refers to 1961/I–1973/IV.

Table 3.5. Cochrane–Orcutt iterative estimates (period of fit: 1961/I–1976/IV). The estimated standard error of each coefficient is given in parentheses.

City	b_{i0}	b_{i1}	b_{i2}	b_{i3}	b_{i4}	$\bar{R}^2$	DW	Standard error	AR(1)	Number of observations
Albany–Schenectady–Troy	2·56 (1·87)	0·13 (0·44)	0·53** (0·17)	0·03 (0·32)	-0·20** (0·10)	0·61	1·88	1·57	0·50	64
Utica–Rome	3·28* (1·65)	-0·19 (0·25)	0·67** (0·13)	0·15 (0·23)	-0·14** (0·06)	0·82	1·99	1·02	0·59	64
Syracuse	3·07** (1·37)	-0·09 (0·28)	0·69** (0·12)	0·41* (0·24)	-0·49** (0·12)	0·83	1·81	0·96	0·70	64
Rochester	5·07** (0·97)	-0·77** (0·28)	0·91** [a] (0·11)	0·04 (0·21)	-0·26** (0·12)	0·81	1·84	1·04	0·44	64
Buffalo	-0·20 (1·53)	0·54** (0·21)	0·59** (0·26)	0·54** (0·28)	-0·31** (0·13)	0·87	2·11	1·03	0·71	64
Erie	2·13** (0·98)	-0·13 (0·18)	0·81** [a] (0·12)	0·44* (0·24)	-0·15 (0·13)	0·85	1·85	1·02	0·65	64
Youngstown–Warren	0·02 (1·62)	0·18 (0·30)	1·04** [a] (0·20)	0·81** (0·36)	-0·47** (0·12)	0·80	1·89	1·16	0·73	64
Cleveland	2·14* (1·17)	-0·09 (0·24)	0·89** [a] (0·13)	0·43** (0·19)	-0·24** (0·11)	0·89	1·93	1·35	0·42	64
Akron	3·16** (1·38)	-0·20 (0·30)	0·59** (0·15)	0·36 (0·29)	-0·25* (0·13)	0·66	1·79	1·23	0·51	64
Canton	3·18* (1·81)	-0·23 (0·31)	0·86** [a] (0·19)	0·08 (0·31)	-0·27** (0·13)	0·65	1·77	1·35	0·47	64
Wheeling	0·02 (2·00)	0·16 (0·18)	0·92** [a] (0·14)	0·33 (0·25)	-0·48** (0·14)	0·72	1·84	1·42	0·48	64

Pittsburgh	1·52 (1·76)	−0·09 (0·23)	0·83** [a] (0·18)	0·61** (0·22)	−0·31** (0·15)	0·89	1·67	0·95	0·53	64
Johnstown	0·47 (2·31)	0·12 (0·14)	1·13** [a] (0·31)	0·87** (0·39)	−0·45** (0·11)	0·67	1·79	2·61	0·46	64
Altoona	3·77 (2·65)	−0·14 (0·28)	0·72** [a] (0·21)	0·43 (0·42)	−0·51** (0·12)	0·53	1·82	2·41	0·47	64
Harrisburg	4·66** (1·34)	−0·48 (0·32)	0·67** (0·12)	0·37 (0·26)	−0·52** (0·26)	0·76	1·97	1·28	0·58	64
York	5·79** (0·94)	−0·84** (0·21)	0·81** (0·11)	0·08 (0·22)	−0·07 (0·09)	0·83	1·99	1·05	0·51	64
Lancaster	4·03** (1·16)	0·13 (0·30)	0·34** (0·15)	0·42* (0·23)	−0·06 (0·07)	0·78	2·03	1·00	0·68	64
Reading	5·13** (1·47)	−0·60* (0·35)	0·67** (0·16)	0·16* (0·31)	−0·47** (0·22)	0·70	1·79	1·49	0·62	64
Philadelphia	2·74** (1·05)	−0·14 (0·17)	0·82** (0·10)	0·31* (0·19)	−0·58** (0·12)	0·85	1·97	1·05	0·72	64
Allentown–Bethlehem–Easton	5·00** (1·79)	−0·56* (0·33)	0·70** [a] (0·21)	0·03 (0·19)	−0·43** (0·11)	0·72	1·76	1·20	0·63	64
Wilkes–Barre–Hazleton	3·40 (2·55)	−0·10 (0·24)	0·83** [a] (0·31)	0·32 (0·29)	−0·27* (0·15)	0·75	1·95	1·16	0·41	52 [b]
Scranton	3·18 (2·10)	−0·11 (0·19)	0·94** [a] (0·28)	−0·30 (0·35)	−0·31** (0·13)	0·68	2·01	1·48	0·35	52 [b]
Binghamton	4·07** (1·50)	−0·44 (0·38)	0·65** (0·13)	0·62** (0·28)	−0·17* (0·10)	0·71	2·00	1·47	0·49	64

* Significant at 90% level. ** Significant at 95% level.
[a] Insignificantly different from unity at 95% level.
[b] Period of fit refers to 1961/I–1973/IV.

The results (and interpretations) obtained thus far are based on the maintained hypothesis of stable coefficients. Table 3.6 summarises our examination of this hypothesis with use of the various tests based on residuals obtained from recursive least squares estimation of equation (3.67). Even if these tests are regarded more as 'diagnostic checks' rather than exact procedures, the results in table 3.6 suggest that for the majority of markets the assumption of constant parameters is highly questionable. In these cases, almost all of the tests show significant results. As I have already noted, departures from constancy may show themselves in different ways, and the tests may not be equally powerful against the particular kind(s) of departure encountered. The 'cusum' test is significant in less than half of the regressions, whereas the more exact ψ test is significant in thirteen of the twenty-three equations. Harvey and Phillips (1977) have shown that the ψ statistic has greater power than the 'cusum' test, and provides a good test against random-walk parameter variation. Taking the increasing variance property of recursive residuals into account, one-sided

Table 3.6. Tests for parametric instability, based on recursive least squares regression.

City	Cusum	Cusum of squares [a]	ψ	H [a]	VNR [b]	HS
Albany–Schenectady–Troy	0·58	0·16	0·59	1·90	2·06	1·76
Utica–Rome	0·41	0·48**	0·97	4·40**	1·92	5·98**
Syracuse	0·62	0·45**	1·10	3·50**	1·88	4·77**
Rochester	0·92	0·29**	3·38**	6·72**	1·03**	6·92**
Buffalo	0·70	0·43**	0·94	14·18**	1·33**	7·12**
Erie	1·03*	0·26**	2·96**	2·52*	0·92**	14·22**
Youngstown–Warren	1·18**	0·33**	3·99**	9·75**	1·16**	18·76**
Cleveland	0·62	0·31**	1·87	2·80*	1·53*	9·10**
Akron	1·12*	0·22	4·64**	2·44*	1·60	8·01**
Canton	1·00	0·36**	5·22**	4·76**	0·84**	16·55**
Wheeling	0·89	0·21	2·51*	1·83	2·11	7·66**
Pittsburgh	1·29**	0·37**	3·73**	12·26**	1·36**	27·05**
Johnstown	0·74	0·29**	5·81**	3·38**	1·59	14·70**
Altoona	0·89	0·21	1·78	2·84*	2·24	1·69
Harrisburg	1·01*	0·34**	4·18**	2·57*	0·86**	18·05**
York	0·81	0·30**	1·12	2·31*	1·62	10·19**
Lancaster	1·37**	0·40**	5·89**	11·98**	1·07**	20·09**
Reading	0·62	0·38**	0·92	6·72**	1·46*	9·21**
Philadelphia	0·39	0·24*	0·35	1·89	1·90	4·40**
Allentown–Bethlehem–Easton	0·65	0·42**	2·02*	10·81**	1·56*	7·81**
Wilkes–Barre–Hazleton	0·81	0·15	2·37*	1·40	2·01	2·38*
Scranton	0·66	0·21	2·45*	2·57*	1·98	3·09*
Binghamton	1·66**	0·12	1·46	2·19*	1·12**	9·34**

* Significant at 95% level. ** Significant at 99% level.

[a] One-sided test.

[b] One-tailed test against positive autocorrelation; critical values for the 'modified' VNR are given in Theil (1971, pages 728–729).

tests were formulated for the heteroscedasticity or H test and the 'cusum' of squares test. The results are significant in almost all of the regressions and give further evidence for the presence of random-walk parameters, or at least some form of stochastic variation in the coefficient estimates. These two tests tend to be more effective when the explanatory variables in the model show a steady increase over time, which is the case here.

According to Harvey and Phillips, the VNR based on recursive residuals has a high power against models in which the parameters follow random-walk processes. Further, looking at the VNR in combination with the heteroscedasticity test can help to distinguish between random-walk parameters and serial correlation to a certain extent. A significant value of the VNR statistic together with a nonsignificant H test is indicative of a fixed parameter model with serial correlation. None of the markets shows this particular combination. On the other hand, significant values both of the VNR and heteroscedasticity tests indicates either random-walk coefficients or very strong serial correlation. This combined rejection of the null hypothesis of parametric constancy is found in every regression for which the VNR is significant.

The 'homogeneity' test for constancy, given in the last column, was calculated from a moving regression of length ten. Under the null hypothesis of parametric constancy, the homogeneity statistic is distributed as $F(25, 34)$; in all but two of the markets the assumption of a stable regression equation is rejected.

It would appear, therefore, that the estimated wage inflation model is characterised by marked parametric instability and by heteroscedasticity. Consequently it is not possible to put too much confidence in interpretations based on results obtained with a fixed parameter formulation. To be fully acceptable a wage equation must satisfy the condition that it is stable; otherwise it cannot be used either in an explanatory role or as a reliable basis for prediction. We have found clear evidence of statistical instability in the estimated wage model. The difficult issue that remains, however, is that of relating this statistical instability to economic instability. In the context of Phillips curve models, there are at least five possible sources of parametric instability: (1) misspecification of the model in the sense of excluding important explanatory variables; (2) misspecification of the functional form, particularly the substitution of linear for nonlinear relationships; (3) lack of correspondence between theoretical variables and their empirical proxies; (4) problems of aggregation; and (5) inherent flexibility in the behaviour of market participants, for example, in the formation of expectations, and in response to or in anticipation of disequilibrating shocks and controls. To the extent that the results of our limited analyses are accepted they imply that existing studies of local Phillips curves should be treated with considerable caution.

3.6 Summary

This chapter has been concerned with the specification and estimation of an extended, expectations-augmented Phillips curve model of wage inflation in subregional (urban) labour markets. The model differs from most existing studies of local wage change relationships in the inclusion of a relative wage factor designed to capture 'institutional' intermarket comparison effects in the wage determination process. A number of econometric and statistical difficulties arise in models of this sort. One problem is the specification and modelling of inflationary expectations; a 'convergence' or 'return to normality' hypothesis was used in place of the conventional adaptive expectations scheme. A second issue concerns the validity of the assumed constancy of the estimated wage equation. As far as the author is aware, this is the first study to give explicit attention to this assumption in the analysis of local markets.

The fixed parameter estimates of the model failed to give support for an inverse wage change–unemployment relationship, and the evidence for the 'partial convergence' hypothesis for inflationary expectation was also rather poor. Adjustment to inflationary expectations (as measured by the 'normal rate' of inflation) was found to be complete in only a minority of the markets, whereas partial adjustment to past price inflation seems to be the dominant response. Thus inflation neutrality as postulated in the 'natural rate' or accelerationist theory does not appear to hold universally at the local level. In contrast, the relative wage hypothesis received overwhelming empirical support.

All of these results must be qualified, however, by the considerable temporal instability detected in most of the individual market equations. Several tests based on recursive residuals suggest that this variation follows a random-walk structure. Certainly, in contrast to studies of the stability of the macro Phillips curve, which have tended to test for discrete shifts and pivots in the relationship, these results suggest that significant stochastic variation is likely to be present. At this point in the analysis I put this forward only as a preliminary finding which clearly requires further investigation. The next step must be to model the trajectories of the parameters to determine the true pattern of variation, and then seek structural explanations accordingly. Econometric work of this nature would seem to be an important complement to the general debate on the Phillips curve, with its proliferation of alleged 'intruders' to account for its apparent demise and the theoretical and conceptual considerations invoked to explain why it appears to hold in some periods but not others.

References

Akerhof G A, 1969 "Relative wages and the rate of inflation" *Quarterly Journal of Economics* **83** 353-374

Albrecht W P, 1966 "The relationship between wage changes and unemployment in metropolitan and industrial labour markets" *Yale Economic Essays* **6** 279-341

Albrecht W P, 1970 "Intermarket and intertemporal differences in the relationship between the rate of change of wages and unemployment" *Mississippi Valley Journal of Business and Economics* **6** 51-58

Archibald G C, 1969 "The Phillips curve and the distribution of unemployment" *American Economic Review* **59** 124-134

Bergstrom A R, Catt A J L, Peston M H, Silverstone B D J, 1978 *Stability and Inflation* (John Wiley, New York)

Black S W, Kelejian H H, 1972 "The formulation of the dependent variable in the wage equation" *Review of Economic Studies* **39** 55-59

Brechling F P R, 1967 "Trends and cycles in British regional unemployment" *Oxford Economics Papers (New Series)* **19** 1-21

Brechling F P R, 1973 "Wage inflation and the structure of regional unemployment" *Journal of Money, Credit and Banking* **5** 355-379

Brinner R E, 1977 "The death of the Phillips curve reconsidered" *Quarterly Journal of Economics* **91** 389-418

Brittan S, 1975 *Second Thoughts on Full Employment Policy* Centre for Policy Studies, London

Brown R L, Durbin J, Evans J M, 1975 "Techniques for testing the constancy of regression relationships over time (with discussion)" *Journal of the Royal Statistical Society B* **37** 149-192

Chow G C, 1960 "Tests of equality between subsets of coefficients in two linear regressions" *Econometrica* **28** 591-605

Cooley T F, Prescott E, 1973 "Varying parameter regression: a theory and some applications" *Annals of Economic and Social Measurement* **2** 463-474

Corina J, 1972 *Labour Market Economics: A Short Survey of Recent Theory* (Heinemann, London)

Cowling K, Metcalf D, 1967 "Wage-unemployment relationships: a regional analysis for the UK, 1960-1965" *Bulletin of the Oxford University Institute of Economics and Statistics* **34** 31-39

Crossley J R, 1966 "Collective bargaining, wage structure and the labour market in the United Kingdom" in *Wage-Structure in Theory and Practice* Ed. E M Hugh-Jones (North-Holland, Amsterdam) pp 157-234

Cullen D E, 1956 "The inter-industry wage structure" *American Economic Review* **46** 353-369

Dunlop J T, 1957 "The task of contemporary wage theory" in *The Theory of Wage Determination* Ed. J T Dunlop (Macmillan, London) pp 3-27

Durbin J, 1969 "Tests for serial correlation in regression analysis based on the periodogram of least squares residuals" *Biometrika* **56** 1-15

Eckstein O, Brinner, R, 1972 "The inflation process in the United States" *Joint Economic Committee* **72-171-0** 1-46

Eckstein O, Wilson T A, 1962 "The determination of money wages in American industry" *Quarterly Journal of Economics* **76** 379-414

Farley J U, Hinich M, McGuire T W, 1975 "Some comparisons of tests for a shift in the slopes of a multivariate linear time series model" *Journal of Econometrics* **3** 297-318

Friedman M, 1966 "What price guideposts" in *Guidelines: Informal Contracts and the Market Place* Eds G P Shultz, R Z Aliber (University of Chicago Press, Chicago, Ill.) pp 17-39, 55-61

Friedman M, 1968 "The role of monetary policy" *American Economic Review* **58** 1-17

Friedman M, 1975 *Unemployment versus Inflation?* Institute of Economic Affairs, London

Friedman M, 1977 *Inflation and Unemployment: The New Dimensions of Politics* Institute of Economic Affairs, London

Garbade K, 1977 "Two methods for examining the stability of regression coefficients" *Journal of the American Statistical Association* **72** 54-63

Gordon J J, 1972 "Wage-price controls and the shifting Phillips curve" *Brookings Papers on Economic Activity* **2** 385-430

Gordon J J, 1977 "Recent developments in the theory of unemployment and inflation" in *Inflation Theory and Anti-Inflation Policy* Ed. E Lundberg (Macmillan, London) pp 42-71

Gray M R, Parkin J M, Sumner M T, 1975 *Inflation in the United Kingdom: Causes and Transmission Mechanisms* Inflation Workshop Paper 7518, University of Manchester, Manchester, England (mimeo)

Haddy P, Tolles N A, 1957 "British and American changes in inter-industry wage structure under full employment" *Review of Economics and Statistics* **39** 408-414

Hamermesh D S, 1970 "Wage bargains, threshold effects and the Phillips curve" *Quarterly Journal of Economics* **84** 501-517

Hart R A, MacKay D I, 1977 "Wage inflation, regional policy and the regional earnings structure" *Economics* **44** 267-281

Harvey A C, Collier P, 1975 "Testing for functional misspecification in regression analysis" paper presented at *3rd World Congress of Econometric Society, Toronto* (available from University Library, University of Warwick, England)

Harvey A C, Phillips G D A, 1977 *Testing for Stochastic Parameters in Regression Models* DP-35, Department of Quantitative Social Science, University of Kent, Canterbury, Kent, England

Hicks J R, 1932 *The Theory of Wages* second edition (Macmillan, London)

Hicks J R, 1974 *The Crisis in Keynesian Economics* (Basil Blackwell, Oxford)

Holmes J M, Smyth D J, 1970 "The relationship between unemployment and excess demand for labour: an examination of the theory of the Phillips curve" *Economica* **37** 311-315

Jackson J M, 1972 *The Relation between Unemployment and the Rate of Change of Occupational Wage Rates in Local Labour Markets* unpublished PhD thesis, Department of Economics, University of Georgia, Athens, Ga

Kane E J, Malkiel B G, 1976 "Autoregressive and non-autogressive elements in cross-section forecasts of inflation" *Econometrica* **44** 1-16

Kaun D E, Spiro M H, 1970 "The relation between wages and unemployment in US cities, 1955-1965" *The Manchester School* **38** 1-14

Kerr C, 1957 "Wage relationships—the comparative impact of market and power forces" in *The Theory of Wage Determination* Ed. J T Dunlop (Macmillan, London) pp 173-193

Keynes J M, 1936 *The General Theory of Unemployment, Interest and Money* (Macmillan, London)

King L J, Forster J, 1973 "Wage rate change in urban labour markets and inter-market linkages" *Papers of the Regional Science Association* **34** 183-196

Lahiri K, 1976 "Inflationary expectations: their formation and interest rate effects" *American Economic Review* **66** 124-131

Laidler D E W, Parkin J M, 1975 "Inflation: a survey" *Economic Journal* **85** 741-809

Lerner S W, Marquand J, 1963 "Regional variations in earnings, demand for labour and shop stewards' combine committees in the British engineering industry" *The Manchester School* **31** 261-296

Lipsey R G, 1960 "The relationship between unemployment and the rate of change of money wage rates in the United Kingdom, 1862-1957. A further analysis" *Econometrica* **27** 1-31

MacKay D I, Hart R A, 1975 "Wage inflation and the regional wage structure" in *Contemporary Issues in Economics* Eds M Parkin, A R, Nobay (Manchester University Press, Manchester) pp 88-119

Maher J E, 1961 "The wage pattern in the United States" *Industrial and Labour Relations Review* **15** 1-20

Marcus R G, Reed J D, 1974 "Joint estimation of the determinants of wages in sub-regional labour markets in the United States, 1961-1972" *Journal of Regional Science* **14** 259-267

Martin R L, 1979 *Wage Inflation in Urban Labour Markets: Causes, Structure, and Transmission* (Pion, London) forthcoming

Metcalf D, 1971 "The determinants of earnings changes: a regional analysis for the UK, 1960-1968" *International Economics Review* **12** 273-282

Muth J F, 1960 "Optimal properties of exponentially weighted forecasts" *Journal of the American Statistical Society* **55** 299-306

Muth J F, 1961 "Rational expectations and the theory of price movements" *Econometrica* **29** 315-335

Muth R, 1968 "Differential growth among large US cities" WP-CWR-15, Institute for Urban and Regional Studies, Washington University, Washington

Okun A M, 1973 "Upward mobility in a high-pressure economy" *Brookings Papers on Economic Activity* **1** 207-252

Papola T S, Bharadwaj V P, 1970 "The dynamics of industrial wage structure: an inter-country analysis" *Economic Journal* **80** 72-90

Phelps E S, 1965 "Anticipated inflation and economic welfare" *Journal of Political Economics* **73** 1-17

Phelps E S, 1967 "Phillips curves, expectations of inflation and optimal unemployment over time" *Economica* **34** 254-281

Phelps E S, 1968 "Money wage dynamics and labour market equilibrium" **76** 678-711

Phelps E S, and others, 1970 *The Microeconomic Foundations of Employment and Inflation Theory* (W W Norton, New York)

Phillips A W, 1958 "The relation between unemployment and the rate of change of money wage rates in the United Kingdom, 1861-1957" *Econometrica* **25** 283-299

Quandt R E, 1960 "Tests of the hypothesis that a linear regression system obeys two separate regimes" *Journal of the American Statistical Association* **55** 873-880

Reder M W, 1958 "Wage determination in theory and practice" in *A Decade of Industrial Relations Research* IRRA-19, Eds N Chamberlain, F C Pierson, T Wolfson (Industrial Relations Research Association, New York) pp 64-97

Rees A R, Hamilton M T, 1967 "The wage-price-productivity perplex" *Journal of Political Economics* **75** 63-70

Reynolds L G, 1951 *The Structure of Labour Markets* (Harper and Row, New York)

Reynolds L G, Taft C H, 1956 *The Evolution of Wage Structure* (Yale University Press, New Haven, Conn.)

Rosenberg B, 1973a "A survey of stochastic parameter regression" *Annals of Economic and Social Measurement* **2** 381-397

Rosenberg B, 1973b "The analysis of a cross-section of time series by stochastically convergent parameter regression" *Annals of Economic and Social Measurement* **2** 399-428

Ross A M, 1948 *Trade Union Wage Policy* (University of California Press, Berkeley, Calif.)

Ross S A, Wachter M L, 1973 "Wage determination, inflation and the industrial structure" *American Economic Review* **63** 675-692

Rothschild K W, 1971 "The Phillips curve and all that" *Scottish Journal of Political Economy* **18** 245-280

Rowley J C R, Wilton D A, 1973 "Quarterly models of wage determination: some new efficient estimates" *American Economic Review* **63** 380-389

Sage A, Melsa J, 1971 *Estimation Theory with Applications to Communications and Control* (McGraw-Hill, New York)

Sargent J, 1971 "A note on the 'accelerationist' controversy" *Journal of Money, Credit and Banking* **3** 721-725

Sarris A H, 1973 "A Bayesian approach to estimation of time-varying regression coefficients" *Annals of Economic and Social Measurement* **2** 501-524

Smith V, Patton R, 1971 "Sub-market labour adjustment and economic impulses: a note on the Ohio experience" *Regional Studies* **5** 91-93

Smith V, Smith S, 1972 "A note on municipal Phillips' curves" *The Annals of Regional Science* **6** 79-83

Theil H, 1971 *Principles of Econometrics* (John Wiley, New York)

Theil H, Wage S, 1964 "Some observations on adaptive forecasting" *Management Science* **10** 198-206

Thirlwall A P, 1969 "Demand disequilibrium in the labour market and wage rate inflation in the United Kingdom" *Yorkshire Bulletin of Economic and Social Research* **21** 65-76

Thirlwall A P, 1970 "Regional Phillips curves" *Bulletin of the Oxford University Institute of Economics and Statistics* **32** 19-32

Throop A W, 1968 "The union-non union wage differential and cost-push inflation" *American Economic Review* **58** 79-99

Tobin J, 1972 "Inflation and unemployment" *American Economic Review* **62** 1-18

Trevithick J A, Mulvey C, 1975 *The Economics of Inflation* (Martin Robertson, London)

Turner H A, 1957 "Inflation and wage differentials in Great Britain" in *The Theory of Wage Determination* Ed. J T Dunlop (Macmillan, London) pp 123-135

Turnovsky S J, 1970 "Empirical evidence of the formation of price expectations" *Journal of the American Statistical Association* **65** 1441-1454

Turnovsky S J, Wachter M L, 1972 "A test of the 'expectations hypothesis' using directly observed wage and price expectations" *Review of Economics and Statistics* **54** 47-54

Wachter M L, 1970a "Cyclical variation in the inter-industry wage structure" *American Economic Review* **60** 75-84

Wachter M L, 1970b "Relative wage equations for US manufacturing industries" *Review of Economics and Statistics* **52** 405-410

Webb A E, 1974 "Unemployment, vacancies and the rate of change of earnings: a regional analysis" *Regional Papers III, National Institute of Economics and Social Research* Cambridge University Press, London, pp 1-49

Weissbrod R S P, 1974 *Spatial Diffusion of Relative Wage Inflation* unpublished PhD thesis, Department of Geography, Northwestern University, Evanston, Ill.

The autocorrelation structure of central-place populations in Southern Germany

A C Gatrell

4.1 Introduction

Spatial structure, spatial process, and spatial prediction form a triumvirate that governs much geographical inquiry and is bound together firmly by the key concept of spatial autocorrelation. Since discovering in the 1960s that their traditional focus on spatial interdependence could be formalised in terms of autocorrelation, geographers have made use of it at a descriptive level (Hodge and Gatrell, 1976), at an inferential level (Dacey, 1966a; Cliff and Ord, 1973; Cliff et al, 1975), and at a theoretical level (Curry, 1967; 1977).

In modern time-series analysis (Box and Jenkins, 1970), autocorrelation plays an important part in the statistical description of series structure, in the identification of stochastic generating processes, and hence in the forecasting of the behaviour of future series. Recent attempts have been made to transfer this role into the spatial domain and the present chapter is part of this research tradition (Tobler, 1975). It summarises and extends results from a larger study (Gatrell, 1976) that encompassed structure, process, and prediction (prediction being performed 'instantaneously', as a function of size elsewhere on the map, rather than as a function of historical spatial structure). Results from, and problems encountered in, an empirical investigation of spatial autocorrelation in a population map of Southern Germany are reported. An analysis of spatial structure predominates, with tentative attempts to identify the underlying spatial processes.

4.2 Geographical approaches to spatial autocorrelation [1]

It is perhaps not too invidious to draw a distinction between two traditions of geographical research on autocorrelation. First, there has arisen the formulation of tests for spatial dependence, an approach based initially on contiguity (Dacey, 1965; 1966a) and generalised by Cliff and Ord (1973) to incorporate any *a priori* specification of spatial relationships. This specification is made in terms of a weight matrix W, in which w_{ij} denotes the influence of j on i [in terms of adjacency in a system of areal units, for instance, or perhaps as a function of distance, given a point pattern. Cliff and Ord (1973) themselves illustrate various specifications of W, and

---

[1] A review of nongeographical investigations of spatial autocorrelation may be found in Gatrell (1976, pages 56-58). The discussion in Agterberg (1974) is more accessible, though confined to geological science.

Bannister (1975) uses five alternative schemes in a study of town growth in Southern Ontario].

Autocorrelation statistics have been obtained for all levels of measurement and for nonlattice arrangements as well as lattice data. The distribution theory has been quite fully worked out. Consequently, use of the statistics for inferential purposes has been demonstrated on numerous occasions and they have made a recent entry into introductory textbooks (for example, Ebdon, 1977). Further extensions have enabled estimates of autocorrelation to be obtained for observations separated at different distances, and yield a correlogram when autocorrelation is plotted against lag (Cliff et al, 1975; Hodder and Orton, 1976, pages 179–183).

Use of these statistics may be subject to two constraints. First, given large n, the dimensions of $\mathbf{W}$ may be too great for computer storage. Second, one may be interested in some applications in obtaining directional information about spatial structure. A second research tradition has explored the possibilities of two-dimensional autocorrelation functions and can cope with both problems. Conceptually the function is obtained in a way analogous to that which yields the time-series correlogram, the map being shifted or lagged in all directions. An estimate of autocovariance in a bounded square lattice is provided by

$$c(k, l) = \frac{1}{n-k}\frac{1}{n-l}\sum_{i=1}^{n-k} \sum_{j=1}^{n-l} (z_{ij} - \overline{z})(z_{i+k, j+l} - \overline{z}) \,, \qquad (4.1)$$

where (k, l) denotes the lag and $\{z_{ij}\}$ denotes the spatial series of interest. The autocorrelations, $r(k, l)$, are obtained from division[2] by an estimate of the variance, $s^2 = c(0, 0)$. Periodicities in the data are revealed by peaks and troughs in the autocorrelation function, whereas orientation effects appear as directional biases.

There have been few empirical geographical investigations of spatial structure that use this approach. Nordbeck and Rystedt (1972), taking population data in Southern Sweden, examined the pattern exhibited by the peaks on the autocorrelation function. Unsuccessful attempts were made to discern order in the arrangement of those peaks, periodicities being suspected *a priori* but not confirmed. Haining (1976), largely concerned with specifying plausible theoretical models for spatial data, estimated autocorrelations for short lags ($l, k \leqslant 2$) from crop yield data in an attempt to discriminate between different models. Other geographical studies have dealt with the more common situation of nonlattice data. For instance Sibert (1975) investigated the statistical structure of urban land values, the autocorrelation functions revealing the clear directional biases that one would expect from a variable constrained in its behaviour by the orientation of the street network.

[2] If the maximum lag in both directions is m, this division yields a square autocorrelation matrix of order $(2m+1)$. However, one half is redundant since the matrix is skewwise symmetric.

4.3 Settlement theory

Classical central-place theory (Christaller, 1933; 1966) seeks a deductive, economic explanation of the size and spacing of settlements. Since all settlements act as supply centres for the provision of goods and services to a surrounding market area they are referred to as 'central places'. Given a set of initial conditions or environmental assumptions and a set of laws of consumer and retailer behaviour[3], Christaller derives central-place landscapes (an example of which is shown in figure 4.1) characterised by the hexagonal spatial arrangement of settlements of different size. Small central places, supplying goods required frequently by consumers, are spaced more closely than those offering higher-order goods. A periodic landscape is predicted, with population peaks recurring regularly in space. The two-dimensional spectrum calculated from such a theoretical landscape highlights this periodicity, as one would expect (Rayner and Golledge, 1972).

Christaller attempted in the early 1930s to verify his theoretical predictions in Southern Germany, a field laboratory that has attracted others as a result of his original contribution. There has, however, been no published study that treats his population map as a realisation of a spatial stochastic process and examines the autocorrelation structure, a gap that this chapter attempts to fill.

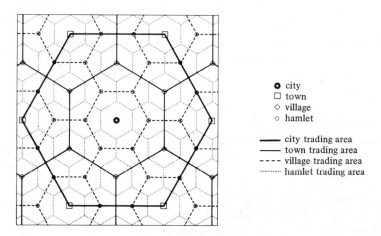

Figure 4.1. A theoretical Christaller central-place landscape.

[3] These environmental assumptions and behavioural postulates cannot be rehearsed here. The reader unfamiliar with them is directed to introductory treatments in Haggett (1975), Kolars and Nystuen (1974), or Berry (1967). I should stress that the brief discussion of central-place theory is intended to set the ensuing empirical analysis in context, no rigorous attempt being made here to veryify the theoretical predictions; this could be done by deriving a theoretical autocorrelation function and comparing the empirically obtained correlation structure with this theoretical model.

Theoretical settlement geography has, in the past fifteen years, undergone a transformation from a deterministic mode to a stochastic one [see, for instance, Dacey (1966a; 1966b), Curry (1964; 1967; 1977), and Hudson (1969)]. In particular, Curry has broken new ground by developing theory in terms of autocovariance functions and spectra (Curry, 1967; 1977). In a recent paper he specifies two distance decay functions, one a 'settlement encouragement' factor, the other describing the extent to which settlements inhibit each others' growth. Given also the mean settlement density and the autocovariance function of settlement density, and by transforming to the frequency domain, Curry obtains the interesting and testable prediction that the settlement spectrum is a square-root function of frequency (Curry, 1977).

4.4 "But first the data analysis!" [4]
Christaller (1933), in an appendix to his seminal work, provided data (from 1925 and 1927 censuses) on the populations of central places in five regional systems (Münich, Nuremberg, Strassburg, Stuttgart, and Frankfurt). Although it proved impossible to locate every central place, 98% of the places were digitised, coordinates being obtained from maps of scale 1 : 50000. One hundred randomly selected settlements were discarded in order that spatial prediction could be performed on this sample of 'missing' observations. This left 1581 central places for study. These varied in size from small villages of four-hundred people to regional capitals with populations of several hundred thousand.

Spatial stationarity is a requirement if the autocorrelation function is to be used for prediction of population throughout the study area. Thus an investigation should be made to see if the mean, variance, and higher-order moments are positionally invariant. However, stationarity per se cannot be examined in a single realisation of a stochastic process, so the tests described below deal, strictly speaking, with self-stationarity.

Given the substantial size variation among settlements, a double logarithmic transformation was performed prior to testing for stationarity. The study area was partitioned into square cells and the mean, variance, and covariance estimated for cells containing at least twenty-five observations. The covariance was estimated with the use of a generalised Moran statistic (Cliff and Ord, 1973) with weights specified in terms of an inverse function of distance. Given the spatial distribution of the three statistics, a further contiguity ratio was computed in order to test for spatial association in these distributions. Intuitively, in order to sustain the hypothesis of stationarity, there should be a random spatial distribution of cell statistics. If high or low values of the mean, variance, and covariance cluster spatially, nonstationarity is indicated. For example, at a cell size of 425 coordinate units the mean of the undifferenced data

[4] This was Besag's closing plea in a recent essay on the statistical spatial analysis of irregularly distributed data (Besag, 1975, page 194).

shows a clear visual trend (figure 4.2) and this is confirmed by the contiguity test statistic, I_1 (table 4.1). [Although the map (figure 4.2) codes the means into two classes, this is a cartographic convenience and it should be stressed that the contiguity statistic was based on the interval data provided by the numerical estimates. Note too the related test of spatial stationarity (homogeneity) provided by Naidu (1970). This requires a large number of cells and a large sample of observations in each and could not be used here.]

It is important to test for stationarity at a variety of spatial scales, since the data may be stationary at one scale but not at another. Ten cell sizes were examined and the results indicated nonstationarity in the mean at all but the finest resolution levels. This reflects the concentration of population in the Rhine Valley. The variance is generally homogeneous spatially but the covariance is strongly nonstationary at virtually all scales (table 4.1).

Spatial differencing is one method of removing a trend in the mean. The procedure is to form an average of observations in the local neighbourhood of point i and to subtract that average from the value at i to obtain a 'first spatial difference':

$$\nabla z_i = z_i - \sum_{j \in J} w_{ij} z_j \,, \tag{4.2}$$

where J denotes the set of neighbours. The choice of neighbourhood is flexible for nonlattice data, there being several ways of specifying neighbours (Tobler, 1975). An unweighted average of the six nearest neighbours, where $w_{ij} = \frac{1}{6}$, was used here, this representing the obvious definition of neighbourhood in a central-place framework [5].

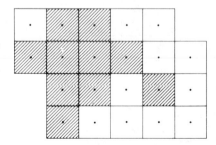

Figure 4.2. Spatial distribution of cell means, undifferenced data (shading denotes that cell mean exceeds median of all values).

[5] It would be worth investigating other schemes in order to estimate how the selection of neighbourhood influences the resulting pattern of differences. Spatial differencing has been employed in other contexts; see, for example, work by Martin (1974) on its use in regression analysis, and Lébart (1969) on factor analysis. Professor Tobler, in conversation (April, 1976), indicated the similarity between spatial differencing and the Laplacian operator.

Table 4.1. Results of the stationarity tests.

Cell size	N	Undifferenced data						First differences					
		I_1	z_1	I_2	z_2	I_3	z_3	I_1	z_1	I_2	z_2	I_3	z_3
275	38	0·00021	1·834	−0·00241	1·657	0·02021	3·180**	−0·05079	−1·600	−0·00222	1·670	0·00318	2·034*
300	37	0·00016	1·846	−0·00386	1·580	0·01077	2·547*	−0·03172	−0·260	−0·00029	1·816	−0·00809	1·301
325	32	0·00017	1·941	−0·00146	1·844	0·05029	4·942**	−0·03750	−0·314	0·00090	1·985*	0·00197	2·049*
350	28	0·00020	1·984*	−0·00465	1·726	0·01201	2·613**	−0·05586	−1·003	0·00054	2·002*	0·00304	2·135*
375	28	0·00018	1·983*	−0·00968	1·458	0·06038	5·190**	−0·04948	−0·663	−0·00094	1·923	−0·00382	1·770
400	26	0·00009	2·025*	−0·00849	1·592	−0·04342	−0·173	−0·02027	0·997	0·00121	2·083*	−0·00120	1·960*
425	21	0·00020	2·081*	−0·01945	1·266	0·01805	2·820*	−0·08352	−1·389	−0·00288	1·953	−0·00007	2·069*
450	18	0·00011	2·240*	−0·00388	2·088*	0·02223	3·080**	0·00941	2·593**	0·00006	2·238*	−0·00358	2·099*
475	19	0·00015	2·225*	−0·01216	1·734	0·02428	3·189**	−0·03286	−0·906	−0·00155	2·157*	0·00203	2·300*
500	17	0·00013	2·272*	−0·00593	2·053*	0·04401	3·864**	−0·06628	−0·137	−0·00063	2·245*	0·00289	2·372*

Notes: N denotes the number of cells at each scale.
I_1, I_2, and I_3 denote the generalised Moran statistics calculated on the cell means, variances, and covariances, respectively; z_1, z_2, and z_3 are the corresponding standard normal deviates.
* denotes significance at 0·05 level; ** denotes significance at 0·01 level.

Once the differences were computed, the hypothesis of stationarity in the mean and variance was retested. Visual inspection of a map of cell means (figure 4.3) suggests that differencing has removed the trend in the mean, and the Cliff–Ord test confirms this at all scales but one (table 4.1). Although the data can be considered stationary in the mean, the variances, curiously, show evidence of autocorrelation at several scales. The individual cell covariances, although still nonstationary, are less autocorrelated than they were when the undifferenced data was considered.

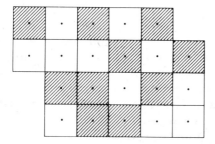

Figure 4.3. Spatial distribution of cell means, first differences.

4.5 The estimation of a spatial autocorrelation function from nonlattice data

Granger (1969; 1975) and Agterberg (1970) have both described procedures for obtaining autocorrelation functions from irregularly distributed spatial data. Sibert (1975) has provided an efficient computer routine which can handle large data sets (for instance, his series comprised over 20000 observations). Briefly, the routine sorts the data by increasing values of the U coordinate and by increasing V values in the case of tied Us. A sampling interval and maximum lag are specified. Differences in the U coordinates and V coordinates for the first pair of observations are computed and if they are within the maximum lag distance the cross product of the variate values (mean removed) is added to the appropriate cell of a cross-products matrix. Differences in U and V are checked for successive pairs of observations; if the V difference exceeds the maximum lag the cross product for that pair is ignored, whereas if the U difference exceeds the maximum lag the routine proceeds to the next observation. A record is kept of the number, $n(k, l)$, of cross products used at any lag (k, l) and the cross products are divided by $n(k, l)$ to yield an auto-covariance $c(k, l)$. Division by the variance yields $r(k, l)$.

Because of the irregular spatial arrangement, $r(0, 0)$ is unlikely to be unity. It is obtained from the cross products of all pairs of observations that lie within one sampling unit of each other. Thus the extent to which $r(0, 0)$ is less than unity will be in part a function of the sampling interval,

Δx [6]. As Δx increases, observations at greater distances will be included in the autocorrelation at lag $(0, 0)$ and, since spatial association generally declines with distance, $r(0, 0)$ will probably decrease.

Selection of the sampling interval is critical, since this determines what variations in the data are resolved. A sampling interval of Δx units means that variations occurring at intervals greater than or equal to $2\Delta x$ will be resolved. Choice of sampling interval was dictated here by the theoretical distance, 7 km, between nearest neighbours (Christaller, 1966, pages 161–162). Central-place theory would not lead us to expect any useful information with $\Delta x < 3 \cdot 5$ km. The sampling interval was subsequently increased to resolve variations in population size relating to orders of central places further up the hierarchy. However, increasing the sampling interval acts as a smoothing operation, which filters out information about high-frequency oscillations. These may be important if one is aiming for optimal spatial prediction of the population of small central places.

Other methodological comments concern the graphical representation of the autocorrelation functions, the question of statistical significance, and the treatment of directional variations in spatial dependence. Spatial autocorrelation functions do not lend themselves to visual presentation as readily as correlograms and the approach here is to depict only statistically significant coefficients. Transects may also be drawn along any axis.

The sampling theory of autocorrelation coefficients in one dimension has been treated in some detail (Bartlett, 1946; Kendall, 1976) and it is possible to place confidence limits around the estimates. There is no distribution theory as yet for spatial estimates and this is unfortunate since we need to be able to say with confidence that peaks in the autocorrelation function are significant and do not merely represent sampling fluctuations. The main problem with obtaining an expression for the variance of r_k is that it depends on the theoretical autocorrelations and so the underlying stochastic process has to be assumed. However, the variance tends to $1/n$ as the autocorrelations tend to zero and as n increases. If we assume a normal distribution and that ρ_k (the theoretical autocorrelation coefficient) is zero, then $r_k/s(r_k)$ may be evaluated as a standard normal deviate (Box and Jenkins, 1970, page 178).

The approach taken here was to write

$$\mathrm{var}[r(k, l)] = \frac{1}{n(k, l)} \tag{4.3}$$

and to obtain the z score

$$z(k, l) = r(k, l)n(k, l)^{\frac{1}{2}} . \tag{4.4}$$

---

[6] This is called a 'nugget effect' in the *Theory of Regionalised Variables* (Matheron, 1971). It "may reflect high discontinuities or micro-regionalisations at a scale smaller than that of the sampling grid. More often it represents an integration of the errors of measurement" (Huijbregts, 1975, page 44). Sayn-Wittgenstein (1970, page 250) presents examples of spatial correlograms, some showing a clear nugget effect.

This assumes a normal sampling distribution for $r(k, l)$ and an expected value of zero for the coefficient at any lag. It further assumes, unrealistically, that there is no covariance among the autocorrelation estimates at neighbouring lags (it is well-known in time-series analysis that successive values of r_k may be correlated, thus complicating the identification of the underlying stochastic process from the empirical auto-correlation function). These are strong assumptions and there are serious risks of a Type-one error, since the true variance of the estimate is likely to be higher than that used here.

A second question concerning the spatial autocorrelation function is the existence of directional bias. If this can be put down to sampling fluctuation we can assume that the autocorrelation function is isotropic, a function only of distance and not direction. If not, however, directional information should be retained since it will be important in spatial prediction.

As a rough guide to directional dependence, transects were drawn through the two-dimensional autocorrelation function and a test made of the null hypothesis that the transects come from identical 'populations'. A one-way analysis of variance looked at the ratio of variation between transects to variation within any transect. This was a rather crude test, since assumptions concerning homogeneity of variance and independence of observations [the $r(k, l)$] could not be completely justified. At the chosen sampling interval, the alternative hypothesis of anisotropy was accepted ($F = 5 \cdot 30$, $\alpha < 0 \cdot 01$) but when the first differencing filter was applied the directional bias was removed ($F = 2 \cdot 00$, $\alpha > 0 \cdot 05$).

4.6 The empirical autocorrelation structure

Any shaded cells in the representation of the autocorrelation function (figure 4.4) indicate that the coefficients at those lags are significant at the $0 \cdot 05$ level [7]. Clearly there is a 5% probability that any of these coefficients is significant purely on a chance basis. In fact, we would expect that 24 of the 481 distinct coefficients (with a maximum of fifteen lags in North–South and East–West directions) would be significant simply because of random fluctuation (fifteen lags were chosen as the maximum, since further lags contained too few cross products to give stable estimates). A similar argument is voiced by Nelson (1973, pages 74–75) who cautions not to impute too much meaning to each individual peak or trough in the autocorrelation function.

That 74 or 15% of the coefficients are statistically significant for the undifferenced data leads us to reject the null hypothesis that, at this scale, the population of Southern Germany is a white-noise spatial series. There is significant structure in the map, dependence persisting more in a

[7] Copies of the autocorrelation functions showing the numerical estimates may be obtained from the author.

North–South direction than in other directions. Transects, representing correlograms along different directions, clarify this (figure 4.5). No obvious periodicity is present, the pattern being one of weak, damped oscillations with the exception of the North–South direction. The transects have little in common other than a significant correlation at lag 1, which corresponds to a distance of between 3·5 km and 10 km.

Averaging the coefficients at similar lags (figure 4.6) masks directional variations but indicates the general behaviour of autocorrelation with increasing lag. An approximately exponential decline is evident. The same is true when the sampling interval is increased to filter out variations that occur at short distances. The sampling interval was increased by $3^{1/2}$

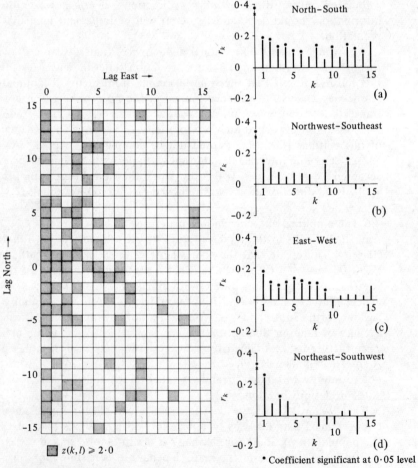

Figure 4.4. Spatial autocorrelation function: undifferenced data, sampling interval twenty-one units.

Figure 4.5. Transects through figure 4.4.

• Coefficient significant at 0·05 level

since central places at the next level of the hierarchy are separated from each other by $3^{1/2}d$, where d denotes the spacing of lowest-order places. A similar pattern of exponential decline was revealed [figure 4.6(b)]. With a further increase in Δx by a factor of $3^{1/2}$ the pattern remained the same [figure 4.6(c)] but the correlogram fails to highlight the band of significant negative coefficients, at lags corresponding to a spacing of about 125 km, that appeared in the two-dimensional autocorrelation function (Gatrell, 1976, pages 124–126).

The autocorrelation function for first spatial differences was estimated only at the lowest scale. The removal of the trend in the mean has created a very different structure (figure 4.7), one characterised by a cluster of significant negative coefficients at short lags, together with a

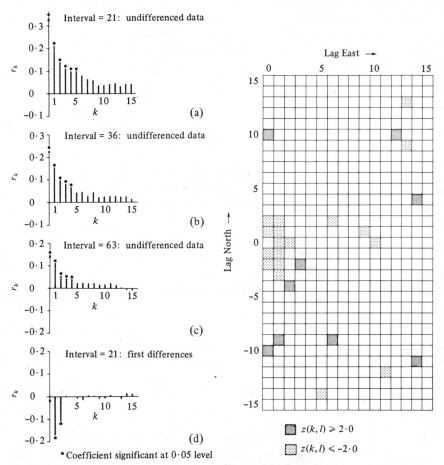

Figure 4.6. Averaged autocorrelation functions (correlograms).

Figure 4.7. Spatial autocorrelation function: first differences, sampling interval twenty-one units.

few outliers. The negative correlation, corresponding to a distance of between $3 \cdot 5$ km and $12 \cdot 5$ km, is highlighted in the averaged correlogram [figure 4.6(d)].

The presence of negative autocorrelations at short distances is of great interest in the context of central-place theory, since these lags correspond approximately to the spacing of low-order central places. A competitive effect is uncovered which was hidden in the undifferenced data. Effects such as this are common in ecological situations where plants compete for territory (Sayn-Wittgenstein, 1970, pages 247-249). Ecological theory rules out the existence of a very large plant next to another large plant, and similarly central-place theory rules out the existence of a large town next to another large town. Constraints operate in both situations to produce the structure found in the autocorrelation function.

Mention should be made of an alternative interpretation of the negative coefficients at short lags. The overdifferencing of a white-noise time series produces a correlogram which may have a structure similar to that observed here [Anderson (1976, chapter 11) gives several examples. It is shown readily that differencing white noise leads to $\rho_1 = -0 \cdot 5$, where ρ_1 denotes the theoretical autocorrelation at lag 1]. The fact that the trend was not overwhelming supports the possibility of overdifferencing. Nonetheless, the stationarity tests indicated that the trend was significant. Anderson (1976, page 110) suggests that an increase in the variance of a series after differencing is an indication of overdifferencing. However, the estimated variance of the undifferenced data was $s_z^2 = 0 \cdot 12155 \times 10^{-2}$, whereas that of the spatially differenced series was $s_{\triangledown z}^2 = 0 \cdot 11887 \times 10^{-2}$. This suggests that overdifferencing is not a problem here.

4.7 Tentative process identification
There are two possible routes that may be followed in the search for stochastic generating processes. A deductive approach specifies a model and obtains a theoretical spectrum or autocovariance function, which is then evaluated in the light of empirical findings. The time-series approach of Box and Jenkins (1970), on the other hand, identifies a model or generating process from the empirical correlation structure and proceeds to estimate parameters, check the model, and make forecasts.

The first approach has not been given detailed attention in central-place studies, although Gatrell (1976) has sketched brief outlines of three models that may be expected to yield periodic spatial series. Clearly it is simpler to write such models for central places spaced along a line (Dacey et al, 1974) although there are problems even with this (Haining, 1977b). These difficulties notwithstanding, the summation of m harmonics describes a simple central-place system with m orders in the hierarchy and yields an undamped correlogram. A second-order autoregressive model may also yield an oscillating correlogram, although theoretically this damps to zero (Nelson, 1973, page 45). Perhaps most intriguing, and worth a detailed look in a spatial context, is the Slutsky–Yule process, which interprets periodic behaviour in terms of the successive summation and differencing of white noise (Slutsky, 1937). It is not too difficult to interpret the appropriate filter in terms of a migration field, with weights declining as some function of distance (Tobler, 1969; Curry et al, 1976, Smith, 1976).

There are, then, plausible models yielding correlograms which oscillate as might be expected *a priori* in a central-place system. However, none of the correlograms estimated from the Southern Germany data suggests any periodicities and an attempt is made instead to identify possible generating processes from the empirical autocorrelation functions.

Part of the identification procedure in time-series analysis comprises an inspection of the autocorrelation function for a cutoff; that is, the point at which autocorrelation drops to zero or a nonsignificant value[8]. Since, in two dimensions, the cutoff will vary with direction, the identification of processes in different directions is implied. Geographically, the directional variation in dependence may suggest directional biases in ease of movement. For instance, with reference to the undifferenced data the longer lag dependencies in a North–South direction imply a lower friction of distance than along other axes, this reflecting perhaps the orientation of the Rhine Valley.

Comments are more readily directed to the averaged correlograms (figure 4.6). However, because of sampling fluctuation and the covariance between neighbouring estimates, significance should not be attached readily to every detail. Simulated realisations from theoretical processes whose properties are well-known often yield correlograms that are interpreted mistakenly as the outcome of some other process (Nelson, 1973, pages 72–73). Bearing in mind this problem, we note the generally exponential decline in dependence for the undifferenced data at all three scales. This suggests a first-order autoregressive model, the population of a central place being a function of that of surrounding neighbours.

The averaged correlogram for first differences indicates some other stochastic process to be at work. The sharp cutoff after lag 2 [figure 4.6(d)] suggests perhaps a moving-average model of order one or two. This writes an observation, z_i, in terms of 'disturbances' occurring elsewhere on the map (Haining, 1976). After differencing, however, the model may be expressed as an exponentially weighted sum of the observations at successive lags. To this extent there is system-wide dependence or what we might call 'spatial memory' (cf Smith, 1976, page 363).

In the absence of estimated partial autocorrelation functions and an assuredly correct sampling distribution for the autocorrelation coefficients

[8] The other aspect of identification concerns the estimation of partial autocorrelation functions. Spatial versions of these are not discussed here, although they are of course essential for a more complete identification.

it would be invidious to attempt anything more than the tentative identification of processes offered here. Haining (1976; 1977a) has suggested two complementary approaches to model identification, one based on the relative magnitudes of low-order autocorrelations, the other a likelihood-ratio statistic that tests various hypotheses of spatial dependence. The latter seems suitable only for relatively small n (say, $n \leqslant 100$), whereas the former requires a very fine sampling interval[9] which may yield more noise than signal, rendering it rather unsuitable for nonlattice data.

4.8 Concluding remarks
There can be little doubt that autocorrelation plays a pivotal role in geography. In the context of the present study, there are several reasons why an examination of autocorrelation in Christaller's data is a useful task to undertake. Most important of these concerns verification of the theoretical spatial structure he developed. Some of the foundations for this task have been prepared here but it needs more rigorous attention. Theoretical models for dependence in central-place systems need elaborating and testing against null models of no spatial dependence, as indicated by Haining (1977a).

Of remaining problems, most serious is that of the sampling distribution of spatial autocorrelation coefficients. One possible solution is to randomise the data and to estimate an autocorrelation function for each permutation but this would be computer time-consuming. Another problem raised above but not resolved entirely satisfactorily concerns directional variation in the autocorrelation function. A test is required of the null hypothesis that there is no significant difference in distance dependence among transects in the autocorrelation function.

Much has rightly been made of the stationarity assumption for spatial series. The differencing procedure employed here was useful in trend removal and in revealing an interesting structure. However, the meaning of the assumption that the covariance between two observations separated by distance d is the same for all pairs d units apart, was not considered. The assumption has been questioned on geographical grounds by Granger (1969) and Berry (1971). It may make more sense to locate the observations in a transformed space and to estimate an autocorrelation function based on separation in this new space.

This was attempted for central places belonging to the Frankfurt regional system (Gatrell, 1976). A simple gravity model predicted 'interactions' between places and these were treated as 'similarities' in a multidimensional scaling algorithm. However, the space created was not substantially different from real geographical space and the relatively small

---

[9] Otherwise we "risk integrating out important information concerning the behaviour of the autocorrelation coefficient around lag zero" (Haining, 1977a, page 120).

number of observations led to autocorrelation functions that were difficult to interpret. The idea of structural analysis (and optimal prediction) in transformed spaces deserves further investigation, with particular attention being given to the type of transformation employed.

References

Agterberg F, 1970 "Autocorrelation functions in geology" in *Geostatistics* Ed. D Merriam (Plenum, New York) pp 113-141

Agterberg F, 1974 *Geomathematics* (Elsevier, Amsterdam)

Anderson O D, 1976 *Time Series Analysis and Forecasting: The Box-Jenkins Approach* (Butterworths, London)

Bannister G, 1975 "Population change in Southern Ontario" *Annals of the Association of American Geographers* **65** 177-188

Bartlett M S, 1946 "On the theoretical specification of sampling properties of autocorrelated time series" *Journal of the Royal Statistical Society Series B* **8** 27-41

Berry B J L, 1967 *Geography of Market Centres and Retail Distribution* (Prentice-Hall, Englewood Cliffs, NJ)

Berry B J L, 1971 "Problems of data organization and analytical methods in geography" *Journal of the American Statistical Association* **66** 510-523

Besag J, 1975 "Statistical analysis of non-lattice data" *The Statistician* **23** 179-195

Box G E P, Jenkins G M, 1970 *Time Series Analysis Forecasting and Control* (Holden-Day, San Francisco)

Christaller W, 1933 *Die Zentralen Orte in Süddeutschland* (Gustav Fischer, Jena)

Christaller W, 1966 *Central Places in Southern Germany* translated by C Baskin (Prentice-Hall, Englewood Cliffs, NJ)

Cliff A D, Haggett P, Ord J K, Bassett K, Davies R, 1975 *Elements of Spatial Structure* (Cambridge University Press, London)

Cliff A D, Ord J K, 1973 *Spatial Autocorrelation* (Pion, London)

Curry L, 1964 "The random spatial economy: an exploration in settlement theory" *Annals of the Association of American Geographers* **54** 138-146

Curry L, 1967 "Central places in the random spatial economy" *Journal of Regional Science* **7** 217-238

Curry L, 1977 "Stochastic spatial distributions in equilibrium: settlement theory" in *Man, Settlement and Culture: Essays in Honour of Professor R L Singh* (forthcoming)

Curry L, Griffith D, Sheppard E, 1976 "Those gravity parameters again" *Regional Studies* **9** 289-296

Dacey M, 1965 "A review on measures of contiguity for two and *K*-colour maps" TR-2. *Spatial Diffusion Study* (Department of Geography, Northwestern University, Evanston, Ill.)

Dacey M, 1966a "A county-seat model for the areal pattern of an urban system" *Geographical Review* **56** 527-542

Dacey M, 1966b "A probability model for central place locations" *Annals of the Association of American Geographers* **56** 550-568

Dacey M, Davies O, Flowerdew R, Huff J, Ko A, Pipkin J, 1974 *Studies in Geography 21. One Dimensional Central Place Theory* Northwestern University, Evanston, Ill.

Ebdon D, 1977 *Statistics in Geography: A Practical Approach* (Blackwell, Oxford)

Gatrell A C, 1976 *Structure, Process and Optimal Prediction in Absolute and Relative Space: Central Place Populations in Southern Germany* unpublished PhD thesis, Department of Geography, The Pennsylvania State University, University Park, Pa

Granger C W J, 1969 "Spatial data and time series analysis" in *London Papers in Regional Science 1. Studies in Regional Science* Ed. A J Scott (Pion, London) pp 1–24

Granger C W J, 1975 "Aspects of the analysis and interpretation of temporal and spatial data" *The Statistician* 23 197–210

Haggett P, 1975 *Geography: A Modern Synthesis* (Harper and Row, New York)

Haining R P, 1976 "The moving average model of dependence for a rectangular plane lattice" abstract in *Advances in Applied Probability* 8 654–655

Haining R P, 1977a "Model specification in stationary random fields" *Geographical Analysis* 9 107–129

Haining R P, 1977b "Markov processes and slope series: the scale problem, a comment" *Geographical Analysis* 9 94–99

Hodder I, Orton C R, 1976 *Spatial Analysis in Archaeology* (Cambridge University Press, London)

Hodge D, Gatrell A C, 1976 "Spatial constraint and the location of urban public facilities" *Environment and Planning A* 8 215–230

Hudson J C, 1969 "A location theory for rural settlement" *Annals of the Association of American Geographers* 59 365–381

Huijbregts C J, 1975 "Regionalised variables and quantitative analysis of spatial data" in *Display and Analysis of Spatial Data* Eds J C Davis, M J McCullagh (John Wiley, New York) pp 38–53

Kendall M G, 1976 *Time Series* (Charles Griffin, London)

Kolars J, Nystuen J, 1974 *Human Geography: Spatial Design in World Society* (McGraw-Hill, New York)

Lébart L, 1969 "Analyse statistique de la contiguité" *Publications de l'Institut de Statistique de l'Universite de Paris* 18 81–112

Martin R L, 1974 "On autocorrelation, bias and the use of first spatial differences in regression analysis" *Area* 6 185–194

Matheron G, 1971 *The Theory of Regionalised Variables* (Cahiers du Centre de Morphologie Mathematique de Fontainebleau No.5, Ecole Nationale Superieure des Mines de Paris, Paris)

Naidu P, 1970 "Statistical structure of aeromagnetic field" *Geophysics* 35 279–292

Nelson C R, 1973 *Applied Time Series Analysis for Managerial Forecasting* (Holden-Day, San Francisco)

Nordbeck S, Rystedt B, 1972 *Computer Cartography* (Studentlitteratur, Lund)

Rayner J N, Golledge R G, 1972 "Spectral analysis of settlement patterns in diverse physical and economic environments" *Environment and Planning A* 4 347–371

Sayn-Wittgenstein L, 1970 "Patterns of spatial variation in forests and other natural populations" *Pattern Recognition* 2 245–253

Sibert J, 1975 *Spatial Autocorrelation and the Optimal Prediction of Assessed Values* (Michigan Geographical Publications 14, University of Michigan, Ann Arbor, Mich.)

Slutsky E, 1937 "The summation of random causes as the source of cyclic processes" *Econometrica* 5 105–146

Smith T R, 1976 "Set-determined process and the growth of spatial structure" *Geographical Analysis* 8 354–375

Tobler W, 1969 "Geographical filters and their inverses" *Geographical Analysis* 1 234–253

Tobler W, 1975 "Linear operators applied to areal data" in *Display and Analysis of Spatial Data* Eds J C Davis, M J McCullagh (John Wiley, New York) pp 14–37

A million or so correlation coefficients: three experiments on the modifiable areal unit problem

S Openshaw, P J Taylor

5.1 Introduction

Although geography and statistics shared a common intellectual heritage of local societies in the nineteenth century (Berry and Marble, 1968), they have diverged somewhat during the twentieth century (Taylor and Goddard, 1974). This divergence has not been reversed as much as might have been expected by the recent rise of a quantitative geography with its emphasis upon statistical analysis. The reason seems to be that the discipline of statistics have moved away from the original interest in the study of published data sources—archival data—so that the modern statistician typically researches in the theoretical mathematics of probabilities with carefully controlled experimental data. Hence where empirical data is employed, it tends to be specially collected and avoids many of the pitfalls that confront the user of 'dirty' archival data. In contrast the majority of geographical studies use archival sources for their data (Haggett, 1965, chapter 7).

The fact that geographical data is typically far removed either from the ideals of classical statistical inference or the assumptions of stochastic modelling has been often noted (Cliff and Ord, 1975). This has led to particular geographical interest in nonparametric statistics (French, 1971) and problems of spatial dependence (Cliff and Ord, 1973). However, one topic which has been relatively neglected has been the arbitrary nature of areal units. Since the area over which data is collected is continuous, it follows that there will be numerous alternative ways in which it can be partitioned to form areal units for reporting the data. There will, in theory, be an infinite number of ways in which a study area can be areally divided, although data will normally be presented for only one particular set of areal units. These units themselves, however, may be combined to form new larger units at a new scale in a large number of alternative ways. This is the modifiable areal unit problem, identified by Yule and Kendall (1950) who point out that "we must emphasise the necessity, in this type of work, of not losing sight of the fact that our results depend on our units" (page 313). Despite this warning there has been relatively little concern for this fundamental problem in spatial analysis. This chapter follows several previous considerations of the problem in geography (Robinson, 1956; Thomas and Anderson, 1965; Curry, 1966, Clark and Avery, 1976) but provides a far more comprehensive empirical study of the topic than has hitherto been attempted.

5.2 Modifiable areal unit problems

The modifiable areal unit problem is in reality two separate but interrelated problems. Statisticians (Gehlke and Biehl, 1934; Neprash, 1934; Yule and Kendall, 1950) have generally discussed what we shall term the *scale problem*. This is defined simply as the variation in results that may be obtained when the same areal data are combined into sets of increasingly larger areal units of analysis. For instance, analysis of the same census data at scales ranging from enumeration districts, wards, local authorities, up to standard regions will almost certainly provide alternative results and possibly interpretations. It should be noted at this point that while this problem may be closely related to making inferences about individuals from aggregate data—in economics the micro/macroanalysis problem (Cramer, 1964) and in sociology the ecological fallacy problem (Robinson, 1950)—in this paper we shall remain interested only in combinations of the data that have already been spatially aggregated. Notice that this is where this study differs in purpose from the recent researches of Williams (1976; 1977a; 1977b).

Although scale differences are the most obvious manifestation of the modifiable areal unit problem there is also the problem of alternative combinations of base units at equal or similar scales. For instance Tooze (1976) and Kirby and Taylor (1976) have both divided up Britain at almost the same scale but have created very different units for analysis. Any variations in results due to alternative units of analysis where n, the number of units, is constant will be termed the *aggregation problem*.

In addition we can distinguish between two different forms of areal arrangement. In the regional taxonomy literature, contiguous areal arrangements are termed regions and noncontiguous areal arrangements are referred to as regional types (Spence and Taylor, 1970). Here we shall use the simpler terminology of *zoning system* for contiguous arrangement and *grouping system* for the noncontiguous case. Alternatively a zoning system may be considered a special case of a grouping system which incorporates a contiguity constraint in its formation. Since most spatial analyses use zoning systems we concentrate on them in what follows although simple groupings are considered for comparative purposes.

Finally Yule and Kendall's (1950) discussion of the problem was originally illustrated by using correlation coefficients and other researchers have followed suit. We continue this tradition because this statistic is easily interpretable and hence both scale and aggregation effects may be simply identified and described. This paper reports the results from three closely related experiments on variations in the correlation coefficient under different spatial and statistical conditions.

5.3 The first experiment: areal associations in Iowa

The first experiments on the correlation coefficient are carried out with the use of a set of data describing Iowa, USA. The data has been chosen partly because of its availability and manageability. The basic areal units are the ninety-nine counties of Iowa. For each unit we have two measures—the dependent variable is the percentage vote for Republican candidates in the congressional election of 1968 and the independent variable is the percentage of population over sixty-years old recorded in the 1970 US census[1]. Notice that we have no need to imply any individual-level correlations here. The percentage old people variable is interpreted as an index of demographic history, largely differential in- and out-migration, reflecting an economic environment which we hypothesise to be positively related to Republican voting. In fact the variables are correlated at $+0.3466$ over the set of ninety-nine counties. These data are regarded as our population of base units.

From Yule and Kendall's (1950) discussion we know that the correlation ($r = +0.3466$) is specific to the particular set of areal units used. In table 5.1 alternative areal arrangements have been employed to compute the correlation to produce five sets of results for the same relationship. In each example the ninety-nine counties have been combined into just six new areal units[2]. Hence differences between these correlations and the original 'county' correlation reflect the scale problem, whereas differences in correlations between the five alternative aggregations illustrate the aggregation problem. The only 'official' aggregation, the congressional districts, produces a correlation below the original r value but the other correlations are, as is generally expected, higher than the base unit value.

Table 5.1. Some effects on the correlation coefficient of different areal arrangements of the Iowa counties into six zones.

Alternative combinations of counties	r
6 Republican-proposed congressional districts	0·4823
6 Democrat-proposed congressional districts	0·6274
6 Congressional districts	0·2651
6 Urban/rural regional types	0·8624
6 Functional regions	0·7128
99 Iowa counties	0·3466

[1] Problems caused by the fact that these measures define closed number sets are not critical for our subsequent analyses since no observations approach the limits of 0% and 100%.

[2] It should be noted that the variables were each aggregated by summing the county percentages and dividing by the number of counties in the aggregation. This does not produce the 'true' percentage values for the new aggregated units since counties vary in population size. This approach was adopted to facilitate computation in the subsequent experiments and will have no systematic effect on the experimental results.

If the researcher is interested in finding high correlations between his variables he will lament the final choice of congressional districts and will wish the Democrat proposal had won the political battle. Still higher correlations can be produced, however, by using specially designed 'geographical' arrangements—either urban/rural types based upon the largest urban area in a county or of which the county is part, or functional regions based upon the counties being allocated to the six largest urban zones by distance. Clearly the correlation between Republican voting and percentage old people is not an easy thing to measure.

5.3.1 Identification of the limits of the scale and aggregation problem
The dimensions of the modifiable areal unit problem for these Iowa data can be determined by finding the limits of the aggregation effects at different scales by applying an automatic zoning algorithm. This method has been developed to identify zonings or groupings of data that approximately optimise any general function defined in terms of the aggregated data (Openshaw, 1977a; 1977b; 1977c). Thus we can identify zoning or grouping systems of Iowa counties which approximately represent the limits of negative and positive correlation. These have been derived for various scales and are presented in table 5.2.

At the aggregation scale both for six zones and six groups the limits show that it is possible to produce the whole range of correlations. As the aggregation scale decreases the range of possible correlations declines especially for zoning systems where contiguity constraints are involved. We can conclude, however, that for all scales there is a relatively wide range of correlations which would certainly involve alternative substantive interpretations. Table 5.2, therefore, specifies the universe of alternatives

Table 5.2. Maximum and minimum values of the correlation coefficient.

Number of zones or groups	Zoning systems [a]		Grouping without contiguity	
	minimum r	maximum r	minimum r	maximum r
6	−0·999	0·999	−0·999	0·999
12	−0·984	0·999	−0·999	0·999
18	−0·936	0·996	−0·977	0·999
24	−0·811	0·979	−0·994	0·999
30	−0·770	0·968	−0·989	0·999
36	−0·745	0·949	−0·987	0·998
42	−0·613	0·891	−0·980	0·996
48	−0·548	0·886	−0·967	0·995
54	−0·405	0·823	−0·892	0·983
60	−0·379	0·777	−0·787	0·983
66	−0·180	0·709	−0·698	0·953
72	−0·059	0·703	−0·579	0·927

[a] Best from fifteen different random zoning systems used as starting points for the automatic zoning algorithm.

within which our experiments on the correlation coefficient are to be carried out.

5.3.2 Zoning and grouping distributions of correlation coefficients

The maximum and minimum correlation coefficients illustrated in table 5.2 may be interpreted as representing the known limits of the distribution of correlation coefficients for alternative aggregations at each scale. To produce the intervening distributions it is necessary to use a random zoning and grouping system generator (Openshaw, 1977d). By using this algorithm, random samples of ten-thousand alternative arrangements into six, twelve, eighteen, twenty-four, thirty, thirty-six, forty-two, forty-eight, fifty-four, sixty, sixty-six, and seventy-two zoning and grouping systems have been produced. Figure 5.1 shows the frequency distributions of correlation coefficients for each sample at class intervals of 0.1. In figure 5.1(a) zoning system correlations are shown and we may term these distributions the *zoning distributions* of correlation coefficients for aggregates of Iowa counties at different scales. Figure 5.1(b) shows the equivalent *grouping distributions* of correlation coefficients, which are sometimes considered as analogous to the sampling distribution of correlation coefficients in standard statistical theory (Williams, 1977a).

Interpretation of figure 5.1 is relatively straightforward. Both sets of distributions show a decrease in spread as the scale decreases (that is, more zones or groups) as would be expected. Furthermore the degree of bias in the estimators compared with the county-level correlation also declines with scale. In terms of our previous discussion this can be interpreted as the aggregation effect being illustrated by the horizontal spread and the scale effect, tending to increase the correlation, which is summarised by the vertical changes in the modes of the distributions. Differences between the two sets of distributions are more interesting. The zoning distributions visually exhibit less spread and less bias than the grouping distributions. The most obvious explanation for this phenomenon would seem to be the interaction between the contiguity constraint of the zoning systems and the positive spatial autocorrelation of the variables. In fact both variables are positively autocorrelated, on using Moran's I statistic $I_x = +0.37$ and

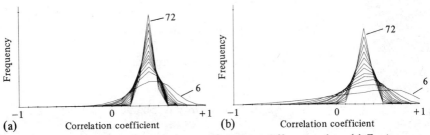

Figure 5.1. Frequencies of correlation coefficients at different scales. (a) Zoning distributions, (b) grouping distributions.

$I_y = +0 \cdot 43$ (Cliff and Ord, 1973). This hypothesis is formally tested in the next section and in this discussion we shall briefly consider other effects of this interaction between pattern and boundary.

The above visual interpretation is confirmed as far as the precision of the correlation estimates are concerned in table 5.3. This shows the spread of four zoning and four grouping distributions as measured by their standard deviations. The other feature that this table shows is that as the number of zones/groups increases, precision also increases. This brings us back to the analogy with sampling theory referred to above. This analogy is explicitly illustrated in table 5.3 by inclusion of the standard deviations for equivalent sampling distributions of the correlation coefficient. It should be noted that the analogy holds for grouping distributions but is highly unsatisfactory for zoning distributions.

Table 5.3. Precision of alternative correlation estimates for Iowa.

Number of zones	Standard deviation [a]	Number of groups	Standard deviation [a]	Size of sample [b]	Standard deviation [a]
6	0·218	6	0·408	6	0·429
12	0·161	12	0·273	12	0·273
24	0·122	24	0·189	24	0·172
72	0·051	72	0·066	72	0·057

[a] All standard deviations are based upon 10000 correlation coefficients derived from zones, groups, or samples.
[b] Samples generated by randomly selecting samples of Iowa counties without replacement, 10000 times.

5.3.3 Correlation coefficients and loss of variation
A random zoning system placed upon a positively autocorrelated variable will tend to have the effect of locating similar base units together. This will have a direct effect upon the inevitable loss in variation in the variable as a result of the aggregation of the base units. The loss can be described in terms of the sums-of-squares equality

$$S^{\mathrm{n}} = S^{\mathrm{B}} + S^{\mathrm{W}} , \qquad (5.1)$$

where

$$S^{\mathrm{n}} = \sum_i \sum_j (x_{ij} - \bar{x})^2 , \qquad \text{total sum of squares,}$$

$$S^{\mathrm{B}} = \sum_j [(\bar{x}_j - \bar{x})^2 n_j] , \qquad \text{between group/zone sum of squares,}$$

and

$$S^{\mathrm{W}} = \sum_i \sum_j (x_{ij} - \bar{x}_j)^2 , \qquad \text{within group/zone sum of squares,}$$

where x_{ij} is the value for the ith unit in the jth group/zone which has n_j base units. Since a sum-of-squares expression cannot be less than zero it follows that $S^B \leqslant S^{\Pi}$. For any particular group/zone the degree of loss of variation in a variable can therefore be measured by

$$\Delta S^B = \frac{S^{\Pi} - S^B}{S^{\Pi}} . \qquad (5.2)$$

At the ninety-nine-zone level each county is its own zone so that $S^{\Pi} = S^B$ and $\Delta S^B = 0$. When the counties are aggregated into groups/zones, variation is lost until we are left with one zone (Iowa) when $S^B = 0$ so that $\Delta S^B = 1$. ΔS^B is therefore a simple measure of loss of variation ranging from zero, or no loss, to unity, or total loss of variation. [Although equation (5.2) actually defines the proportion of within group/zone variation, the interpretation in terms of loss of between group/zone variation is more useful for the present discussion.]

ΔS^B can be computed for either variable, of course, although theoretical work in related areas suggests that loss of variation in the independent variable is the more interesting. Cramer (1964), for instance, has shown that the most efficient grouping of individuals in a linear regression equation for aggregate data is where the loss of variation in the independent variable is minimised. He thus employs income classes based upon individual income returns.

Since positive autocorrelation may be regarded as the 'norm' in most spatial analyses, it follows that aggregation by zoning will have a similar effect on the independent variable as 'efficient grouping' has in econometrics. This would account for the more precise correlation estimates given by the zoning systems as indicated by the smaller spread in figure 5.1(a). A similar result has been found by Williams (1976) in a more closely related study. He contrasted random and homogeneously grouped data and found that the latter provided much more precise estimators in regression and correlation analyses.

Given the above studies, we have chosen to characterise all of our zoning and grouping systems in terms of ΔS^B for the independent variable. In figure 5.2 one-thousand correlation coefficients are arranged against ΔS^B for zoning and grouping systems at each of the aggregations into six, twelve, and twenty-four units. Notice, first of all, that the vertical spread of observations is the same as the horizontal spread for six, twelve, and twenty-four groups/zones in figure 5.1. Hence the precision and bias in the correlation estimates as the number of groups/zones increases from left to right in figure 5.2, is repeated here. These diagrams also show a systematic variation in ΔS^B with scale. As would be expected, as the number of groups/zones increases, loss of variation is less noticeable. In figure 5.2(b) the grouping systems for a six-unit scale lost nearly all of their variation, only a few arrangements lose less than 90% ($\Delta S^B = 0 \cdot 9$). In contrast at the scale of twenty-four almost all grouping systems have lost less than 90% of the original sums of squares and typically about three-quarters of the variation is lost ($\Delta S^B = 0 \cdot 75$).

Similar patterns are found in the zoning systems on figure 5.2(a). The contrasts with figure 5.2(b) are interesting, however. Brought over from figure 5.1 is the smaller spread and bias on the vertical (correlation) scale. Differences along the ΔS^B scale are even more clear-cut, however. As our previous discussion has implied, there is far less loss of variation in the zoning systems at each of the three scales. In fact the level of ΔS^B for six zones is approximately equivalent to ΔS^B for twenty-four groups. Only about half of the original variation is lost when zoning systems of twenty-four are produced. Figure 5.2 clearly shows that choice of a zoning system has a very profound effect on the type of aggregation produced when the data are autocorrelated. Very similar results have been produced for arrays of r against ΔS^B for the dependent variable.

Before leaving figure 5.2 it is worth noting where the five aggregations of six from table 5.1 occur among these randomly generated systems. The four zoning systems occur within the distribution of random zoning systems but this is not true of the one grouping system in table 5.1, the urban/rural types. This grouping preserves much more variation than any random grouping and is similar in fact to the zoning systems. The particular criteria for the grouping seem to be equivalent to zoning constraints in terms of ΔS^B. In contrast the criteria employed for the four zoning systems seem to have had no additional effect beyond contiguity constrained random zoning.

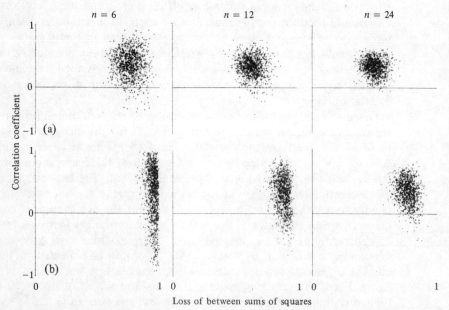

Figure 5.2. Graph plots of one-thousand correlation coefficients against loss of between sums of squares for the independent variable. (a) Zoning systems, (b) grouping systems.

5.3.4 Correlation coefficients and differential loss of variation

One feature of figure 5.2 is that there is no apparent relationship between r and ΔS^B. Whatever degree of loss of variation that occurs, a wide range of correlations can result. There is nothing in the literature to suggest otherwise. Blalock (1964), however, does propose that differential loss of variation between independent and dependent variables will have a systematic effect on the correlation coefficient. His argument is simply that if the dependent variable, y, is caused by several independent variables, then if the variation of one independent variable, x, is particularly maintained in an aggregation, this will enable this particular variable to account for more of the remaining variation in y, that is, increase the correlation. This is because, of the total variation of y that is lost in the aggregation, less of that part of the variation of y caused by x will be lost than that part caused by the other independent variables whose own variation is not being maintained. This proposed effect may be measured by

$$\Delta S^D = \Delta S_x^B - \Delta S_y^B \, , \tag{5.3}$$

where the subscripts x and y refer to independent and dependent variable sums of squares respectively. ΔS^D may be termed the differential loss of the sums-of-squares term. This measure ranges from -1 when there has been no loss in the variation of x and total loss of the variation of y to $+1$ when total loss of variation of x coincides with no loss in variation of y. These two limiting cases afford a theoretical addition to Blalock's hypothesis despite the fact that the correlation coefficient itself is indeterminate in each case owing to the zero variation in one of the variables. If we briefly revert to a regression format, however, we can see that the case where $\Delta S^D = -1$ corresponds to a horizontal line and $\Delta S^D = +1$ to a vertical line. Maximum loss in the sum of squares of y with respect to the sum of squares of x ($\Delta S^D = -1$) corresponds to perfect prediction of y from x, whereas maximum loss of the sum of squares of x with respect to the sum of squares of y ($\Delta S^D = +1$) corresponds to no prediction of y from x and variations in y depend solely on other variables. These are limiting cases, however. In general we expect that as ΔS^D gets larger, r will decline since variables other than x are maintaining that part of the variability of y not dependent on x at the expense of that part dependent on x.

Figure 5.3 shows scatters of one-thousand zoning and grouping systems at scales of six, twelve, and twenty-four units with r plotted against ΔS^D. The two sequences of scatters are very different from one another with respect to the changes in the spread of ΔS^D. With aggregations into just six units the spread of ΔS^D is far greater for zoning systems than for simple random groupings. However, the difference is eliminated by the twenty-four-unit scale. In effect we find two different scale effects—in zoning systems the spread of ΔS^D declines with scale whereas with grouping systems the range increases. The explanation for this phenomenon is not readily apparent. Presumably the particular scale of six units interacts with the particular spatial autocorrelation pattern in these variables to counteract the expected decline in ΔS^D as illustrated in the grouping systems and is even strong enough to slightly increase the range of ΔS^D. We present an informal test of this idea in the next section.

A second difference between the zoning and grouping systems in figure 5.3 concerns the modal values of ΔS^D. In the grouping scatters, the mode is at zero, which is expected, that is, random grouping is as likely to cause loss of variation in x as in y. In the zoning systems, however, the mode of ΔS^D is positive for all three data plots. This suggests that zones tend to cause more loss of variation in x than y. The explanation of this may be presumed to lie in the interaction of the contiguity constraint and the particular autocorrelation patterns of the two variables. Once again we informally test this assertion in the next section.

Having commented on the nature of ΔS^D in the data plots in figure 5.3 we can now turn to the hypothesised relationship between it and the correlations. A cursory glance at the six diagrams indicates no tendency towards $r = 0$ for increasing ΔS^D as we had previously suggested.

Finally before we leave these plots we can note once again where the five six-unit aggregations from table 5.1 fit onto the diagrams. In this case all five grouping/zoning systems lie within the scatter of aggregations. Although they all lie away from the ΔS^D modes on the two diagrams, there is no particular pattern to this divergence.

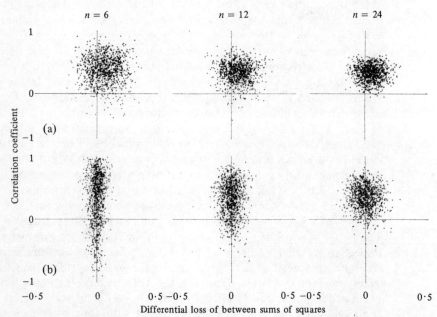

Figure 5.3. Graph plots of one-thousand correlation coefficients against differential loss of between sums of squares. (a) Zoning systems, (b) grouping systems.

We can summarise these experiments on the Iowa data in two general statements.
(1) There seem to be very distinct differences between zoning and grouping systems in many situations and these seem to be caused by the interaction of the contiguity in the zoning with the spatial autocorrelation in the data.
(2) There seem to be no systematic relationships between the sums-of-squares terms and the correlation coefficients.

We have designed two further experiments to explore both statements more fully.

5.4 The second experiment: spatial autocorrelation
We investigate the effects of spatial autocorrelation on the correlation coefficient and sums-of-squares terms by employing specially constructed artificial data. The experiments carried out upon the Iowa data are then repeated on the artificial data.

5.4.1 A data generator
We require to produce new variables, the properties of which we can carefully control, for our set of ninety-nine Iowa counties. A suitable data generator can be based upon the quadratic loss function $F(x^*, y^*)$, which tends to zero as data with the required properties are produced. Thus our objective function to be minimised is

$$F(x^*, y^*) = w_1 (r_{yx}^* - r_{yx})^2$$
$$+ w_2 (S_x^* - S_x)^2 + w_3 (S_y^* - S_y)^2 + w_4 (K_x^* - K_x)^2$$
$$+ w_5 (K_y^* - K_y)^2 + w_6 (I_x^* - I_x)^2 + w_7 (I_y^* - I_y)^2 , \qquad (5.4)$$

where x^* and y^* are 198 undetermined parameters (two for each county) which are estimated to minimise F; r_{yx}^*, S_x^*, S_y^*, K_x^*, K_y^*, I_x^*, and I_y^* are values of correlation, skewness, kurtosis, and spatial autocorrelation for any given vectors x^* and y^*; r_{yx}, S_x, S_y, K_x, K_y, I_x, and I_y are desired values which x^* and y^* should possess; and $w_1, ..., w_7$ are a set of weights that indicate the relative importance associated with each characteristic of the data. The spatial autocorrelation measure used is Moran's I statistic based on first-order contiguity relations (Cliff and Ord, 1973).

The function $F(x^*, y^*)$ is minimised by a quasi-Newtonian optimisation procedure available as a Harwell subroutine VA10AD (Fletcher, 1972). This data generator proved remarkably successful albeit expensive in computer time with a typical run requiring 920 seconds of central processing unit time on an IBM 370/168.

Two sets of data were generated for our experiment. In both sets our target r_{yx} was $+0.3466$ for comparability with the Iowa data. Furthermore we defined normal data so that $S_x = S_y = K_x = K_y = 0$ were targets. The data sets were designed to differ only in terms of spatial autocorrelation.

In data set A the targets were $I_x = I_y = 0$ to produce no spatial auto-correlation and in data set B the targets were $I_x = I_y = 1$ to produce maximum positive spatial autocorrelation. The targets are summarised in table 5.4 together with the properties of the generated data.

Table 5.4. Properties of generated data.

Data set	Targets					Actual					
	r	S	K	I	r	S_x^*	S_y^*	K_x^*	K_y^*	I_x^*	I_y^*
A	0·34	0	0	0	0·34	−0·00	0·00	−0·00	0·00	0·00	0·00
B	0·34	0	0	1	0·34	0·01	0·00	−0·00	−0·00	0·82	0·92

5.4.2 Zoning and grouping distributions of correlation coefficients

The zoning and grouping distributions for both artificial data sets are shown in figure 5.4. If we consider data set A first, we can see that with no autocorrelation there are no major differences between the zoning and grouping distributions. In contrast, for data set B, differences between the two sets of distributions are very clearly illustrated. Data set B replicates our findings from the Iowa data with the zoning distributions showing less spread and less bias than the grouping distributions. We may conclude that spatial autocorrelation and zoning does interact in the way suggested for the Iowa data.

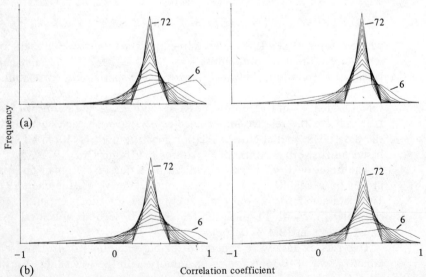

Figure 5.4. Frequencies of correlation coefficients at different scales for the artificial data. (a) Zoning distributions, (b) grouping distributions.

5.4.3 Correlation coefficients and the sums-of-squares terms

Let us turn to the sums-of-squares analysis. With one exception, the scale effects noted in the Iowa data are repeated for both sets of artificial data and so we shall just consider one scale here, that of twelve zones/groups. The exception will be briefly commented on below.

The scatter of results for the correlation coefficient against ΔS^B is shown in figure 5.5. Clearly the distinctive feature is the reduced loss of sums of squares for the zoning distributions in data set B. Hence this data set has again replicated the result found for the Iowa data and we therefore confirm the hypothesis of the interaction of contiguity constraint and auto-correlation, with reduction in loss of variation in a variable with aggregation.

The four scatter plots of correlation coefficients against ΔS^D are shown in figure 5.6. Once again it is the zoning system with data set B which has the most distinctive pattern. Here, however, replication of the Iowa data is not exact. Although autocorrelation and zoning interact to produce a wide spread of ΔS^D values, the bias in the values is the opposite to that of the Iowa data. This is inconsequential because specification of x and y as dependent and independent variables is wholly arbitrary in our artificial data. These results do suggest that the particular bias in the Iowa data set is specific to that data set. Furthermore the peculiar scale effect for the

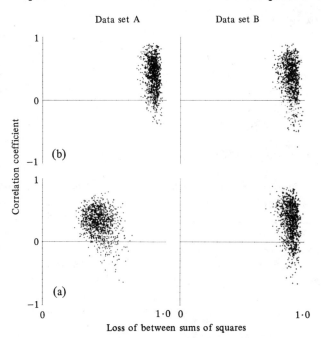

Figure 5.5. Graph plots of one-thousand correlation coefficients against loss of between sums of squares for the independent variable in the artificial data. (a) Zoning systems, (b) grouping systems.

zoning system with ΔS^D for Iowa that was found in figure 5.4 is not repeated for data set B, which further suggests an effect specific to the Iowa data.

The conclusions to be drawn from these second experiments are that zoning and spatial autocorrelation do interact in quite predictable ways and that this interaction explains much of the variety of results previously obtained from the original autocorrelated Iowa data.

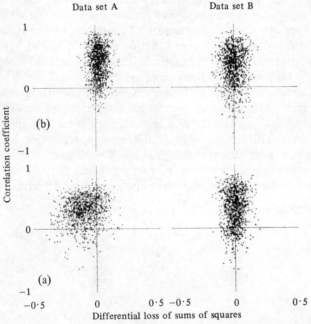

Figure 5.6. Graph plots of one-thousand correlation coefficients against differential loss of sums of squares for the artificial data. (a) Zoning systems, (b) grouping systems.

5.5 The third experiment: target sampling of Iowa zoning distributions

The second experiment produced quite expected results, which generally support hypotheses generated by the Iowa data. These positive findings from the Iowa data contrast sharply with the negative findings relating correlation coefficients to the sums-of-squares terms. Our strategy for investigating this situation needs to go beyond the random generation of zoning and grouping systems. We simply ask the question: if the proposed relationships are not discernible in the pattern of one-thousand observations, are the relationships confined to 'rare' cases beyond the distributions identified so far?

5.5.1 Target sampling

Target sampling involves defining a desired value of correlation or sums-of-squares term and producing a set of aggregations that have the desired value. For instance we may wish to know the pattern of correlation coefficients for zoning systems at a specific scale for which $\Delta S^D = +0.5$ or -0.5. Such aggregations can be produced by using the automatic zoning algorithm employed in section 5.3 to define maximum or minimum correlation coefficients. In this case we define an aggregation of x and y which minimises the quadratic loss function $Z(x^1, y^1)$

$$Z(x^1, y^1) = (\Delta S^D - \text{target})^2 \ . \tag{5.5}$$

For any one target we can often generate several different solutions by starting from different initial random zoning systems. This is possible both because of the large number of alternative solutions in terms of equation (5.5) and because of the suboptimal nature of the automatic zoning algorithm. In effect we are sampling vertical strips in the distribution represented by the data plots of one thousand previously described (figure 5.6). If in equation (5.5) we replace ΔS^D by r_{xy} we can sample horizontal strips in the same way.

5.5.2 Target sampling experiments

The purpose of this exercise is to obtain a more thorough understanding of the sum-of-squares/correlation relationship than can be obtained from random arrangements. The problem with the latter is that they fail to emphasise the limits and extremes of a distribution. Our target sampling will enable us to push beyond the known distribution patterns previously reported. In these experiments we use the original Iowa data and concentrate on zoning distributions. We generate one-hundred solutions of twelve zones both for ΔS^D and correlation targets, although we are not always completely successful. Figure 5.7 shows both sets of target samples and in both graphs there is some scatter of results around extreme targets. Elsewhere the algorithm has been successful in producing specific vertical and horizontal samples through the distribution. Both graphs should be

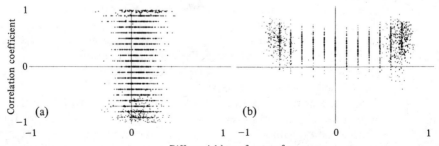

Differential loss of sums of squares

Figure 5.7. Target sampling of correlation coefficients against differential loss of sums of squares. (a) Correlation targets, (b) sums-of-squares targets.

compared with the random twelve-zone distribution for the Iowa data in figure 5.3(a). This comparison clearly illustrates how the target sampling provides information for a much wider range of r and ΔS^D.

If we consider the correlation target results first [figure 5.7(a)], we notice that although the whole range of possible r values have been generated, no systematic relationship emerges. Since we have previously argued that the size of correlation is likely to be dependent upon the sum-of-squares term then we can expect a higher probability of discerning a relationship when it is ΔS^D which is varied. This is achieved by the target sampling for ΔS^D illustrated in figure 5.7(b). In this graph there is again no systematic trend in the main central portion of the plot but the extreme ΔS^D samples do suggest some effects on the correlation coefficient. For high negative values of ΔS^D (that is, when much of the variation in x is preserved relative to y) at less than -0.4, there is a slight tendency for a relatively large number of higher correlation coefficients to be produced. This is consistent with the previous arguments of section 5.3. At the other extreme, for high positive values of ΔS^D (that is, when most of the variation in x is lost relative to y) at greater than $+0.4$, a relatively large quantity of high correlations are again produced suggesting a shallow parabola. In this case, however, the high correlations are accompanied by a set of weak negative correlations. This suggests that since high ΔS^D values eliminate most of the variation in x we are approaching the 'vertical' regression situation discussed in section 5.3, where the underlying relationship will be highly unstable.

5.6 Conclusions
The purpose of these experiments has been to increase our understanding of the modifiable areal unit problem both from statistical and geographical perspectives. Three particular sets of findings are likely to be of some general interest. In section 5.3 the relationship between zoning, grouping, and also sampling distributions was illustrated in terms of correlation coefficients. In particular it is clearly shown that the simple zoning/ grouping/sampling analogy is misleading since zoning distributions exhibit less bias and more precision. Secondly, in section 5.4, we were able to demonstrate that the interaction between spatial autocorrelation and the zoning procedure with its contiguity constraints directly affected resulting statistics. Finally in section 5.5 the expected relationship between the sums-of-squares term and the correlation coefficient was found to be much more elusive than initially expected.

Our general feeling is that the modifiable areal unit problem is much more complex than has previously been believed. The fact that there seems to be no simple solution does not mean that the problem is any less important. We *have* been able to find a very wide range of correlations. We simply do not know why we have found them. Hence we can make no general statements about variations in correlation coefficients so that each areal unit problem must be treated individually for any specific piece of research.

In many ways the use of any particular zoning or grouping system may not be such a bad thing if it brings the problem of the geographical individual back into consideration in spatial analysis. Hannan (1971) has suggested that the substantive way to overcome scale problems is to develop cross-level theory. The aggregation problem of the geographer seems to be much simpler. No special theory linking scales is required; all that is needed is agreement upon what constitute the objects of the geographical enquiry. This question has long been a thorny one for geographers and seems to have been avoided more than it has been explicitly faced. *The question is simply what objects at what scales do we wish to investigate?* In contrast to our previous statistical perspective, this is an essentially geographical problem and answers to the question should form the basic geographical contribution to, and solution of, the modifiable areal unit problem. If all researchers in a field of geographical enquiry agree on their objects of interest, and these objects can be defined in a nonarbitrary manner, then this constitutes a unique set of units and the problem disappears. Of course the situation is not, and is never likely to be, that simple and is discussed further in Openshaw and Taylor (1978).

Furthermore, even if it is possible to agree on a unique set of areal units on geographical grounds, statistical variations, such as those illustrated in this paper, will continue to be of interest in order to place a set of results into a meaningful statistical and spatial perspective. For as Williams (1976, page 16) has so neatly put it "No self-respecting statistician would take just any selection of individuals as his sample in a study and give it no further thought. Likewise we would hope that the days are numbered for urban and regional scientists who produce zoning systems, as it were, out of a hat and proceed to use them, blissfully unaware of the effects the grouping might have on any subsequent empirical investigations they carry out".

References
Berry B J L, Marble D F, 1968 *Spatial Analysis* (Prentice Hall, Englewood Cliffs, NJ) pp 1–9
Blalock H M, 1964 *Causal Influences in Non Experimental Research* (University of North Carolina Press, Chapel Hill, NC)
Clark W A V, Avery K L, 1976 "The effects of data aggregation in statistical analysis" *Geographical Analysis* 8 428–438
Cliff A D, Ord J K, 1973 *Spatial Autocorrelation* (Pion, London)
Cliff A D, Ord J K, 1975 "Model building and the analysis of spatial patterns in human geography" *Journal of the Royal Statistical Society B* 37 297–348
Cramer J S, 1964 "Efficient grouping, regression and correlation in Engel curve analysis" *Journal of the American Statistical Association* 59 233–250
Curry L, 1966 "A note on spatial association" *Professional Geographer* 18 97–99
Fletcher R, 1972 "FORTRAN subroutines for minimization by quasiNewton methods" *AERE* R7125 (HMSO, London)

French H M, 1971 "Quantitative methods and non-parametric statistics" in *Quantitative and Qualitative Geography* Eds H M French, J B Racine (University of Ottawa Press, Ottawa) pp 119–128

Gehlke C E, Biehl K, 1934 "Certain effects of grouping upon the size of the correlation coefficient in census tract material" *Journal of the American Statistical Association, Supplement* **29** 169–170

Haggett P, 1965 *Locational Analysis in Human Geography* (Edward Arnold, London)

Hannan M T, 1971 *Aggregation and Disaggregation in Sociology* (Lexington Books, D C Heath, Lexington, Mass)

Kirby A M, Taylor P J, 1976 "A geographical analysis of voting patterns in the EEC Referendum" *Regional Studies* **10** 183–191

Neprash J A, 1934 "Some problems in the correlation of spatially distributed variables" *Journal of the American Statistical Association, Supplement* **29** 167–168

Openshaw S, 1977a "An optimal zoning approach to the study of spatially aggregated data" in *Spatial Representation and Spatial Interaction* Eds I Masser, P J B Brown (Marturius Nijhoff, Leiden)

Openshaw S, 1977b "A geographical solution to scale and aggregation problems in region-building, partitioning and spatial modelling" *Transactions of the Institute of British Geographers, New Series* **2** 459–472

Openshaw S, 1977c "Optimal zoning systems for spatial interaction models" *Environment and Planning A* **9** 169–184

Openshaw S, 1977d "Algorithm 3: A procedure to generate pseudo-random aggregations of N zones into M zones, where M is less than N" *Environment and Planning A* **9** 1423–1428

Openshaw S, Taylor P J, 1978 "Some statistical and geographical aspects of the modifiable areal unit problem" data organisation for British Cities, Working Note 2, Centre for Urban and Regional Development Studies, University of Newcastle upon Tyne, England

Robinson A H, 1956 "The necessity of weighting values in correlation of areal data" *Annals, Association of American Geographers* **46** 233–236

Robinson W S, 1950 "Ecological correlations and the behaviour of individuals" *American Sociological Review* **15** 351–357

Spence N A, Taylor P J, 1970 "Quantitative methods in regional taxonomy" *Progress in Geography* **2** 1–63

Taylor P J, Goddard J, 1974 "Geography and statistics: an introduction" *The Statistician* **23** 149–155

Thomas E N, Anderson D L, 1965 "Additional comments on weighting values in correlation analysis of areal data" *Annals, Association of American Geographers* **55** 492–505

Tooze M J, 1976 "Regional elasticities of substitution of the United Kingdom in 1968" *Urban Studies* **13** 35–44

Williams I N, 1976 "Optimistic theory validation from spatially grouped regression: theoretical aspects" *Transactions of the Martin Centre* **1** 113–145

Williams I N, 1977a "Some implications of the use of spatially grouped data" in *Dynamic Models for Urban and Regional Systems* Eds R J Bennett, N J Thrift, R L Martin (Pion, London) pp 53–64

Williams I N, 1977b "Estimating individual correlation coefficients with grouped data" Martin Centre, Cambridge University, Cambridge (mimeo)

Yule G U, Kendall M G, 1950 *An Introduction to the Theory of Statistics* (Charles Griffin, London)

Relationships between Great Britain census variables at the 1 km aggregate level

I S Evans

6.1 Introduction

Although many studies (Rees, 1971) have analysed the spatial structure of cities by correlating census data for small areas, this has not yet been done for a whole country the size of Britain. A country-wide study is desirable to provide a datum against which individual cities or regions can be compared. Furthermore, a country is a less open system with boundaries that are less arbitrary than those of a city (however defined); hence it provides a very useful spectrum of variation over which relationships between census variables such as unemployment and overcrowding can be observed. Although correlations at the national scale gloss over regional variations and could be quite different from the average of correlations within each region or within each city, there is a general consistency between the national correlations reported here and the correlations within conurbation cores and outer rings reported by Coulter (1977).

Previous work at the national scale has dealt with large units such as administrative towns [boroughs of England and Wales; Moser and Scott (1961)] or post-1974 districts [classified by Webber and Craig (1976)]. It is now computationally feasible to handle even larger data sets. Data for all 1 km national grid squares throughout Great Britain (this excludes Northern Ireland, the Isle of Man, and the Channel Islands) which meet confidentiality requirements have been made available by the Office of Population Censuses and Surveys (OPCS) to the Census Research Unit (CRU), Department of Geography, University of Durham, for the production of a national census atlas (CRU, 1979). The need to select a limited number of variables for inclusion in this atlas, from a long list of candidates, provided a further incentive to the measurement of relationships between census variables.

Correlation can be used as a measure of information redundancy: if a number of variables are closely related, it is necessary to map only one, from which the others can be predicted. The study of correlation structures also enhances understanding of the variables to be mapped, and aids the selection of one or two to represent each cluster of variables. Hence a matrix of correlations between 102 socioeconomic ratio variables was constructed. This chapter provides a discussion of the production and interpretation of this matrix, and of how such results can be presented.

Although the more interesting relationships between pairs and within clusters of variables are discussed, emphasis is placed upon methodological considerations here. If product-moment correlation coefficients are to be reasonable measures of association between variables, it is necessary to choose appropriate transformations with care, and to weight correlation by population per grid square. Even so, correlations between ratio variables within a single classification (for example, proportions in various socioeconomic groups) are biased by the closure effect. Solutions to this include use of the dissimilarity index, and the correlation of absolute numbers. These are shown to be comparable and useful, but differ from ratio correlations in that there is no control for relationships with respect to total population per grid square. Correlations between absolute numbers and total population were useful in deciding which variables could be mapped by absolute numbers, and which were so closely related to population that they should be mapped in relation to population.

Relationships between variables at the finest scale available for the whole country (1 km) are of particular interest. They are also the first step in studying how relationships vary with areal scale, which will be reported in later publications. In the present chapter an attempt is made to establish the methodology required and to experiment with the graphical and tabular presentation of sets of relationships, highlighting the main structures and revealing interesting details.

6.2 Data

The results presented here would be routine were it not for the exceptional nature of the data on which they are based. For the first time, a high-resolution multivariate socioeconomic spatial series is available for the whole country. The areal divisions are equal in size and shape and are reasonably compact (square). Hence a large part of the indeterminacy and arbitrary nature of areal aggregate correlations, discussed by Openshaw and Taylor (chapter 5 this volume), is avoided. Distortions due to varying area and shape of divisions (Robinson and Hsu, 1970; Robertson, 1969) are avoided, and it can be specified clearly that the calculated correlation relates to the 1 km scale, or 1 km level of aggregation: we are dealing with areal *units*, not with irregular areal divisions. If there were few grid squares, we would be concerned with the incidence of their arbitrary boundaries on the irregularities of the underlying distribution; by displacing the grid mesh by half a unit, four estimates of the correlation at that scale could be calculated to assess this effect. With such a fine mesh as 1 km for all Great Britain, the incidence effect can be assumed negligible.

For the 1971 census, the Office of Population Censuses and Surveys (OPCS) provides 1571 numbers (small area statistics) for each 1 km square which passes all the suppression criteria imposed to preserve confidentiality. These criteria vary between the four records: 100% population, 100% household, and (two) 10% sample records. Numbers of households, of people, of males, and of females are not suppressed at all. Often the degree of cross-classification given by OPCS is justifiable only for squares with large populations; many of the 1571 numbers are usually zero, for example, 'born in Ceylon', and especially 'married women aged 14 or younger'.

By judicious selection and aggregation, most of the information that is useful for small populations can be expressed in 102 ratio variables [Rhind et al (1977); see also Hakim (1975)]. Table 6.1 gives examples of definition of these ratio variables, and table 6.2 gives a complete list of their abbreviations (six or fewer characters each), which are used throughout this chapter. The list of variables is an aggregate of several alternative lists: for example, age divisions are given at two levels of detail. In fact a further purpose in calculating the correlations is to shed light on differences between alternative definitions.

Table 6.1. Examples of operational definition of ratios.

Number	Abbreviation	Definition
VP03	SEXMAL	Sex ratio: males per 10000 of total population
VP05	YFERT	Young fertility: number of children ever born to married women 16–29 years in PHH[a], per 1000 such women
VP08	WORKMF	Economically active married females aged 15–59 years per 10000 married females aged 15–59 years
VP15	UNEMMA	Unemployed males per 10000 economically active males
VY20	YOUTHS	Persons aged 15–24 years per 10000 of total population
VE40	NC1+2	New Commonwealth parentage: both parents, or self and one parent, born in New Commonwealth, per 10000 present residents
VH42	OWNOCC	Owner-occupied households per 10000 PHH
VH50	WCDEF	Households which share or lack an inside wc, per 10000 PHH
VS54	HHONE	One-person households, per 10000 PHH
VS61	P/RM	Persons per 100 rooms
VS62	1+P/RM	Households with more than 1·0 persons per room, per 10000 PHH
VE69	SCHOOL	Employed persons with ONC, HSC, or A level per 10000 employed
VG71	MANAG	Employers and managers (socioeconomic groups 1, 2, and 13) per 10000 economically active persons in stated civilian employment
VG76	SEMSKL	Semiskilled personal service and farm workers (socioeconomic groups 7, 10, and 15) per 10000 economically active persons in stated civilian employment
VO84	MANUF	Employed in manufacturing, per 10000 employed persons
VT89	CARTW	Persons travelling to work by car, per 10000 employed persons
VS101	LONPAR	Lone-parent families with dependent children, per 10000 families with dependent children

[a] PHH = private households.

Table 6.2. Effect of transformation and weighting on skewness (moment-based) of 102 ratio variables.

Unweighted			Weighted			
original skew	skew after transformation	variable number	population weighted mean (%)	original skew	skew after transformation	transform
Other population variables						
5·62	0·76	VP01 POP†	4135·00		−1·65	log
5·09	1·43	VP02 PNOTPH	2·98	6·34	0·12	log
0·46	0·83	VP03 SEXMAL	48·50	2·97	4·00	ang
−0·09	−0·07	VP04 MARFEM	51·29	0·11	0·19	ang
1·72	−0·62	VP05 YFERT*	1·216	1·56	−1·57	sqrt
2·75	−0·60	VP06 MFERT*	2·258	2·80	−0·72	sqrt
2·97	−0·51	VP07 CFERT*	1·988	3·04	−0·51	sqrt
−0·02	−0·42	VP08 WORKMF	49·13	−0·78	−1·35	ang
−0·12	−0·18	VP09 WORKFE	54·83	−0·95	−1·09	ang
−2·79	−0·75	VP10 WORKMA	91·54	−6·14	−3·24	ang
6·17	−0·34	VP11 VISIT	2·55	7·85	−1·11	log
2·48	−0·10	VP12 STUDNT	3·04	3·89	0·58	ang
4·65	−0·45	VP13 UNEMPL	4·08	6·09	−3·62	log
5·67	0·20	VP14 UNEMFE	3·84	9·48	−3·00	log
4·64	−0·07	VP15 UNEMMA	4·25	4·95	−2·82	log
8·58	0·59	VP16 SICK	1·21	15·84	−1·76	log
Age variables						
0·92	−0·30	VY17 RETAGE	16·02	1·01	0·00	ang
1·05	−0·68	VY18 INFANT	8·07	1·06	−0·38	ang
1·23	−0·43	VY19 SCHCHL	15·91	0·86	−0·37	ang
2·83	0·43	VY20 YOUTHS	14·50	5·05	2·73	ang
0·19	−0·60	VY21 YOADLT	24·10	0·71	0·25	ang
0·23	−0·62	VY22 MIDAGE	24·18	−0·50	−1·22	ang
0·95	−0·56	VY23 OLD	8·53	1·06	−0·22	ang
3·33	0·19	VY24 VOLD	4·71	3·00	0·46	ang
0·61	−0·42	VY25 CHILDR	23·98	0·27	−0·47	ang
0·78	0·69	VY26 YOPEOP	38·60	1·44	1·61	ang
1·33	−0·17	VY27 AGED	13·24	1·39	0·19	ang
0·13	−0·34	VY28 DEPAGE	40·32	−0·27	−1·20	ang
10·48	−2·45	VY29 AGINDX*	0·671	18·69	−1·53	log

* Open ratio: mean in units, not %. † absolute numbers: mean relates to many more grid squares than for other variables.
Transformations used are logarithm (log), angular (ang), square root (sqrt), and reciprocal (recip). Variable numbers and abbreviations are given; those for age and household size and composition are defined in table 6.9, others are defined where relevant in the text or tables, and a full set of definitions is given in Rhind et al (1977, pages 31–40).

Table 6.2 (continued).

Unweighted			Weighted			
original skew	skew after transformation	variable number	population weighted mean (%)	original skew	skew after transformation	transform
Ethnic variables						
−1·57	−1·46	VE30 ENGL	78·10	−1·81	−1·72	ang
3·27	2·91	VE31 WELSH	5·29	4·26	3·97	ang
2·55	2·42	VE32 SCOTS	10·07	2·78	2·72	ang
10·23	0·87	VE33 ULSTER	0·45	17·95	−1·35	log
8·34	0·48	VE34 EIRE	1·34	3·55	−1·30	log
6·64	0·48	VE35 NCBORN	2·17	4·24	−0·57	log
9·37	−0·33	VE36 OTHBOR	2·51	9·36	−1·68	log
5·79	−0·77	VE37 FORBOR	6·02	3·34	−1·66	log
12·67	2·88	VE38 2NCPA	0·62	4·46	0·61	log
12·30	1·78	VE39 NCIMM	1·64	4·56	0·08	log
10·43	1·25	VE40 NC1+2	2·40	4·35	−0·16	log
Household tenancy and amenity variables						
6·29	0·93	VH41 HOUSEH†	13·95	1·96	−1·63	log
−0·26	−0·28	VH42 OWNOCC	47·58	−0·10	−0·27	ang
1·47	0·99	VH43 COUNCL	32·10	0·80	0·55	ang
1·39	1·19	VH44 UNFRPR	15·86	1·77	1·07	ang
5·95	0·38	VH45 FURNPR	4·32	4·82	−0·77	log
7·66	1·62	VH46 SHARE	3·37	3·55	−0·18	log
−1·36	−0·47	VH47 ALAMEX	83·25	−1·29	−0·63	ang
2·33	1·02	VH48 HWDEF	7·97	1·98	0·81	ang
1·96	0·89	VH49 BATHDF	11·54	1·73	0·86	ang
1·74	0·80	VH50 WCDEF	13·95	1·51	0·67	ang
0·40	−0·27	VH51 NOCAR	48·99	−0·19	−0·35	ang
1·51	0·60	VH52 MLTCAR	8·58	2·65	1·33	ang
Household composition, size, and density						
2·61	0·71	VS53 P/HH*	2·914	0·97	0·71	recip
1·07	−0·38	VS54 HHONE	17·64	0·82	−0·13	ang
0·73	0·01	VS55 HHTWO	31·19	0·32	−0·34	ang
2·08	0·35	VS56 HHLARG	6·27	2·00	0·31	ang
0·88	−0·37	VS57 PENSHH	19·52	0·59	−0·52	ang
−0·18	0·80	VS58 NOCHIL	63·49	−0·58	−0·15	ang
1·43	−0·43	VS59 1CHIL	14·37	1·13	−0·77	ang
1·16	−0·16	VS60 2+CHIL	22·19	1·05	0·31	ang
2·75	0·60	VS61 P/RM*	0·617	1·68	−0·05	recip
2·90	0·93	VS62 1+P/RM	7·51	2·51	0·98	ang
0·57	0·46	VS63 SPACUS	32·13	0·05	−0·28	ang
2·34	1·08	VS64 RM/HH*	4·813	0·19	2·03	recip
2·24	0·89	VS65 COMPCT	16·56	1·79	1·09	ang
1·36	0·36	VS66 ROOMY	8·14	2·45	1·02	ang

Table 6.2 (continued).

Unweighted			Weighted			
original skew	skew after transform-ation	variable number	population weighted mean (%)	original skew	skew after transform-ation	transform

Migration variables

2·75	1·50	VM67 1MIGRT	1·12	3·31	−3·54	log
0·77	0·46	VM68 5MIGRT	3·35	1·18	0·64	ang

Education variables

2·36	−0·40	VE69 SCHOOL	11·31	2·23	−2·86	log
2·60	−0·20	VE70 GRAD	8·99	2·50	−2·10	log

Socioeconomic groups

2·29	−0·60	VG71 MANAG	9·80	3·00	−2·96	log
4·73	0·47	VG72 PROFES	3·79	4·56	−1·25	log
3·34	−0·08	VG73 INTNOM	8·07	3·34	−2·71	log
1·31	−0·75	VG74 JUNNOM	22·05	0·33	−4·90	log
1·21	−0·87	VG75 FORSKL	24·83	0·32	−4·62	log
1·57	−1·18	VG76 SEMSKL	19·48	1·62	−4·86	log
3·67	0·16	VG77 UNSKL	7·93	2·04	−2·17	log
3·45	0·08	VG78 SELFNP	4·05	7·62	−1·86	log
8·70	2·69	VG79 DEFCE	0·97	11·88	1·14	log
1·76	−0·87	VG80 URCLS	13·59	2·16	−3·20	log
0·69	0·31	VG81 MIDCLS	34·17	0·33	−0·23	ang

Industrial occupations

2·31	0·43	VO82 FARMRS	1·72	8·44	0·78	log
6·39	2·17	VO83 MINERS	1·78	5·35	1·15	log
0·74	0·26	VO84 MANUF	34·83	0·10	−0·63	ang
3·46	−0·13	VO85 CONSTR	7·21	4·62	−2·97	log
3·85	0·05	VO86 TRPTUT	8·23	2·69	−2·78	log
0·44	0·22	VO87 SERVIC	39·57	0·55	0·29	ang
3·95	0·12	VO88 GOVDEF	6·66	6·15	−2·38	log

Travel to work

0·28	0·20	VT89 CARTW	35·94	0·73	0·65	ang
1·48	0·71	VT90 BUSTW	25·18	0·54	−0·23	ang
4·77	1·22	VT91 TRANTW	6·17	2·05	−0·31	log
7·21	1·18	VT92 MCTW	1·46	8·90	−0·57	log
4·14	0·54	VT93 PCTW	4·16	3·64	−0·90	log
1·58	−0·72	VT94 FONOTW	20·95	1·19	−3·67	log

Additional variables

3·33	−0·40	VP95 UNEMPH	4·01	2·45	−3·63	log
5·42	0·63	VP96 SICKPH	1·11	4·26	−1·77	log
1·81	0·86	VH97 NOINWC	11·07	1·98	0·91	ang
6·34	0·90	VS98 OVCRD	1·96	4·43	−0·92	log
6·93	0·79	VG99 UNCLASS	2·64	6·70	−1·35	log
1·81	−0·03	VP100 RETRD2	10·14	1·88	0·34	ang
3·93	2·64	VS101 LONPAR	9·34	2·82	0·40	ang
3·40	2·39	VP102 WKMUM	9·94	3·55	0·87	ang

Figure 6.1 represents the size of this data matrix in terms of the number of unsuppressed squares. In detail the number of squares available for ratio variables varies from denominator to denominator: a denominator of zero gives an indeterminate ratio, which should not be confused with a zero ratio.

Each correlation is based upon the intersection of unsuppressed squares for the two variables involved: this varies from 152 440 squares for POP versus SEXMAL, which covers 100% of the population, through 67 546 squares for pairs of age variables and 68 422 for 'private household' variables to 53 277 squares for 'travel-to-work' variables. The use of 'pairwise deletion' for missing values was convenient since many computer runs were necessary to produce the (102 × 102) correlation matrix [in strips of (102 × 3)]; casewise deletion would have involved further processing.

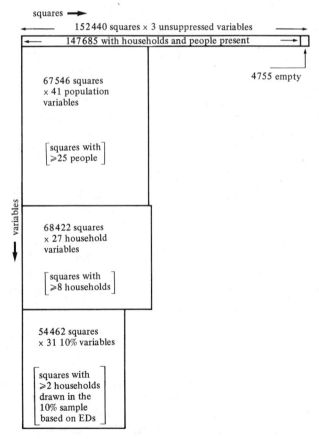

Figure 6.1. Size of data base, after compression to 102 'ratio' variables. The height of each box is proportional to the number of variables in that group, and the length is proportional to the number of squares that have escaped suppression. The 10% sample was drawn for each enumeration district (ED), so the percentage sampled varies between grid squares.

6.3 Transformation

Two procedures are strongly advisable before *r* values are calculated from data such as these. First the skewness of many variables exceeds $\pm 1 \cdot 0$ (the most extreme is $+12 \cdot 7$), and should be reduced by transformation (reexpression). This usually permits a closer approximation to the requirements of the General Linear Model, by reducing dependence of variance on mean, nonnormality of error variances, and curvilinearity of relationships, and by improving additivity of different effects. Second, the weight given to each grid square in calculation of a correlation coefficient should be proportional to its population (Gilbert, 1973).

The 102 variables studied here are of three types: (1) absolute numbers, POP and HOUSEH; (2) ratios with no upper limit, YFERT, MFERT, CFERT, AGINDX, P/RM, P/HH, and RM/HH; of these the last two have a lower limit of one by definition, whereas the others have lower limits of zero; (3) closed ratios, proportions varying from zero to one because the numerator is a subset of the denominator and cannot exceed it. Ninety-three of the variables are closed ratios, as indeed are most statistical variables in social geography.

The three types of variable have different frequency distributions and no single transformation would be appropriate. For each variable a transformation is required that will increase the probability of coming close to the requirements of the General Linear Model, for a series of relationships with other variables: it would be undesirable to use different transforms for the same variable in different relationships. There are grounds for believing that transforms which minimise skewness of the univariate frequency distribution are most likely to have the desired all-round effect (Evans et al, 1975).

The specific transformations listed in table 6.2 are applied. Absolute numbers per grid square have J shaped distributions (many squares with few, few squares with many) and a logarithmic transformation reduces but does not eliminate the positive skew. Square-root transformation would leave greater skew. The fertility variables YFERT, MFERT, and CFERT are moderately skewed and a square-root transformation seems adequate; reciprocal transformation is too severe. For the three ratios between numbers of rooms, of people, and of households, however, reciprocal transformation is useful in reducing skew. This is tantamount to redefinition, for instance 'persons per room' becomes 'rooms per person'.

For all closed ratios, there are grounds for applying the angular transformation (the arcsine of the square root of the proportion) to stabilise variance for binomial ratios (Bartlett, 1947). This transformation compresses the middle part of the scale (around 50%), and stretches both ends, where small differences (1% versus 2%, 98% versus 99%) are of greater importance. Hence it has little effect on variables such as VP03, VP04, VP08, VP09, VY28, and VH42 which have means near 50% (table 6.2); it reduces the negative skew of majority variables such as

VP10 and VH47; and it reduces the positive skew of minority variables such as VP12 and VH48–VH50. For the complete set of closed ratios that deal with age, household size, and composition (VY17–VY28 and VS54–VS60, defined in table 6.9), skewness is reduced to less than ±1·0. (A skewness of ±1·0 is still troublesome and markedly nonnormal, but is unlikely to cause great distortions in further analysis.) The angular transformation is relatively gentle, and only produces small changes in relationships.

For internal birthplace variables (VE30–VE32), however, frequency distributions are bimodal, with most values well above 50% for the country concerned (England, Wales, or Scotland), and well below 50% for the rest of Britain. No simple transformation can (or should) remove this feature, and the correlations of such variables must be interpreted with caution in any analysis of the whole of Great Britain. The angular transformation is nevertheless applied, not because of the small reduction in skewness it produces but rather for consistency with the other closed ratios, all of which are transformed.

Unfortunately the angular transformation is inadequate to remove the positive skew of many closed ratios, especially those for small minorities (figure 6.2). Where there is a tendency for spatial concentration or segregation from the rest of the population, as for immigrants (VE33–VE40), defence workers (VG79), farmers (VO82), and miners (VO83), a logarithmic transformation seems appropriate and is more effective than the angular transformation in reducing skewness. Transformations were

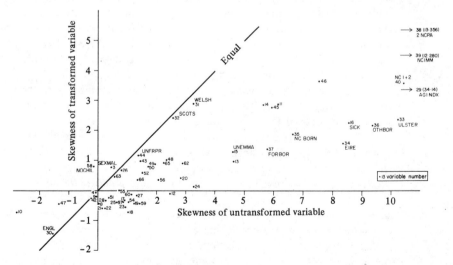

Figure 6.2. The effect of angular transformation on skewness, with weighting by population per grid square both for transformed and untransformed variables. Reduction is general, but skewness for birthplace and a number of other 'minority' variables remains disturbing.

selected on the basis of histograms and degree of skewness without
population weighting, as given in the left-hand columns of table 6.2.
Unfortunately the right-hand columns of this table show that weighting
and transformation interact, and many 10% variables (for example,
socioeconomic groups) which have low skew when *unweighted* after
logarithmic transformation, give a marked negative skew when also
weighted for population. For these, an angular transformation might be
more appropriate to a weighted analysis.

The problem of choice of transformations may be caused by a peculiar
characteristic of many of the frequency distributions, a separate secondary
(or primary) mode at zero [figure 6.3(a)]. This is not eliminated on
transformation [figure 6.3(b)], but it decreases in importance as attention
is confined to increasingly populous squares [figure 6.3(c)]. It seems, then,
to be due to the small number of discrete possibilities for ratios that involve
small numbers; for example with a mean of 5%, 0/10 is more likely than
1/10 and intermediate values are impossible. This confirms the importance
of weighting each grid square by its population, otherwise squares with
very low populations are given undue prominence, as in the first two
histograms: in a histogram, each grid square is taken as an equal unit.

If several areas are to be compared, it is useful to apply the same
transformations in each. The work of Coulter (1977) dealt with population
and thirty-six variables related to deprivation, for fourteen areas defined
as metropolitan cores and rings by LSE (1974). Coulter considered the
effect of a number of transformations on skewness and kurtosis, and
selected those given in table 6.3: the variables are a subset of those in
table 6.2, but the transformations selected sometimes differ, for example
for INFANT and ALAMEX. The results provide important information
on the suitability of the same transformation for a series of different
areas. Almost all these transformations produced great or considerable
reductions in skew, for almost all the cores and ring areas.

The right-hand columns summarise the skews remaining after both
transformation and weighting for population. They show that the process
has been successful for most variables, but that several (MLTCAR and
5MIGRT) retain consistent positive skews, many of which are serious.
For these, the angular transformation is too gentle, whereas for YFERT
the logarithmic transformation is slightly too strong because it consistently
produces small negative skews. The household amenity variables HWDEF,
BATHDF, WCDEF, and NOINWC are undertransformed, but the residual
skew is serious only in ring areas; likewise use of the square of ALAMEX
is a poor transform for every ring area.

Attempts to choose one suitable transform for a variable, regardless of
the area, represent a somewhat different approach to that of Evans et al
(1975), who minimised skewness separately for each area. To facilitate
comparison within a set of comparably-defined areas, it seems worthwhile
to tolerate some increase in skewness. Furthermore, the use of fractional

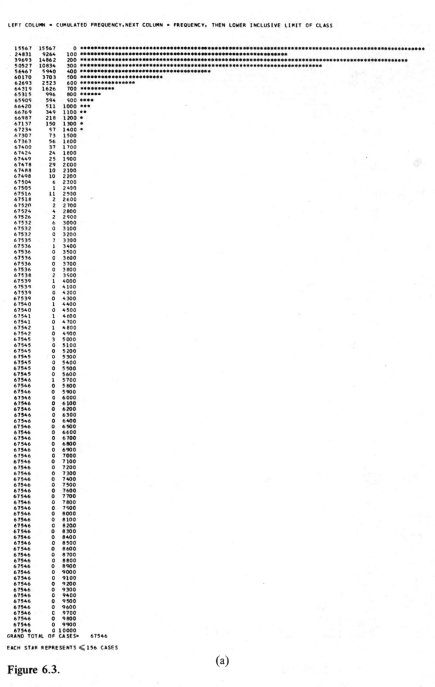

LEFT COLUMN = CUMULATED FREQUENCY,NEXT COLUMN = FREQUENCY, THEN LOWER INCLUSIVE LIMIT OF CLASS

EACH STAR REPRESENTS ⩽ 156 CASES

(a)

Figure 6.3.

LEFT COLUMN = CUMULATED FREQUENCY,NEXT COLUMN = FREQUENCY, THEN LOWER INCLUSIVE LIMIT OF CLASS

```
14093   14093      0  ****************************************************************************************
14227     134     100  *
15567    1340     200  *********
22172    6605     300  ***********************************************
36779   14607     400  ***********************************************************************************
50527   13748     500  ******************************************************************
58159    7632     600  *******************************************
62693    4534     700  *******************************
65187    2494     800  **************
66372    1185     900  ********
66958     586    1000  ****
67249     291    1100  **
67390     141    1200  *
67467      77    1300  *
67498      31    1400
67515      17    1500
67524       9    1600
67532       8    1700
67536       4    1800
67536       0    1900
67539       3    2000
67540       1    2100
67542       2    2200
67545       3    2300
67545       0    2400
67546       1    2500
67546       0    2600
67546       0    2700
67546       0    2800
67546       0    2900
67546       0    3000
67546       0    3100
67546       0    3200
67546       0    3300
67546       0    3400
67546       0    3500
67546       0    3600
67546       0    3700
67546       0    3800
67546       0    3900
67546       0    4000
67546       0    4100
67546       0    4200
67546       0    4300
67546       0    4400
67546       0    4500
67546       0    4600
67546       0    4700
67546       0    4800
67546       0    4900
67546       0    5000
67546       0    5100
67546       0    5200
67546       0    5300
67546       0    5400
67546       0    5500
67546       0    5600
67546       0    5700
67546       0    5800
67546       0    5900
67546       0    6000
67546       0    6100
67546       0    6200
67546       0    6300
67546       0    6400
67546       0    6500
67546       0    6600
67546       0    6700
67546       0    6800
67546       0    6900
67546       0    7000
67546       0    7100
67546       0    7200
67546       0    7300
67546       0    7400
67546       0    7500
67546       0    7600
67546       0    7700
67546       0    7800
67546       0    7900
67546       0    8000
67546       0    8100
67546       0    8200
67546       0    8300
67546       0    8400
67546       0    8500
67546       0    8600
67546       0    8700
67546       0    8800
67546       0    8900
67546       0    9000
67546       0    9100
67546       0    9200
67546       0    9300
67546       0    9400
67546       0    9500
67546       0    9600
67546       0    9700
67546       0    9800
67546       0    9900
67546       0   10000
GRAND TOTAL OF CASES=   67546
```

EACH STAR REPRESENTS ≤146 CASES

(b)

Figure 6.3 (continued)

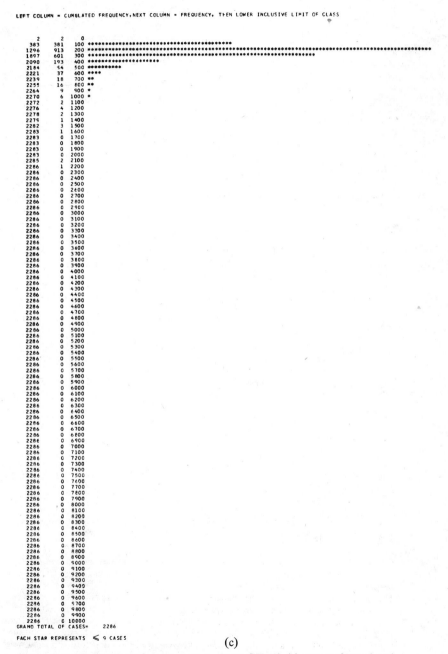

Figure 6.3. Histograms of percentage students (VP12) (a) untransformed and
(b) angular transformed. The sharp mode at zero relates mainly to squares with low
populations. It is absent from histograms for squares with 5000 or more people,
shown (c) untransformed.

Table 6.3. Effect of weighting and transformation on skewness of thirty-six variables for seven metropolitan 'rings' and seven 'cores' (Birmingham, Glasgow, Leeds, Liverpool, London, Manchester, and Newcastle), summarised from tables of skewness in Coulter (1977, pages 121–126).

Skew						Skew						
improvement on weighting:		improvement on transformation:					number of negative:		greatest remaining:		number exceeding ±1·0:	
rings	cores	rings	cores				rings	cores	rings	cores	rings	cores
O	G	G	W	log	YFERT	−	7	7	−0·51	−1·06	0	1
W	C	−	−	no	WORKMF	−	6	5	−1·02	−1·15	2	2
O	W	G	C	ang	STUDNT		2	0	−0·66	2·11	0	5
W	O	G	G	log	UNEMPL		5	4	0·67	0·61	0	0
W	C	G	C	log	UNEMFE	(−)	5	6	−0·84	−0·88	0	0
W	O	G	C	log	UNEMMA	(−)	4	7	−0·83	−0·74	0	0
W	O	G	G	log	SICK		5	7	0·97	−0·97	0	0
O	O	C	S	sqrt	INFANT		6	5	−1·15	1·24	1	1
S	C	G	G	ang	VOLD	+	0	2	1·19	0·38	1	0
S	O	−	−	no	CHILDR	(+)	1	2	0·86	0·78	0	0
O	C	G	G	log	ULSTER		4	5	−0·53	−1·12	0	1
W	C	G	G	log	EIRE		3	4	−0·30	0·74	0	0
W	S	G	G	log	NCBORN	(+)	2	1	0·47	0·75	0	0
O	O	O	S	ang	OWNOCC		5	2	0·75	±0·39	0	0
S	O	S	C	ang	COUNCL		3	1	0·87	0·79	0	0
W	C	C	G	ang	UNFRPR	+	0	1	4·03	0·52	6	0
W	C	G	G	log	FURNPR		1	2	0·73	−0·59	0	0
C	C	G	G	log	SHARE	(+)	0	2	0·88	0·90	0	0
W	C	C	C	square	ALAMEX		7	5	−3·78	0·82	7	0
W	C	G	C	ang	HWDEF	+	0	2	2·14	1·06	4	1
O	C	G	C	ang	BATHDF	+	0	1	2·21	1·01	4	1
O	C	G	G	ang	WCDEF	+	0	1	2·16	0·85	3	0
O	O	−	−	no	NOCAR	−	6	7	−0·80	−1·07	0	2
W	O	C	G	ang	MLTCAR	+	0	0	2·12	1·95	7	6
G	C	G	S	log	P/HH		1	3	0·94	−0·58	0	0
S	O	G	G	ang	HHLARG	(+)	2	0	0·71	0·90	0	0
G	O	−	−	no	PENSHH		3	7	0·59	−1·00	0	1
C	C	C	C	log	P/RM		3	3	0·64	0·47	0	0
C	S	G	G	sqrt	1+P/RM		4	2	0·65	−0·38	0	0
O	C	G	G	log	1MIGRT		6	6	−0·71	0·66	0	0
W	O	C	S	ang	5MIGRT	+	0	0	1·32	2·46	3	4
S	O	G	G	log	SCHOOL		6	4	−0·80	±0·32	0	0
O	C	G	C	sqrt	UNSKL		6	7	−0·68	−0·56	0	0
W	O	S	C	ang	CARTW	(+)	0	0	0·72	0·79	0	0
W	C	G	G	ang	NOINWC	+	0	0	2·43	1·00	3	1
W	C	G	G	log	OVCRD		3	7	−0·89	−1·13	0	2

G = great improvement (most skews halved, or better).   O = no consistent change.
C = considerable improvement.   W = worse (most skews increased).
S = small but consistent improvement.
The left-hand columns describe the changes in skewness owing to weighting by population, followed by the *further* changes owing to the transformation given in the fifth column. The transformations were chosen by Coulter for this specific data set, and differ from those used for Great Britain. The sixth column gives the abbreviation for each variable, and the right-hand columns all describe the remaining skewness after both transformation and weighting for population. The number of negative skews is out of seven, so six or seven indicate consistent negative skew whereas zero or one indicate consistent positive skew. If the results for rings and cores are consistent, and the greatest remaining skew agrees in sign, a + or − sign is given in the seventh column: this is bracketed if no skews are serious (stronger than ±1·0).

powers such as $0\cdot63$ or $1\cdot81$, for example in Box–Cox transforms, has
been found to give only small improvement over powers such as $0\cdot5$
(square root) or $2\cdot0$ (square), and the increased difficulty in interpretation
does not seem worthwhile. Likewise, the specific transformations
(logarithms after addition of a fitted constant) used by Lindqvist (1976)
produce low skewness at the expense of ignoring the special characteristics
of closed ratio data: hence the need for clumsy 'mirror transformations'
for negatively skewed variables.

The need for transformations (Mosteller and Tukey, 1977, chapter 5)
is not yet accepted by all geographers. Where coefficients of variation
are lower than in the present study, the effects of transformation are
correspondingly reduced: nevertheless, it is safer to test them.

Roff (1977) found that transformation had little effect on the strength
of principal components and on the frequency distribution of correlation
coefficients. What is more important, however, is the *composition* of
principal components, and the values of individual correlation coefficients.
Evans et al (1975) found that a number of the latter were seriously
affected, and this changed the correlation structure for Durham County.
After transformation, an 'age' cluster of variables separated clearly from
a 'deprivation' cluster, including (negatively) ALAMEX (exclusive use of
all amenities), OVCRD (overcrowding, more than $1\cdot5$ persons per room),
and POP (population density). Without transformation, the correlations
of the three latter variables, each of which had a highly skewed frequency
distribution, were grossly understated and they were excluded from the
'deprivation' cluster.

The correlation of persons per household (P/HH, in the 'age' cluster)
and rooms per household (RM/HH, in the 'deprivation' cluster) was
overstated without transformation, giving a spurious 'bridge' between the
two clusters of variables. It must be admitted, however, that these
correlations were not weighted for population: grid squares with 25 or
with 10000 people were treated as equal. The (pre-1974) administrative
districts considered by Roff (1977) varied from 1200 to 1 100 000 people
and the larger populations involved probably gave lower ranges for the
various ratios, thus explaining the possibly reduced effect of transformations.

Figure 6.4 shows the importance of transformation for population.
The strengths of almost all correlations are increased when the logarithm
of population is taken, together with appropriate transformations of the
other variables. Without transformations, population is shown to be
related to New Commonwealth variables and to low car ownership, but
only after transformations are the relationships with members of the
population who are unemployed, sick, Ulster-born, or visitors, and
furnished private–rented households clear. The strong positive skew of
population per grid square, and of some of these variables, reduces
correlation: logarithmic transformation has a powerful effect and gives a

clearer picture of relationships. With such data, transformation is not only desirable, it is essential.

In further work, the logit transform [favoured by Wrigley (1976)] was tested in the hope that it would be suitable for all closed ratios. Unfortunately it overtransforms many variables, such as age and household composition variables, giving strong negative skewness both with and without weighting. On the other hand it provides a modest reduction in skew for variables log-transformed in this study.

My recommendation for future analyses, then, is that only transformations such as angular and logit, symmetrical about 50%, should be used for closed ratios, and logarithmic or power transformations should never be used for such variables. For the 102 variables studied here, logit should be used for birthplace, ethnic, tenancy, and travel-to-work variables, and for PNOTPH, VISIT, DEFCE, FARMRS, MINERS, and OVCRD. Angular should replace log for unemployment, education, migration, industrial occupation, and socioeconomic groups except DEFCE, FARMRS, and MINERS. Angular transformation should be retained for age; household

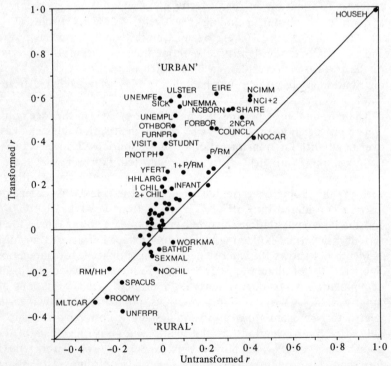

Figure 6.4. Effect of transformation on unweighted correlation coefficients for **POPULATION** per 1 km square, for census variables in the 100% records. Almost all correlations are strengthened after application of the transformations listed in table 6.2.

size, composition, and density; amenities; car ownership; work variables, and STUDNT. Other difficult cases are OWNOCC, COUNCL, BUSTW, and FONOTW, which are better served by angular transforms, unlike other variables in their sets. For MARFEM, MLTCAR, SELFNP, and GOVDEF, which are marginal as between angular and logit transforms, the angular is preferred since it is a weaker transformation and requires no special treatment for zero values. For weighted analyses, POP and HOUSEH can be square-root transformed, like the fertility variables. RM/HH is better untransformed, but reciprocal transformation is useful for P/HH and P/RM.

These changes in transformation do have considerable effects on weak and moderate correlations, emphasising the need to transform and to make a careful choice of transformations. The logarithmic transformation tends to give a positive bias to correlation coefficients.

6.4 Weighting

The case for weighting for population has already been mentioned: when the absolute numbers involved have been lost through ratio formation, weighting by the denominator is highly desirable unless the denominator is near constant. This is readily achieved when sums are cumulated, by multiplying each contribution to the sum by its denominator. The formula is comparable to that for correlation with grouped data, and the result is *not* the same as correlating absolute numbers. The denominators of all ratios used here correlate strongly with total population ($r > +0 \cdot 86$), so rather than take weighted means of the two denominators involved, total population is used to weight all ratio correlations.

In a small data set with much of the population concentrated in one square, it would be necessary to guard against this having a dominating influence on any correlation coefficient. But the total population of some fifty-four million in our study is over 2000 times the largest individual weight. If we had a sample of 2000 independent observations, correlations of $\pm0 \cdot 04$ would be significant at the 95% confidence level. Although values for adjacent grid squares are positively autocorrelated, interpretable correlations are well in excess of $\pm0 \cdot 04$, and each is based on at least 53 000 differently-weighted squares. It is hoped that this gives sufficient leeway that problems of statistical significance—always dubious for geographical aggregates (Gould, 1970)—can be ignored; the following interpretation therefore concentrates on the magnitude of correlations.

For grid squares, the variation in population per square is so great that weighting by population is almost forced upon the researcher. It might be thought that since enumeration districts are defined to have approximately 160 households (in urban areas), their population would be approximately equal. However, data presented by Holtermann (1975) for 85 578 enumeration districts (those in administratively rural districts, 'special' enumeration districts, and those with under fifty people in private households having been excluded), show that their population varies considerably even for urban areas. The mean population (in 1971) was 470, standard deviation 159, and some populations exceeded 1200.

In addition to reducing the 'spikiness' (discreteness) of histograms, especially the mode at zero, population weighting might be expected to improve the analysis in several other respects. Figure 6.5 shows that variability is usually reduced, because part of the unweighted variability is caused by differences between squares with different populations, and variability (of ratios) is usually greater for grid squares with lower populations. The exceptions are New Commonwealth variables and 'shared dwellings', which are so preferentially concentrated in urban areas that their variability is greater after population weighting. Correlations are most often increased by population weighting, though the patterns of correlation are broadly consistent. However, as table 6.2 shows, skewness is increased by weighting almost as often as it is decreased: likewise table 6.3 shows no consistent improvement. Hence we must look to transformation to reduce skewness.

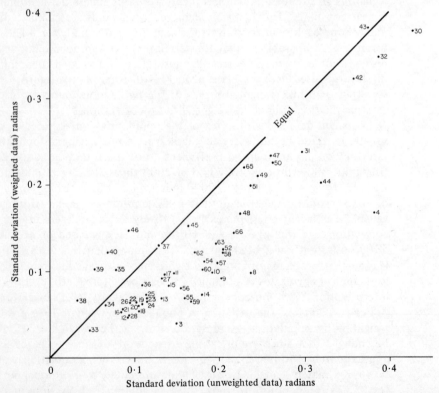

Figure 6.5. Effects of weighting by population on the standard deviation of closed ratio variables, which have been angular transformed throughout. The reduction is not as strong or as consistent as might be expected. Variables are numbered as in table 6.2.

Weighted correlations were calculated by using the OSIRIS (1973) package, which proved rather faster than SPSS (1975) and produced comparable results.

6.5 Geographical correlations

Transformation and population weighting lead to product-moment correlation coefficients which better describe relationships, and the remaining problems are those of interpretation and presentation of these results. The relationships described relate to areas—grid squares, not individual people or households. Sociologists call them 'ecological' correlations, but they might be called 'geographical' correlations in that they are based on geographical individuals—places 1 km x 1 km in area. Households without a car (NOCAR) are spatially associated with crowded $(1 + P/RM)$ and with one-person households (HHONE); a greater proportion of one implies a greater proportion of the other, but the households with each of these characteristics may be different: they are just in the same grid square.

Most of the relationships at the 1 km aggregate level make sense in terms of associations within individual households, for example, pensioner households (PENSHH) are likely to be childless households (NOCHIL), hence there is a strong positive correlation. But a few do not, for example, one of the stronger correlations $(+0·35)$ of lone-parent families (LONPAR) is with one-person households (HHONE), though the two are mutually exclusive types of household. Presumably they are spatially associated in rooming-house areas, or areas with small or cheap housing units. Hence we must avoid the 'ecological fallacy' of assuming individual relationships on the basis of spatial aggregate relationships: likewise the converse, and various other scale-transference fallacies, should be avoided (Alker, 1969). Nevertheless, contradictory relationships such as LONPAR : HHONE are not common, and are more likely for minorities such as these, than for larger groups where strong individual relationships are likely to produce aggregate relationships.

The correlation coefficient is an aspatial statistic in that once the areas (grid squares) are defined, their relative location is ignored. A very strong correlation $(>±0·9)$ does imply that the spatial distributions are very similar, but a very weak correlation $(<±0·3)$ may nevertheless be accompanied by some complex relationship at the 1 km scale, or by a stronger linear relationship at a broader scale. Furthermore, there could be contradictory relationships in different regions, weakening the overall relationship.

In interpreting the correlations between 102 variables, it is essential to simplify in various ways, rather than just to present the whole (102 x 102) correlation matrix. A valuable initial step (initial in terms of presentation, but formulated *after* many of the following analyses) is to distinguish clusters of mutually correlated variables within which all correlations are strong and most exceed ±0·7: these are defined in table 6.4. The largest clusters are those related to children and to the aged, each of which contains a number of household composition variables as well as age variables: a minus sign indicates a negative relationship, for example children are rarely found in one-person households and this individual relationship leads to a strong negative spatial association. These two clusters overlap in a fringe of quite strongly related variables.

Then there is a cluster of five (or seven) immigrant variables, and others of four variables related to household, five related to amenities, four related to car ownership and to travel to work, and a number of smaller clusters involving pairs of variables that overlap either in definition (SICK, SICKPH), or owing to strongly exclusive categories (ENGL and SCOTS, $r = -0·74$; OWNOCC and COUNCL, $r = -0·78$). The strongest relationship, which again is not surprising, is between numbers of people and of households. Relationships between variables in *different* clusters tend to be of greater interest.

Table 6.4. Clusters of variables as defined by correlations stronger than ±0·7. The main clusters are confirmed in terms of Duncan's dissimilarity index for the numerators.

Children:	CHILDR, SCHCHL, 2+CHIL, −NOCHIL, P/HH (weaker) −HHONE, 1CHIL, INFANT
Aged:	AGED, RETAGE, RETRD2, OLD, VOLD, PENSHH (weaker) −YOPEOP, and main children variables
Immigrants:	NC1+2, NCIMM, 2NCPA, NCBORN, FORBOR (weaker) OTHBOR, EIRE
Crowding:	P/RM, 1+P/RM, OVCRD, −SPACUS
Unemployment:	UNEMMA, UNEMPH, UNEMPL but *not* UNEMFE
Amenities:	−ALAMEX, WCDEF, HWDEF, BATHDF, NOINWC
Cars:	−NOCAR, MLTCAR, CARTW
Numbers:	POP, HOUSEH ($r = 0·987$)
Tenure:	OWNOCC, −COUNCL
Birthplace:	ENGL, −SCOTS
Rooms:	RM/HH, −COMPCT but *not* ROOMY
Hotels:	PNOTPH, VISIT
Sick:	SICK, SICKPH
Working females:	WORKFE, WORKMF
Employers and managers:	URCLS, MANAG

Note: Fertility variables YFERT, MFERT, CFERT do *not* cluster.

6.6 Correlates of population density, unemployment, and overcrowding
Another useful step is to take the correlations of a single variable and to arrange them in rank order. Table 6.5 does this for population per grid square, which (when arbitrary pieces of the sea for coastal squares are included) is roughly equivalent to population density. Brackets are used for the abbreviations of variables in the same cluster (that is, HOUSEH),

and also for second and later variables in other clusters, for example, NCIMM is the first of the 'immigrants' cluster, so the other three New Commonwealth variables, and the total foreign-born, are bracketed. The intention here is to focus interest on the unbracketed correlations.

Table 6.5. Population-*weighted* correlation coefficients of *transformed* ratio variables with log POP (total population present, in a 1 km square), in rank order, for 1 km x 1 km squares in Great Britain. The first column gives correlations of corresponding log *numerators* with log POP (see also table 6.10).

0·984	(HOUSEH	0·987)	Number of households (in 1 km square)
0·800	1MIGRT	0·629	Changed address in last year
0·826	EIRE	0·622	Born in Republic of Ireland
0·715	UNCLASS	0·582	Occupation (SEG) unclassified
0·801	SICK	0·580	Sick economically active persons
0·820	TRPTUT	0·579	Transport and utilities employees
0·704	OVCRD	0·577	Overcrowded households (>1·5 persons per room)
0·792	ULSTER	0·575	Born in Northern Ireland
0·644	NCIMM	0·573	New Commonwealth born, parents likewise
0·868	UNEMMA	0·560	Unemployed economically active males
0·693	(NC1+2	0·555)	Both parents, or one parent and self born in New Commonwealth
0·616	SHARE	0·554	Households in shared dwellings
0·867	UNEMFE	0·549	Unemployed economically active females
0·784	UNSKL	0·542	Unskilled workers
0·946	NOCAR	0·533	Households with no car
0·778	(NCBORN	0·525)	Born in New Commonwealth
0·864	(MLTCAR	−0·515)	Households with ⩾2 cars
0·519	(2NCPA	0·503)	Born in Great Britain, both parents born in New Commonwealth
0·847	BUSTW	0·503	Travelling to work by bus
0·904	(UNEMPL	0·501)	Unemployed economically active persons
0·787	GOVDEF	0·483	Government and defence employees
0·866	JUNNOM	0·479	Junior nonmanual workers
0·890	(FORBOR	0·475)	Born outside UK
0·804	CONSTR	0·470	Construction workers
0·600	TRANTW	0·459	Travelling to work by train
0·742	LONPAR	0·440	Lone-parent families with dependent children
0·807	INTNOM	0·440	Intermediate nonmanual workers
0·968	WORKFE	0·426	Economically active females, of those aged 15–59 years
0·801	FONOTW	0·418	Travelling to work on foot, or no travel
0·854	FORSKL	0·411	Foremen and skilled workers
0·610	MCTW	0·404	Travelling to work by motor cycle
0·963	(WORKMF	0·404)	Economically active married females, of those aged 15–59 years
0·864	OTHBOR	0·403	Others born abroad
0·855	(CARTW	−0·399)	Travelling to work by car
0·853	COMPCT	0·398	Compact households (1, 2, or 3 rooms)
0·691	FURNPR	0·387	Furnished private-rented households
0·652	SELFNP	0·384	Self-employed nonprofessionals

Variables which correlate strongly with population density are of great
interest in that they show strong urban bias (if positive) or rural bias (if
negative: only MLTCAR has a considerable rural bias). Proportions of
migrants, sick, unemployed, and unskilled workers, of overcrowded,
shared, or carless households are greater in more densely populated areas.
More people there travel to work by bus, train, or other means, rather
than by car, and there are larger proportions of Irish, New Commonwealth,
and other immigrants. Some relationships with total population may be
curvilinear even after transformation.

Table 6.6. Population-*weighted* correlation coefficients of *transformed* variables with
log UNEMMA (*male unemployment* per economically active males aged 15–64 years),
in rank order, for 1 km × 1 km squares in Great Britain. The first column gives
corresponding figures for log UNEMFE.

Female		Male	
0·708	(UNEMPL	0·849)	Unemployed economically active persons
0·548	HOUSEH	0·562	Number of households (in 1 km square)
0·549	(POP	0·560)	Total population present (per 1 km square)
0·321	NOCAR	0·531	Households with no car
0·413	SICK	0·511	Sick economically active persons
−0·349	(MLTCAR	−0·493)	Households with ≥2 cars
0·373	OVCRD	0·482	Households with >1·5 persons per room
1·000	(UNEMFE	0·473)	Unemployed economically active females
0·359	UNSKL	0·457	Unskilled workers
0·294	BUSTW	0·440	Travelling to work by bus
0·398	UNCLASS	0·389	Occupation unclassified
0·246	(1+P/RM	0·388)	Households with >1 person per room
0·432	1MIGRT	0·376	Changed address within last year
0·329	FORSKL	0·376	Foremen and skilled workers
0·372	TRPTUT	0·374	Transport and utilities employees
0·345	CONSTR	0·369	Construction employees
−0·182	(CARTW	−0·338)	Travelling to work by car
0·230	LONPAR	0·337	Lone-parent families with dependent children
0·209	COMPCT	0·335	Compact households (1, 2, or 3 rooms)
0·276	SEMSKL	0·326	Semiskilled workers
0·364	ULSTER	0·322	Born in Northern Ireland
0·343	EIRE	0·310	Born in Republic of Ireland
0·186	COUNCL	0·310	Council-rented households
0·303	FONOTW	0·302	Travelling to work on foot, or no travel
0·212	P/RM	0·287	Persons per room
−0·206	ROOMY	−0·286	Roomy households (≥7 rooms)
−0·167	RM/HH	−0·283	Rooms per household
−0·145	OWNOCC	−0·279	Owner-occupied households
0·341	JUNNOM	0·277	Junior nonmanual workers
0·199	HHLARG	0·274	Large households (≥6 persons)
0·255	SHARE	0·267	Households in shared dwellings
0·127	HHONE	0·258	One-person households
0·327	GOVDEF	0·250	Government and defence employees

If we look in turn at one of these important variables, namely male unemployment (table 6.6), we note first that its correlation with the female unemployment rate is only 0·47 (or 0·30 for untransformed variables), which causes reservations about combining the two in UNEMPL. Its strongest substantive correlation is with population (or household) density, followed by NOCAR, SICK, OVCRD, and UNSKL. These four are, like unemployment, commonly considered to be aspects of deprivation, and their association makes sense both for individual households and for spatial aggregates: each is associated with lack of resources and opportunity. Likewise, although it improves accessibility to social facilities and opportunity for social interaction, high population density may be viewed as deprivation of space. However, these correlations are only moderate ($r = 0·46$ to $0·56$, roughly one quarter of their variability is

Table 6.7. Population-*weighted* correlation coefficients of *transformed* ratio variables with (arcsine square root) $1 + $P/RM (*crowded* households, those with more than one person per room), in rank order, for 1 km x 1 km squares in Great Britain. The first column gives corresponding figures for log OVCRD.

OVCRD		1 + P/RM	
0·548	(P/RM	0·825)	Persons per room
1·000	(OVCRD	0·708)	Overcrowded households (>1·5 persons per room)
−0·447	(SPACUS	−0·701)	Spacious households (<0·5 persons per room)
−0·483	OWNOCC	−0·629	Owner-occupied households
0·416	HHLARG	0·628	Large households (≥6 persons)
0·587	COMPCT	0·608	Compact households (1, 2, or 3 rooms)
0·634	NOCAR	0·587	Households with no car
−0·574	(MLTCAR	−0·557)	Households with ≥2 cars
0·382	(COUNCL	0·553)	Council-rented households
−0·491	RM/HH	−0·551	Rooms per household
0·493	BUSTW	0·531	Travelling to work by bus
0·302	SCOTS	0·531	Born in Scotland
−0·388	ROOMY	−0·519	Roomy households (≥7 rooms)
−0·506	(CARTW	−0·492)	Travelling to work by car
0·309	MFERT	0·484	Middle fertility (of married women 30 years– 44 years)
−0·298	(ENGL	−0·473)	Born in England
−0·266	HHTWO	−0·471	Two-person households
0·265	CFERT	0·440	Completed fertility (married women 45 years– 64 years)
0·487	UNSKL	0·394	Unskilled workers
0·482	UNEMMA	0·388	Unemployed economically active males
0·484	SICK	0·369	Sick economically active persons
0·216	YFERT	0·367	Young fertility (married women 16 years–29 years)
0·436	(UNEMPL	0·366)	Unemployed economically active persons
0·577	POP	0·345	Total population present (per 1 km square)
0·113	CHILDR	0·350	Children (persons 0 years–14 years old)
0·102	(SCHCHL	0·332)	Schoolchildren (persons 5 years–14 years old)
−0·091	RETAGE	−0·311	Men ≥65 years old plus women ≥60 years old
0·563	(HOUSEH	0·306)	Number of households (per 1 km square)

shared with UNEMMA), so they cannot be said to form a close-knit cluster. Without weighting or transformation, correlations within this group are even lower. Maps of unemployment show strong regional contrasts between north and south, centre and periphery. Female unemployment [table 6.6, column (a)] has lower correlations with deprivation and household size and space, but equal correlations with POP, TRPTUT, and CONSTR. (Further analyses with angular rather than logarithmic transformations give considerably lower correlations for unemployment variables.)

When the correlates of crowding $(1 + P/RM)$ are considered (table 6.7), the other deprivation variables except for NOCAR come well down the list (see the next section). Crowding relates more to (council) tenure, number of people, (small) number of rooms per household, and to Scotland.

Table 6.8 ranks the correlates of foreign-born. The first five are either partially (NC1 + 2) or totally constituents of FORBOR; 2NCPA has the

Table 6.8. Population-*weighted* correlation coefficients of *transformed* variables with log FORBOR (*born outside UK*), in rank order, for 1 km × 1 km squares in Great Britain.

(OTHBOR	0·843)	Born outside UK, Eire, and New Commonwealth
(NCBORN	0·771)	Born in New Commonwealth
(NC1 + 2	0·700)	Both parents, *or* one parent and self, born in New Commonwealth
(EIRE	0·684)	Born in Republic of Ireland
(NCIMM	0·678)	Born in New Commonwealth, parents likewise
2NCPA	0·603	Born in Great Britain, both parents born in New Commonwealth
FURNPR	0·551	Furnished private-rented households
SHARE	0·489	Households in shared dwellings
POP	0·475	Total population present (per 1 km square)
ULSTER	0·463	Born in Northern Ireland
(HOUSEH	0·457)	Number of households (in 1 km square)
TRANTW	0·398	Travelling to work by train
VISIT	0·386	Visitors
1MIGRT	0·367	Changed address in last year
PNOTPH	0·358	Persons not in private households
HHONE	0·296	One-person households
PROFES	0·295	Professional workers (SEG 3 and 4)
SCHOOL	0·293	Employed persons with ONC, HSC, or 'A' level
STUDNT	0·290	Students (aged ⩾15 years)
INTNOM	0·283	Intermediate nonmanual workers (SEG 5)
WORKMF	0·276	Economically active married females 15–59 years
GRAD	0·273	Employed persons with HNC or degree
SERVIC	0·267	Workers in distribution and service
HWDEF	0·267	Households that share or lack hot water
UNCLASS	0·266	Occupation unclassified
COMPCT	0·263	Compact households (1, 2, or 3 rooms)
(WORKFE	0·256)	Economically active females, 15–59 years)
OVCRD	0·253	Overcrowded households (>1·5 persons per room)
MINERS	−0·251	Employed in mining

next strongest correlation, presumably because of the individual relationship between these people born in Great Britain and their NCBORN parents. The foreign-born are concentrated in urban areas, especially those areas with much furnished private–rented accommodation and/or shared dwellings. ULSTER immigrants are not a constituent of FORBOR, but they do seem to associate with EIRE. Foreign-born are often visitors and recent migrants, but the correlations with TRANTW and 1MIGRT may be owing to their spatial association in urban areas, especially London. They are also associated with qualified and professional people.

In terms of relations between ethnic variables, there are very close relations between the four 'New Commonwealth' definitions. 2NCPA is the most dissimilar from other variables, whereas NCBORN is the most similar to ENGL. EIRE and ULSTER are correlated; both these and OTHBOR relate to ENGL rather than to WELSH, SCOTS, or New Commonwealth groups.

6.7 Comparative correlations

The next step is to compare the correlations of two related variables, for example crowding, defined as the proportion of households with over one person per room ($1 + P/RM$), with 'average crowding', the overall number of persons in the grid square divided by the number of rooms. Such comparison is sometimes attempted by plotting two correlation profiles, with the other variables in arbitrary order. Graphic efficiency can easily be improved by plotting variables in rank order of their correlations with one of the two variables being compared, and in the same order for the other variable: but it is much better still to plot one set of correlations against the other, as in figure 6.6 (which excludes 10% sample variables).

Both variables are related to households with many people (HHLARG) and few rooms (COMPCT, $-RM/HH$), to fertility (YFERT, MFERT, CFERT), to Scots, and (negatively) to owner occupation. There are, however, important differences despite the strong correlation of $0·89$ between $1 + P/RM$ and P/RM (untransformed: $0·825$ for transformed). Average crowding relates more strongly to children, to council tenancy (COUNCL), and (negatively) to old people. Crowded households, on the other hand, have a relatively stronger spatial association with deprivation variables such as poor housing (HWDEF, BATHDF, WCDEF, $-ALAMEX$), unemployment, and lack of a car, and with New Commonwealth parentage. If the threshold is raised to $1·5$ persons per room giving OVCRD, these changes are emphasised further [table 6.7, left-hand column]. Correlations with age variables almost disappear, and those with fertility, SCOTS, P/RM, HHLARG versus HHTWO, and COUNCL versus OWNOCC are greatly reduced. Correlations with NOCAR, POP, UNSKL, SICK, and UNEMMA are increased to $0·634$, $0·577$, $0·487$, $0·484$, and $0·482$ respectively. Only the correlation with COMPCT is stable, at around $0·6$. OVCRD also correlates $0·373$ with UNEMFE (unemployed females), $0·370$ with ULSTER, $0·361$ with EIRE, $0·352$ with SHARE, $0·348$ with 2NCPA, and 0·343 with WORKFE, but less than 0·3 with poor housing. Clearly OVCRD is a deprivation variable, whereas P/RM is more influenced by children increasing household size, and 1 + P/RM is a compromise. 1 + P/RM involves 7·5% of the households in unsuppressed squares, whereas OVCRD covers only 2·0%.

Figure 6.7 makes a similar comparison for (angular transformations of) the two main tenure types, owner occupation (OWNOCC), and council renting (COUNCL): they form respectively 48% and 32% of households in this study. They are mutually exclusive and their overall proportions alone would give a closure correlation of some −0·65: the actual correlation of −0·78 shows a small additional tendency to segregate. On this diagram two sets of quadrants are labelled. In the top right both are positive, in the bottom left neither is positive, in the top left OWNOCC only, and in the bottom right COUNCL only. If diagonal dividing lines are taken, in the top and bottom quadrants OWNOCC has the stronger correlations, whereas COUNCL has the stronger correlations in the left and right quadrants. Because of the strong negative correlation between OWNOCC and COUNCL, their correlations with other variables also have a strong negative relationship, with most points falling in the top left and bottom right quadrants.

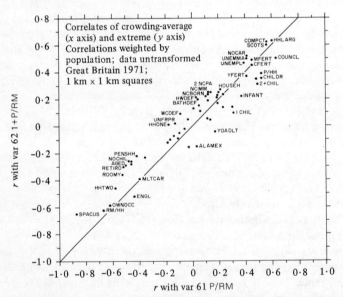

Figure 6.6. Correlates of crowding: a comparison of those for average degree of crowding (overall persons per room, VS61) with those for the proportion of households with over one person per room (VS62, which represents the upper tail of the frequency distribution of crowding per household), after population weighting but without transformation. The correlation between these two variables is 0·89 (0·83 after transformation).

OWNOCC has stronger (positive) correlations with car ownership and use, with average rooms per household, with managerial and professional classes and graduates, and (negative) with youths and lone-parent families. COUNCL has stronger correlations with children and fertility (factors involved in the allocation of council housing), and with MANUF and FORSKL. COUNCL has weak positive correlations with household amenities (ALAMEX), whereas OWNOCC is unrelated to this and related variables: this is a reminder of the heterogeneity of owner-occupied property, which includes much old housing as well as modern suburban estates. Negative correlations between COUNCL and the other two tenure types (FURNPR and UNFRPR), which are weakly related to OWNOCC despite its greater proportion of households (and thus greater 'closure'

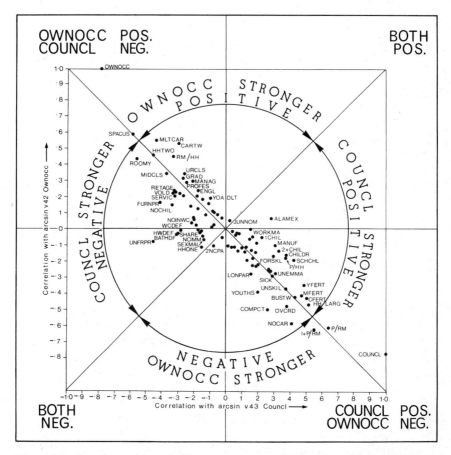

Figure 6.7. Correlates of household tenure; a comparison of those for proportion owner-occupied (VH42) with those for proportion council rented (VH43), after angular transformation and weighting for population. The correlation between these two variables is −0·78.

bias towards negative correlation), show that council renting is the most
segregated housing type. In fact, there is a positive association—an
intermixing—of OWNOCC and FURNPR (furnished private-rented households). Finally, the associations of COUNCL positively with unemployed,
unskilled, and crowding, and negatively with spacious (SPACUS) and
ROOMY households are balanced by the converse associations of OWNOCC.

These two examples show that, even where two variables are closely
related, differences in their correlation with third variables are informative.
Such information is lost when a cluster of variables is represented by one
of its number, and when a cluster is compressed into one dimension, a
'component' or 'factor'. This reinforces Meyer's (1971) reservations
about factor analysis compared with the careful study of correlations.
Each variable has been defined because it is of some substantive interest:
its identity should not be submerged in a 'factor'.

6.8 Correlation structure: age and household size
In addition to highlighting subtle differences in the correlations of related
variables, it is important to examine the internal structure of a cluster,
that is, the pairwise correlations between all member variables. Figure 6.8
attempts this graphically for the seventeen variables defined in table 6.9,
which form the two clusters related to children and the aged, with related
household size and composition variables (HHTWO and HHLARG have
rather weaker correlations). This is a correlation structure diagram, in
which the $(17 - 1)$-dimensional space where distances between these
variables could be accurately represented is compressed subjectively into
two dimensions. Because distortions due to the loss of fourteen dimensions
are considerable, no attempt is made to approximate the objective
multidimensional scaling solution by relying on distance alone: instead,
both proximity *and* thickness of line are used to represent similarity
between variables. For example, MIDAGE has relatively weak correlations
and in sixteen dimensions it would be some distance off, in a dimension
at right angles to the paper: here it is shown at one side. 'Similarity' is
measured here by strength of correlation, regardless of sign.

Age divisions are made at the years five, fifteen, twenty-five, forty-five,
sixty-five, and seventy-five and again at fifteen, forty-five, and sixty-five to
give broader groups. This produces a high degree of redundancy,
especially between AGED and OLD and, together with RETAGE and
PENSHH, these have considerable overlaps in definition, and form the core
of the 'aged' cluster. Variables involving children are also closely interrelated by definition: INFANT and SCHCHL are subdivisions of CHILDR,
households with 2+CHIL are likely to correlate with overall proportions
of children, and childless households (NOCHIL), which form 63·5% of
private households, not surprisingly have strong negative correlations with
these other variables.

As is often the case then, the strongest correlations arise from our definitions and are of little interest. Their study is, however, salutary in showing how the cores of well-defined 'components' or 'factors' can be produced purely by including sets of variables with overlapping definitions: this is commonly done in factorial ecology. Although there are again subtle differences in the correlations of, say, AGED and OLD, it is reasonable to take only one variable from each cluster for mapping, when only some 30 of the 102 variables can be included in an atlas. We could choose either AGED or RETAGE, plus either CHILDR or 2+CHIL.

The negative correlations of HHONE with 'children' variables are stronger than its positive correlations with 'aged' variables. P/HH correlates with both groups, and most strongly with 2+CHIL. A notable feature of

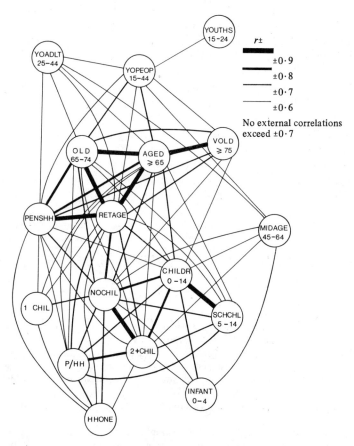

Figure 6.8. Correlation structure within the age and household size and composition cluster, of untransformed variables defined in table 6.9, weighted for population. Variables are positioned in relation to the strength (modulus) of correlation, which is portrayed also by the thickness of line.

this structure is the position of YOADLT, YOUTHS, and their sum, YOPEOP. Their negative correlations with 'aged' variables exceed their positive ones with 'children' variables. Given the strength of negative correlation between 'aged' and 'children' variables, YOADLT, YOUTHS, and YOPEOP must be positioned at the periphery of the structure, and on the 'aged' side.

A clearer view, in this case, may be obtained by considering pure age variables only, and by positioning variables in relation to the *sign* as well as the magnitude of their correlation [figure 6.9(a)]. This permits portrayal of the relatively weak positive correlations between YOADLT, INFANT, and SCHCHL, and between MIDAGE and the two subsets of AGED; OLD, and VOLD. Across the 'great divide' at age 45 years (± <10 years) all correlations are negative, between −0·47 and −0·70 except for YOUTHS. The latter has weak negative correlations with MIDAGE, OLD, and VOLD and is practically uncorrelated with groups adjacent to it in age. Figures 6.8 and 6.9 both portray correlations of untransformed variables. Because most of these groups are large, their frequency distributions are only slightly skewed, and correlations after angular transformation are only slightly different: there is a tendency for negative correlations to be reduced, and positive ones to be slightly increased.

Figure 6.9(b) shows correlations between the more aggregated age variables. Here the negative correlations are reinforced, whereas the positive ones are roughly averaged compared with correlations between constituent groups. This is less informative than the detailed picture, especially since YOPEOP is the sum of two variables, YOADLT and

Table 6.9. Definitions of age group and (below) related variables for size and composition of private households. Age group variables are per 10000 total population.

INFANT	SCHCHL	YOUTHS	YOADLT	MIDAGE	OLD	VOLD	(a) detailed
0	5	15	25	45	65	75 years	
	CHILDR		YOPEOP		MIDAGE	AGED	(b) broad

RETAGE	=	Retirement-aged: males 65 years and over, females 60 years and over, per 10000 persons in private households
PENSHH	=	Pensioner households: 1- and 2-person households composed solely of pensioners, per 10000 private households
NOCHIL	=	Households with no children, per 10000 private households
1CHIL	=	Households with one child, per 10000 private households
2+CHIL	=	Households with two or more children, per 10000 private households
P/HH	=	Persons per household: number of persons per number of private households
HHONE	=	One-person households, per 10000 private households
HHTWO	=	Two-person households, per 10000 private households
HHLARG	=	Households with six or more persons, per 10000 private households

YOUTHS, which correlate −0·146 (−0·141 transformed). The constituents of CHILDR correlate just 0·385 (0·415 transformed), and only those of AGED which correlate 0·728 (0·760 transformed) can really justify amalgamation.

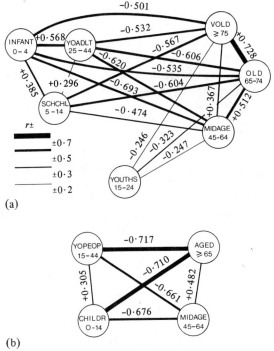

(a)

(b)

Figure 6.9. Correlation structure for untransformed age variables defined in table 6.9, weighted for population: (a) detailed classification, (b) aggregate classification. Note; CHILDR and especially YOPEOP group together two variables with little intercorrelation. The broader aggregation increases the effects of array closure (inbuilt negative correlation).

6.9 The closure problem: comparisons with dissimilarity index

The stronger negative correlations between more aggregated groups are an instructive exemplification of the effect of *closure*. There is an inbuilt negative correlation between proportions that are mutually exclusive constituents of a fixed total (their denominator) (Mosimann, 1962; Chayes, 1971). If 40% of the population in one square are AGED, the proportions in all other groups taken together cannot exceed 60%. The greater the proportions in any one group, the less there are to share among the others. Hence the inbuilt negative correlations are greater for large groups than for minorities: this is a good reason for using detailed subdivisions in any correlation study. Some aggregate groups are useful, however, for mapping, to avoid production of a long series of similar maps.

Several strategies for coping with closure correlation have been proposed elsewhere (Evans, 1978): here only two will be assessed, the use of correlations between absolute numbers and, especially, of Duncan and Duncan's (1955) coefficient of dissimilarity, D. This is widely used in studies of segregation and minority groups, for example in many of the papers reprinted in Schwirian (1974), and it would appear to have broader value in measuring the dissimilarity of two distributions. It is based not on proportions of a grid square total, but on proportions of a national total, that is, it compares *absolute numbers* after scaling. Two groups are expressed as proportions of their respective national totals, and the difference between their proportions is calculated in each grid square, then summed and divided by two:

$$D = \tfrac{1}{2} \sum \left| \frac{x}{\sum x} - \frac{y}{\sum y} \right|,$$

where the summation is over all grid squares in the system. The result is a measure of dissimilarity on a scale from 0 to 1, and D equals the minimum proportion of *one* variable which would have to be moved between grid squares, to make its distribution identical to that of the other variable. Note that like r, D is an aspatial measure in that the relative positions of grid squares are not considered.

Since we are dealing with absolute numbers, not proportions of a variable total, no weighting is required, nor would transformation be appropriate given the way in which D is defined. A flexible program to calculate D, capable of handling the present large data set, has been written by Evans, Visvalingam, and Rhind. It is applied to the file of 121 numerators and denominators of the ratio variables (Rhind et al, 1977, pages 42–49) and employs casewise deletion for missing values. The results are more like correlations between (logarithms of) absolute numbers, than correlations between ratios.

The great disadvantage of D is its sensitivity to the set of boundaries used. Since it expresses proportions which would have to cross boundaries, in order to balance distributions between the areas used, any differences in boundaries or grouping of areas is likely to be reflected in D. With traditional irregular areas, values of D can be compared only for one set of boundaries *in one area*: they cannot be used to compare segregation in London with segregation in Glasgow, for example for enumeration districts, since it cannot be assumed that districts were defined in comparable fashion: local authorities are invited to influence the drawing of boundaries. However, with grid squares D has much greater value since it can be said to relate to a particular *scale* of segregation, in this case the 1 km scale. It would be feasible to compare different cities, so long as each contained a large number of grid squares. (Otherwise the particular incidence of grid boundaries may become important.)

Because D values relate to absolute numbers, and most large groups of the population are strongly positively correlated, D values between large groups tend to be low, showing similarity. For example, between 'males' and 'persons in private households', $D = 0 \cdot 025$: very little change would be needed to bring the two into exact proportion in every grid square. Both of these variables have a similar pattern of D values, and most are below $0 \cdot 5$, whereas for 'persons *not* in private households' most are above $0 \cdot 6$. For 'pensioner households' most D values are between $0 \cdot 1$ and $0 \cdot 5$, and for 'households with over one person per room' most are between $0 \cdot 2$ and $0 \cdot 7$.

Further understanding of values of D may be obtained by comparison with correlations between (logarithms of) absolute numbers.

Table 6.10 shows numerators which are unusual in that their correlations with total population are low; they are ranked in order of increasing correlation. Farmers are uncorrelated with the population as a whole, whereas miners and defence workers both have low correlations, followed

Table 6.10. The most 'unusual' or segregated census variables, ranked in order of increasing correlation of log (numerator) with log (total population present). These variables are considered the least suitable for ratio treatment: they should preferably be analysed as absolute numbers.

FARMRS	$-0 \cdot 044$	Farmers and agricultural employees
DEFCE	$0 \cdot 343$	Members of the armed forces (socioeconomic group 16)
MINERS	$0 \cdot 348$	Mining employees
PNOTPH	$0 \cdot 506$	Persons not in private households
2NCPA	$0 \cdot 519$	Born in Great Britain, both parents born in New Commonwealth
TRANTW	$0 \cdot 600$	Travelling to work by train
MCTW	$0 \cdot 610$	Travelling to work by motor cycle
SHARE	$0 \cdot 616$	Households in shared dwellings
PCTW	$0 \cdot 627$	Travelling to work by pedal cycle
(NCIMM	$0 \cdot 644$)	Born in New Commonwealth, parents likewise
SELFNP	$0 \cdot 652$	Self-employed nonprofessional workers
NOINWC	$0 \cdot 652$	Households with no inside w c
WELSH	$0 \cdot 661$	Born in Wales
SCOTS	$0 \cdot 677$	Born in Scotland
FURNPR	$0 \cdot 691$	Furnished private-rented households
(NC1+2	$0 \cdot 693$)	Both parents, or one parent and self, born in New Commonwealth
PROFES	$0 \cdot 701$	Professional workers (socioeconomic groups 3 and 4)
OVCRD	$0 \cdot 704$	Overcrowded households ($>1 \cdot 5$ persons per room)
HWDEF	$0 \cdot 704$	Households share or lack hot water
BATHDF	$0 \cdot 714$	Households share or lack fixed bath
WCDEF	$0 \cdot 714$	Households share or lack inside w c
UNCLASS	$0 \cdot 715$	Occupation (socioeconomic group) unclassified
WKMUM	$0 \cdot 740$	Working mothers of children <5 years old
LONPAR	$0 \cdot 742$	Lone-parent families with dependent children
MANAG	$0 \cdot 750$	Employers and managers (socioeconomic groups 1, 2, and 13)
GRAD	$0 \cdot 754$	Employed persons with HNC or degree

by persons not in private households and second-generation New Commonwealth immigrants. Numerators with moderate correlation to total population include several travel-to-work minorities, households in shared or furnished private–rented or overcrowded dwellings, households deficient in amenities, self-employed and professional people, Welsh and Scottish people. Those not on this list have strong correlations with total population.

Correlations between logarithms of numerators are not directly comparable to correlations between ratios until their correlation with total population is partialled out. They are, however, comparable with D values (figure 6.10). The latter are relatively low for English-born members of the population, but high for groups concentrated in certain regions, such as Scottish-, Welsh-, and foreign-born: in general, the relationship is close (and negative).

As defined by D, the age structure again has an 'aged' cluster and a 'children' cluster of variables, and is broadly comparable to the correlation

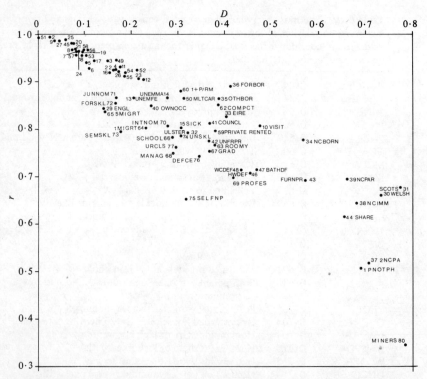

Figure 6.10. A comparison of product-moment correlation coefficients of logarithms (r), and dissimilarity indices (D), between numerator fifty-one (persons in private households) and the first eighty numerators. Numerator numbers are as defined in Rhind et al (1977, pages 42–49); abbreviations (not given for variables most similar to population) are those for the corresponding ratios.

structures *with sign taken into account* (figure 6.9 rather than figure 6.8), but there are some further differences. MIDAGE is now almost as similar to YOUTHS (0·130) and YOADLT (0·135) as it is to OLD (0·128). INFANT and SCHCHL are now as closely related (0·119) as OLD and VOLD, and both relate equally (0·099, 0·098) to YOADLT. It seems that the lower positive correlations of SCHCHL, compared with INFANT and VOLD, are caused by its being a larger proportion of the total population (16% compared with 8% and 5%) which produces greater closure effects, that bias correlations negatively. Closure likewise accounts for most of the negative correlation between MIDAGE, YOADLT, and YOUTHS. Hence if correlations for closed proportions are to be useful, and comparable, variables must be defined not only as small proportions, but also as roughly equal proportions overall, that is, large groups and small groups should not be used in the same classification.

In the more aggregate grouping, the age sequence CHILDR-YOPEOP-MIDAGE-AGED is apparent, with D values of 0·079, 0·121, and 0·139 for adjacent age groups, 0·158 and 0·210 for those two-steps apart, and 0·245 between the extremes, CHILDR-AGED. This seems a big improvement over the picture given by ratio correlations.

If aggregate and detailed classifications are mixed, there are some very low D values because of overlapping definitions, for example, CHILDR : SCHCHL 0·040. Household composition variables are still closely related: 2+ CHIL is in the core of the 'children' cluster, but NOCHIL and HHTWO fall between MIDAGE and RETAGE. HHONE relates to NOCHIL and PENSHH whereas 1CHIL is on the fringe of the 'children' cluster, including YOADLT and YOPEOP. Here again D values are in effect taking the sign of correlation into account. Finally the three fertility variables (YFERT, MFERT, and CFERT) have as numerators 'number of children ever born to married women aged ...' for the age groups 15–29, 30–44, and 45–59. Hence the first two numerators relate to INFANT and to SCHCHL. The second and third relate to YOADLT and MIDAGE respectively, that is, to the age groups of the mothers involved.

It should be noted that MIDAGE and younger groups have D values with large groups (such as 'people', 'males' and the numerators of WORKMA, WORKFE, WORKMF, and ALAMEX) which are often lower than those with other age variables.

A further comparison between D and r values may be made by plotting one against the other for all pairwise relationships between age variables (figure 6.11). This shows much more scatter than figure 6.10, because we are now considering ratio correlations rather than absolute number correlations. However, if variables that are large proportions of the population are distinguished by crosses, we can see that their D values are low for a given r: on the other hand, D values for minorities (cased) are high. This is due to two effects: r values for large, mutually exclusive groups are negatively biased by closure, and D values for large groups are lower because both usually have strong similarity to total population.

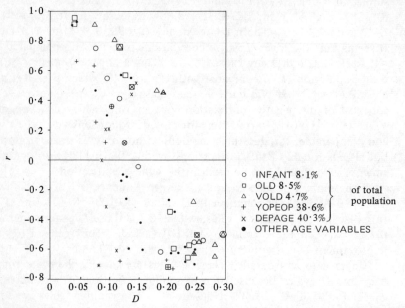

Figure 6.11. A comparison of dissimilarity index (D) between numerators, and product-moment correlation coefficients (r) between corresponding ratios, for age-group variables. Note that correlations involving small groups (INFANT, OLD, VOLD) are high, and those involving large groups (YOPEOP, DEPAGE) are low, relative to D. The other age variables (RETAGE, SCHCHL, YOUTHS, YOADLT, MIDAGE, CHILDR, and AGED) form between $13 \cdot 2\%$ and $24 \cdot 2\%$ of total population, by population-weighted mean values for unsuppressed grid squares.

6.10 Socioeconomic correlation structure: comparison of D and r

Table 6.11 compares values of D and of r for occupations, socioeconomic groups, and qualifications. Correlations are fairly weak—perhaps because these 10% variables have greater 'noise': the highest are $0 \cdot 748$, $0 \cdot 675$, and $0 \cdot 711$ for GRAD with SCHOOL, INTNOM, and PROFES respectively, except for $0 \cdot 914$ between URCLS and its subset MANAG. URCLS ('upper class') is the sum of MANAG and PROFES (managerial and professional) and MIDCLS ('middle class') is the sum of INTNOM, JUNNOM, and SELFNP (intermediate and junior nonmanual workers, and self-employed nonprofessionals): both are dominated by their largest constituents and add little to the picture given by the detailed classification. No close-knit clusters, then, are found among these variables.

Both D and r suggest the same structure for the socioeconomic groups. At one end comes UNSKL (unskilled), closely related to FORSKL and SEMSKL (foremen and skilled, and semiskilled). In the middle comes JUNNOM, then INTNOM and MANAG, with PROFES as the most segregated at the opposite extreme to UNSKL. SELFNP is intermediate but has low correlations and high dissimilarities with other groups. Its strongest link is with JUNNOM; its D values are relatively high (dissimilar) for UNSKL and PROFES, whereas its r value is low for SEMSKL. Defence employees (DEFCE) have high dissimilarities ($0 \cdot 75$ to $0 \cdot 79$) with all other groups, and no correlation exceeds $\pm 0 \cdot 1$ (note, they are included in GOVDEF).

The fact that PROFES rather than MANAG is the extreme group probably reflects the definition of the former being narrow and of the latter, wide, so that they cover $3 \cdot 8\%$ and $9 \cdot 8\%$ of the workforce respectively. Otherwise the relationships are as expected, except that FORSKL is as similar to UNSKL as is SEMSKL, and their D values with JUNNOM are comparable, that is FORSKL and SEMSKL seem to occupy comparable positions in any hierarchy of social segregation. The separation of UNSKL, SEMSKL, and FORSKL ('working class') from others is emphasised by r values with JUNNOM being reduced by the closure constraint, since FORSKL, JUNNOM, and SEMSKL are the three largest groups. The same effect reduces correlations between PROFES, MANAG, INTNOM, and SELFNP less than those with JUNNOM, so that the centrality of JUNNOM is deemphasised in the correlation structure. The structure based on D values is probably more reliable.

There are fewer relationships between industrial occupations, with most correlations being near zero or negative owing to the closure constraint, and rising to $-0 \cdot 575$ between the two largest groups, MANUF and SERVIC (manufacturing and services). As expected, farmers and miners have distributions quite different from other groups (cf table 6.10). Outside this classification, both SERVIC and GOVDEF relate to qualifications (SCHOOL and GRAD) and to nonmanual socioeconomic groups. SERVIC also correlates with VISIT ($0 \cdot 410$), STUDNT ($0 \cdot 404$), OTHBOR ($0 \cdot 349$), and negatively with children.

6.11 Travel to work and car ownership
Six modes of dominant travel to work are distinguished in the 10% sample data. These do not form a cluster of variables, indeed there is a basic dichotomy between use of car (CARTW) which correlates with car ownership and (negatively) with population density, and all the other modes (especially bus) which have the inverse correlations. BUSTW correlates $0 \cdot 623$ with NOCAR, $-0 \cdot 602$ with MLTCAR, $0 \cdot 503$ with POP, $0 \cdot 530$ with $1 + $P/RM, $0 \cdot 469$ with SICK, $-0 \cdot 468$ with CARTW, $0 \cdot 449$ with UNSKL, $0 \cdot 440$ with UNEMMA, and $0 \cdot 438$ with COUNCL. In addition to the inverse correlations, CARTW correlates $0 \cdot 538$ with OWNOCC, $-0 \cdot 465$ with COMPCT (households with one or two rooms), and $0 \cdot 422$ with ALAMEX.

Table 6.11. Comparative values (× 1000) of (*D*) dissimilarity indices between numerators and (*r*) product-moment correlation coefficients between corresponding ratios, for occupational, class, and educational variables. Log transforms were used

GRAD	MANAG	PROFES	INTNOM	JUNNOM	FORSKL	SEMSKL	UNSKL	SELFNP	DEFCE	URCLS
748	*463*	*627*	*541*	*273*	*−469*	*−451*	*−394*	*−037*	*051*	*596*
186	243	259	217	274	415	412	492	424	786	205
	504	*711*	*675*	*194*	*−519*	*−498*	*−466*	*−015*	*026*	*668*
	258	216	213	341	480	476	555	454	793	207
		411	*266*	*138*	*−507*	*−475*	*−467*	*041*	*−000*	*914*
		314	268	273	404	397	492	378	789	087
			399	*237*	*−434*	*−438*	*−404*	*001*	*033*	*709*
			324	382	517	518	590	482	806	227
				219	*−375*	*−394*	*−316*	*−033*	*078*	*330*
				250	381	381	460	395	782	253
					−210	*−346*	*−181*	*−020*	*114*	*143*
					262	270	362	342	781	276
						207	*350*	*−156*	*−051*	*−580*
						178	259	383	800	416
							328	*−121*	*−013*	*−544*
							245	369	798	411
								−107	*045*	*−534*
								445	811	506
									013	*018*
									800	391
										−001
										790

for DEFCE, FARMRS, and MINERS, and angular transforms for all others (unlike those used in table 6.2 and elsewhere in this chapter). Population-weighted mean percentages for unsuppressed grid squares are given on the right.

MIDCLS	FARMRS	MINERS	MANUF	CONSTR	TRPTUT	SERVIC	GOVDEF		
440	−042	−139	−254	−129	−032	441	148	r	SCHOOL
245	769	843	395	385	372	246	358	D	11·3%
471	−018	−129	−300	−174	−093	513	115	r	GRAD
299	770	849	455	443	433	297	400	D	9·0%
218	073	−163	−278	−101	−062	393	058	r	MANAG
247	731	838	385	365	363	230	359	D	9·8%
327	−038	−147	−191	−115	−023	404	119	r	PROFES
360	787	867	486	478	466	357	438	D	3·8%
623	−062	−110	−235	−123	007	465	177	r	INTNOM
193	771	830	364	358	338	213	334	D	8·1%
703	−136	−134	−064	−008	256	266	307	r	JUNNOM
077	775	821	254	265	222	140	281	D	22·0%
−492	−084	319	521	246	141	−521	−168	r	FORSKL
273	770	741	148	241	282	274	385	D	24·8%
−545	095	072	235	044	014	−295	−129	r	SEMSKL
274	735	784	189	256	280	255	380	D	19·5%
−413	−091	135	296	230	185	−286	004	r	UNSKL
374	789	784	286	310	330	355	431	D	7·9%
354	148	−090	−152	137	014	106	−031	r	SELFNP
303	691	832	386	335	373	313	418	D	4·1%
096	018	−046	−132	019	027	022	393	r	DEFCE
778	893	937	806	784	794	775	656	D	1·0%
258	053	−184	−314	−141	−096	456	071	r	URCLS
246	741	843	393	377	374	231	359	D	13·6%
	−039	−211	−328	−085	083	463	263	r	MIDCLS
	758	821	263	262	237	107	278	D	34·2%
		−028	−192	016	−105	−047	−040	r	FARMRS
		900	786	742	782	754	769	D	1·7%
			065	050	−049	−195	−087	r	MINERS
			789	789	823	819	831	D	1·8%
				−028	−017	−575	−287	r	MANUF
				290	311	286	400	D	34·8%
					109	−151	−046	r	CONSTR
					287	260	357	D	7·2%
						−029	019	r	TRPTUT
						248	349	D	8·2%
							−031	r	SERVIC
							294	D	39·6%
									GOVDEF
									6·7%

Other modes have very low correlations. Although MCTW and PCTW (use of motor cycle and pedal cycle, respectively) correlate only 0·375, their other correlations are very similar, starting with POP. Like FONOTW (foot and no travel to work) they have low (0·1 to 0·3) correlations with urban variables such as immigrants, unemployment, working women, and the industries TRPTUT, CONSTR, and MANUF (but not SERVIC). FONOTW has stronger correlations with NOCAR, with manual workers, with HHONE, and with overcrowding, but none exceed 0·45. Some of these may be based on correlations at the individual level (for example, NOCAR-FONOTW), but others such as all positive travel to work versus unemployment relationships must be purely spatial associations, since the variables are mutually exclusive at the individual level.

TRANTW (travelling to work by train) has correlations of 0·46 with POP and between 0·33 and 0·43 with the various immigrant groups (VE33–VE40) and SHARE, but this is probably because all of these concentrate in the London region. Otherwise its highest correlations are 0·33, with TRPTUT and PROFES.

6.12 Broader correlation structure

Having considered relations within some of the main clusters and groups, we can now study the overall correlation structure (figure 6.12) for selected variables, for example, INFANT, SCHCHL, and AGED from the age clusters; all of the tenure variables; and those related to deprivation with ALAMEX and UNEMMA representing 'amenities' and 'unemployment' clusters. These have been chosen after inspecting the ranked correlations of each variable.

The correlations define perhaps five groups of variables, none of which is especially strong or isolated. OWNOCC and COUNCL have strong negative closure correlation, but both are included because although OWNOCC has a stronger negative correlation with NOCAR, COUNCL has stronger relations with demographic variables (as discussed above in 'comparative correlations'). Both relate strongly to crowding, and form a 'crowding and major tenures' group.

The three age variables relate to P/HH and HHONE in an 'age and household size' group, with INFANT on the periphery, related also to MARFEM (the proportion of females aged fifteen to twenty-nine years old who are married). Sex ratio (SEXMAL) relates only to AGED, reflecting the lesser survival of males, and STUDNT correlates with service employment, and negatively with male economic activity and MARFEM. Unfurnished private–rented households (UNFRPR) have a strong negative spatial association with household amenities (ALAMEX), but the latter also relates negatively to HHONE, NOCAR, and SHARE. NOCAR is the only variable which correlates strongly with all the deprivation variables, but they do not form a distinct cluster: ALAMEX has just been considered and crowding relates more to major tenure types.

These two are linked only via NOCAR to the main deprivation cluster of
NOCAR, SICK, UNEMMA, and POP, which intercorrelate between +0·5
and +0·6.

   UNSKL (unskilled) relates mainly to NOCAR and POP (total population),
as does LONPAR (lone-parent families) with weaker correlations.
FORSKL (foremen and skilled workers) relates to UNSKL and POP, and
also to MANUF (manufacturing workers). URCLS (managerial and
professional) has weak correlations, mainly with qualifications, SERVIC
and INTNOM (not included in figure 6.12), OWNOCC and STUDNT.
WKMUM (working married women with children under five years old) is
even less related to other variables: its strongest correlations are 0·290
with POP (0·275 with HOUSEH), 0·238 with INFANT, and −0·233 with
MLTCAR. Its denominator is 'working married women': 'married
women with children under five years old' would have been preferred but
is not available.

   Finally, a distinct 'immigrant housing' group relates the variables
FORBOR (foreign-born) and NC1+2 (New Commonwealth parentage) to

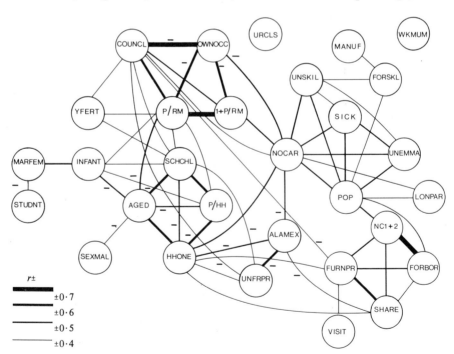

Figure 6.12. Correlation structure for twenty-four transformed census variables for
1 km grid squares, Great Britain 1971: correlations weighted by population. As far
as possible, variables are positioned in relation to all correlations exceeding ±0·3, so as
to be close to strongly correlated variables and far from weakly correlated variables.
Strength of correlation is shown also by the thickness of connecting lines, and negative
relationships are shown by minus signs.

FURNPR (furnished private-rented households) and SHARE (shared dwellings). The first two relate to POP.

Clearly the groups mentioned have interrelationships which involve one or two variables at a time. Important linkages such as these are ignored when orthogonal principal component or factor models are applied. I believe that the correlation structure presents results more clearly and meaningfully. Each variable has a clear meaning, whereas factors and components rarely do.

The low or moderate correlations between 'deprivation' variables bring the concept of 'multiple deprivation' into question. It may be that each type of problem *is* concentrated in certain areas, which are therefore 'multiply deprived', but that each problem occurs also in some other area, and these areas differ for different problems. Poor housing, for example, is found in rural East Anglia and Wales, as well as in city slums: shocking unemployment occurs in the Outer Hebrides, as well as in Sunderland. But Coulter (1977) found that in seven conurbation cores and their rings, poor housing formed a cluster distinct from UNEMPL, UNSKL, NOCAR, and SICK. Only in the cores of London, Birmingham, and Leeds do NCBORN and BATHDF (bath deficient) join in the main deprivation cluster, and FURNPR never does: in the same areas the correlation between NOCAR and UNEMPL falls below $+0\cdot45$. NOCAR and UNSKL, on the other hand, are consistently highly correlated. It seems most likely, then, that several dimensions of deprivation must be recognised. Perhaps action on housing should not be expected to have much effect on the main set of deprivation variables.

6.13 Conclusion

Considerable attention has been paid here to methodological issues, in the hope that more precise descriptions of relationships may thus be obtained. Transformation of variables to reduce skewness, and weighting of ratio correlations (and descriptive statistics) by total population, often make small differences to correlations, but in a number of cases they greatly clarify the situation and provide a truer description. They do not remove the 'closure' problem, and because of this the dissimilarity index, D, provides a better view of relationships within classifications. To compare sets from different classifications, D must be interpreted with great care because it is based on absolute numbers. D values can be compared with correlations between absolute numbers, but because most absolute numbers correlate strongly with total population it might be better to partial this out of such correlations.

Several types of graphical and tabular displays of correlations are presented, to portray correlations both within and between clusters of variables closely related by definition. Since the original variables are of prime interest, and there is little theoretical support for combining these in 'factors', it is maintained that these displays, and especially correlation structure diagrams, are of greater value and less arbitrary than the results of factor or component analysis.

Although these relationships are geographic (ecological), many of them must be based on relationships at the individual household level, especially those for age structure and household size and composition. Some, such as those between GRAD and PROFES or INTNOM, are probably partly individual, and partly due to purely spatial associations. Others, such as UNSKL and UNEMMA, or MANAG and INTNOM, are probably more spatial than individual. However, such comments on the relative importance of relationships at the individual person or household level are speculative; these geographic relationships do not form a satisfactory substitute for cross-tabulations that are not included in the national census tables, but which are required if individual relationships corresponding to these geographic ones are to be obtained.

The strongest intercluster correlations are portrayed in figure 6.12 which shows several groupings of variables. The basic 'deprivation' variables are NOCAR, POP, OVCRD, UNEMMA, SICK, and UNSKL; LONPAR is comparable but has weaker correlations. Crowding $(1 + P/RM)$ is related to this group only via OVCRD and NOCAR.

Poor housing, represented negatively by ALAMEX, relates mainly to unfurnished private renting, and not to 'deprivation' variables other than NOCAR. Sharing correlates geographically with furnished private renting and immigrants, which relate to POP but not to other deprivation variables. Variables not shown here either correlate strongly with one of those shown, or have correlations most of which are weaker than $\pm 0 \cdot 4$.

Acknowledgements. This paper reports on work of the Census Research Unit, University of Durham, funded by the Social Science Research Council. The unique data set was kindly provided by the Office of Population Censuses and Surveys, in connection with our experiments in automated cartography. This project is very much a team effort and I am deeply indebted to David Rhind and Mahes Visvalingam for work on the data base, and to Kate Stanness, Jane Coulter, and Barbara Perry for research assistance. John Clarke and John Dewdney were also involved in planning and consultation. I thank Jerry Donnini for drawing the figures and Joan Dresser for typing, and am grateful to John Dewdney, Catherine Hakim, Alan Townsend, Allan Williams, and the editor for comments on the chapter, but responsibility for its expression and structure remains my own.

References
Alker H R, 1969 "A typology of ecological fallacies" in *Quantitative Ecological Analysis in the Social Sciences* Eds M Dogan, S Rokkan (MIT Press, Cambridge, Mass) pp 69–86
Bartlett M S, 1947 "The use of transformations" *Biometrics* 3 39–52
Chayes F, 1971 *Ratio Correlation: A Manual for Students of Geochemistry and Petrology* (University of Chicago Press, Chicago, Ill.)
Coulter J M M, 1977 *The Use of Grid Square Census Data in a Study of Deprivation in the Conurbations of the United Kingdom* unpublished M Sc dissertation, Department of Geography, University of Durham, England (see also WP-13, Census Research Unit, Department of Geography, University of Durham, 1978)

CRU, 1979 *People in Britain: A Census Atlas* (HMSO, London, for the Office of Population Censuses and Surveys, Census Research Unit)

Duncan O D, Duncan B, 1955 "A methodological analysis of segregation indexes" *American Sociological Review* **20** 210–217

Evans I S, 1978 "Strategies for coping with percentage and ratio correlation" unpublished typescript

Evans I S, Catterall J W, Rhind D W, 1975 "Specific transformations *are* necessary" WP-2, Census Research Unit, Department of Geography, University of Durham

Gilbert N, 1973 *Biometrical Interpretation* (Oxford University Press, London)

Gould P R, 1970 "Is *Statistix inferens* the geographical name for a wild goose?" *Economic Geography* **46** 439–448

Hakim C, 1975 "Social indicators from the census" *Proceedings of the PTRC Summer Annual Meeting: Corporate Planning Organisation and Research and Intelligence Techniques*, July 1975 PTRC, P/112 pp 114–125 (London) (see also Occasional Papers 2 and 5, Office of Population Censuses and Surveys)

Holtermann S, 1975 "Census indicators of urban deprivation" WN-6, ECUR Division, Department of the Environment

Lindqvist L, 1976 "SELLO, a FORTRAN IV program for the transformation of skewed distributions to normality" *Computers and Geosciences* **1** 129–145

LSE, 1974 "Urban change in Britain: 1961–1971. Standard metropolitan labour areas and Metropolitan economic labour areas" WR-1, WR-2, Department of Geography, London School of Economics, London

Meyer D R, 1971 "Factor analysis versus correlation analysis: are substantive interpretations congruent?" *Economic Geography* **47** 336–343

Moser C A, Scott W, 1961 *British Towns: A Statistical Study of Their Social and Economic Differences* (Oliver and Boyd, Edinburgh)

Mosimann J E, 1962 "On the compound multinomial distribution, the multivariate β-distribution, and correlations among proportions" *Biometrika* **49**(1 and 2) 65–82

Mosteller F, Tukey J W, 1977 *Data Analysis and Regression: A Second Course in Statistics* (Addison-Wesley, Reading, Mass)

OSIRIS, 1973 "OSIRIS III: an integrated collection of computer programs for the management and analysis of social science data" 6 volumes, University of Michigan, Ann Arbor

Rees P, 1971 "Factorial ecology: an extended definition, survey and critique of the field" *Economic Geography* **47** 220–233

Rhind D W, Evans I S, Dewdney J C, 1977 "The derivation of new variables from population census data" WP-9, Census Research Unit, Department of Geography, University of Durham

Robertson I M L, 1969 "The census and research: ideals and realities" *Transactions, Institute of British Geographers* **48** 173–187

Robinson A H, Hsu M-L, 1970 *The Fidelity of Isopleth Maps: An Experimental Study* (University of Minnesota Press, Minneapolis)

Roff A E, 1977 "The importance of being normal" *Area* **9** 195–198

Schwirian K P (Ed.), 1974 *Comparative Urban Structure: Studies in the Ecology of Cities* (D C Heath, Lexington, Mass)

SPSS, 1975 *Statistical Package for the Social Sciences* by N H Nie, C H Hull, J G Jenkins, K Steinbrenner, and D M Bent, 2nd edition (McGraw-Hill, New York)

Webber R J, Craig J, 1976 "Which local authorities are alike?" *Population Trends* **5** 13–19 (HMSO, London)

Wrigley N, 1976 "Introduction to the use of logit models in geography" *Concepts and Techniques in Modern Geography* number 10 (Geo Abstracts, Norwich)

Spatial patterns of the past: problems and potentials

I Hodder

7.1 Introduction

The application of mathematics and statistics in archaeology was given an initial stimulus by the development of radiocarbon dating. The obvious importance of this for the study of prehistory forced the archaeologist to accept the value of science to his discipline and demonstrated the need to grapple with numerical and statistical problems. Also important for the early application of numerical methods in archaeology was the joint work by Brainerd (1951), an archaeologist, and Robinson (1951), a mathematician. They showed how seriation—the relative dating of archaeological deposits by organising their contents in a sequence based on similarity—could be studied mathematically.

The archaeologist typically is faced with large amounts of data (fabrics of potsherds, types of bone, measurements of flint tools, etc) from excavated sites. The possibility of applying numerical analysis to these data led to the desire to manipulate large collections of numbers, and to carry out large quantities of calculations. Developments in computer science thus had an obvious attraction for some archaeologists and were early applied to classification (Hodson et al, 1966) and seriation. The use of numerical methods, objectivity, and computers became one of the clarion calls of the so-called 'new archaeology' in the 1960s and early 1970s. For example, *Analytical Archaeology* (Clarke, 1968) contains a bewildering assortment of 'number crunching' methods derived from authors in other disciplines [for example, Sokal and Sneath (1963)], and methods for studying spatial patterns derived from the 'new geography'. The application of explicit models (Clarke, 1972), again reflecting developments in geography, was part of the same desire to bring objectivity and statistical rigour into archaeology.

Perhaps one of the most widely-used concepts derived from other disciplines has been systems theory (Clarke, 1968; Flannery, 1972; Renfrew, 1972). However, there has been a general unwillingness or inability to use this in any rigorous or quantitative form (Doran, 1970). The archaeologist often lacks the data necessary for detailed application of systems theory concepts, and more use has been made of general statements concerning the interrelationship of variables, that is of General Systems Theory. A general concern with the interaction between variables has been usefully involved in computer simulation (see below).

Gradually statistical and mathematical applications in archaeology have become more diverse. They cover a range from Hawkins's (1973) and Thom's (1971) work on the astronomical significance of stone alignments, to studies of the distribution of bone tools on palaeolithic living floors (Whallon, 1973). Measures of entropy have been applied to degrees of organisation in personal rank and status as seen in the artifacts associated with individuals in burials (Tainter, 1977). But, as in the initial work by Brainerd and Robinson, the central concern of mathematics in archaeology has been with relative dating, classification, and typology. The aim has been to reach objective definitions of 'types' of tool or artifact, and types of assemblage of artifacts. It is in this sphere especially, that developments in numerical methods have been made within archaeology (Doran and Hodson, 1975), rather than being simply derived from other disciplines.

The other major area of application of quantitative methods in archaeology has been 'spatial archaeology' (Clarke, 1977). As the review by Hodder and Orton (1976) shows, the techniques used are largely derived from geography, geology, and plant ecology, although developments more directly related to the nature of archaeological data are beginning to be made. 'Spatial archaeology' covers a wide range of activities (Hodder and Orton, 1976). Distributions of artifacts within sites supply information about refuse disposal procedures and activity areas, whereas the distributions of the sites themselves are studied in relation to environmental variables and in order to understand settlement processes. More emphasis is also beginning to be placed on the analysis of artifact distributions, their structures and associations. In particular, the frequency of artifacts (such as neolithic stone axes) from a known source (such as a particular type of rock outcrop) has often been noted to fall off with distance. While this is much what a geographer might have expected, such falloffs have been formulated into a 'Law of Monotonic Decrement' (Renfrew, 1977). However, the most interesting aspect of this type of work, often involving regression analysis, is the reason for variation in the shape of the falloff, as noted in human geography by Taylor (1971).

Again in parallel with the development of 'behavioural geography' and the increasing interest in phenomenology in geography, some archaeologists are beginning to doubt whether the functionalist and systems approach and the associated analytical methods mentioned above are valid [for example, Athens (1977)]. More emphasis is being placed on the need to examine the underlying 'whys' of social, cultural, and economic change in prehistory. It seems probable, however, that prehistoric spatial distributions will, for some time, be studied by making use of quantitative and statistical techniques. The aim of this chapter is to outline some general problems posed by the spatial analysis of archaeological data. An attempt will then be made to explain why the geographer or spatial analyst might be at all interested in these problems and in archaeology and prehistory.

7.2 Problems

The major and immediate problem facing those interested in the analysis of spatial patterning in archaeological data is that little is known about those data. Consider a distribution of settlement sites, fairly evenly spaced, in Europe in about 4000 BC. Over the next 6000 years up to the present there is further habitation of parts of the same area which completely destroys some of the original settlements. In that 6000 years also, the changing courses of rivers cause erosion and removal of more sites, and deposit deep silt on others making them difficult to locate. Modern gravel digging, coal mining, motorways, and urban sprawl further destroy and cover up sites in the area. Finally archaeological teams with insufficient funds and resources are able to carry out adequate surveys in only small parts of the area, usually near museums and universities (Hamond, 1978).

So the archaeologist of today, studying his map of 400 BC settlement, may have only a very small sample to work with—perhaps as small as 10% or less of the sites. In addition, this sample is by no means a random nor a representative sample of the original distribution. It is usually a highly distorted and biased sample.

The first and obvious consequence of this is that, with such data, it is exceptionally difficult to reconstruct the original process of settlement. It is often possible, with small samples, to conceive of large numbers of hypothetical processes which could have produced the data. Because there are so few data it is often impossible to choose between the hypotheses. Conversely, a lack of fit to the data can always be explained away by reference to survival and recovery biasses, and this has been done.

But the effect that survival and recovery factors have on the testing of hypotheses partly depends on the scale of study. For example, it may be possible to build general models of the diffusion of the neolithic across Europe on a European scale (Ammerman and Cavalli-Sforza, 1973) without worrying too much about survival and recovery factors. It is more difficult, however, to discuss detailed patterns of settlement spread within counties or parishes without a very good understanding of the way in which the data have survived (Cunliffe, 1972), though many archaeologists would claim that detailed historical studies of land use, environmental, and ecological change, and thorough, extensive and well-controlled field survey, can lead to an understanding of how spatial patterns at the small scale have been damaged.

So, the nature of the data means that it is especially difficult in archaeology to tackle satisfactorily the problem of inferring process from form. But a second problem concerns the application of statistics to such inference. If it is accepted that the archaeological sample is biassed, the type of statistical analysis that can be used is obviously restricted. The problem would perhaps be less severe if the archaeologist could present some idea of how his data have been distorted. But any precision in this is usually extremely difficult. In most cases the archaeologist has very little idea of the relationship between his sample and the original population of sites and artifacts. Doran (1977) has pointed to the uncertainties associated with the notions of a 'random pattern' and a 'null hypothesis' in an archaeological context [see also Doran and Hodson (1975)], and is tempted by these difficulties 'to move outside the statistical tradition' (Doran, 1977, page 97).

Recent work in archaeology (Mueller, 1975; Redman, 1975) has laid much emphasis on the need to develop more objective sampling designs. Such studies show the way in which the nature of the data limits the use of the techniques with which human geographers are most acquainted. Random or systematic sampling strategies, for example, have limited advantages if the samples are taken from an already biassed sample that survives from an unknown population of sites. In addition, the archaeologist is unlikely to have sufficiently large sets of data to allow the original characteristics of spatial patterns to be approximated adequately. Consequently random sampling procedures are likely to play a less important role in spatial studies in archaeology than in related disciplines.

Another problem posed by the nature of archaeological data is that it is difficult to achieve or justify the use of high levels of measurement. Yet many of the techniques of spatial analysis which have been applied in archaeology assume a high order of measurement on interval scales. As Wrigley (1977b) has recently pointed out, it cannot help to transform categorised measurements into ratios and percentages in an attempt to use available techniques without modification. Modifications can often be applied to the available techniques in order to deal with categorised data directly (Wrigley, 1977a). For example, the usual trend surface model can be modified to a 'probability surface' model which is appropriate for categorised data. Such an approach allows analysis of the large amount of archaeological data which can only be assessed as presence/absence (the presence or absence of a site or a particular artifact type).

The problems posed by the nature of the archaeological data would be better understood if research was carried out into which types of spatial patterns were most robust in the face of distortion. But another approach which can be applied to the understanding of spatial patterns and processes and which answers some of the objections discussed above is computer simulation.

7.3 Simulating the past
Norton (1976) has used the simulation of settlement spread to construct 'abstract worlds of the past'; that is, 'would have been' patterns according to particular hypotheses about the settlement process that occurred. The results of his simulation of the development of agricultural settlement in Canada (1782–1851) are compared with the actual data to see how they fit. Norton suggests that this setting up of a stochastic null hypothesis, this use of the 'what might have happened if ...' situation as a basis for comparison, is growing in historical geography.

Computer simulation seems especially useful in archaeology, partly because with inadequate data the number of hypothetical processes that could have produced a particular pattern is usually immense—much more so than in the typical geographical situation. Simulation on computers ideally allows the rigorous examination and testing of the whole range of alternative hypotheses, and the assessment of their relative fits.

However, simulation is only valuable as an alternative to classical approaches if the likelihood is assessed of getting by chance the observed value of a statistic that summarises the data—'by chance' here means 'as a result of the stochastic component in the simulated model'. The aim, then, should be to see how likely one is to produce the observed data by simulating a particular stochastic process a large number of times. Such tests have not been used as yet in archaeology, even though computer simulation has recently begun to be employed (Chadwick, 1977; Thomas, 1973; Wobst, 1974; Zubrow, 1975).

Two examples of simulation studies in archaeology will be given which demonstrate something of their value in this context. The first concerns the relationship between the rank and size of settlements. Archaeologists can often estimate the sizes of their sites. This is especially the case when walls encircle the sites, but estimates can also be made of the areas covered by pottery spreads or occupation debris in the plough soil above sites. In a recent study (Hodder, 1978), twelve different archaeological data sets with different rank-size graphs were examined. The aim of the work was to determine what sorts of stochastic processes could have produced the different rank-size graphs, and what factors could have caused variation between them.

Many of the twelve data sets were of very poor quality, having been affected and distorted by a wide range of postdepositional factors. For example, it is rarely possible to be certain that all of a sample of sites are exactly contemporary. For early periods our chronologies are insufficiently precise to say whether two sites were occupied during the same years, decades, or even centuries. The rank-size graph, then, may not represent one contemporary situation. In addition, only small parts of a large site may have been inhabited at any one time, while the spread of surface pottery over a site may not be an accurate guide as to the size of that site. Also many sites, and in particular the smaller sites, will have been destroyed or may be difficult to find. So any original rank-size relationship is blurred and distorted by the time it reaches the archaeologist. If, and only if, this blurring and distortion result from a variety of factors, it may be reasonable to equate the aggregate effect with random blurring.

A number of geographers have suggested stochastic processes for the allocation of populations and sizes to settlements [for example, Chapman (1970)]. These processes vary in the number and nature of the constraints that are imposed. The archaeologist can simulate stochastic processes involving different constraints and compare the resulting rank-size graphs with the actual data. When this was done for the twelve data sets it was found that several had rank-size graphs very similar to the end results of very unconstrained stochastic processes [for example, figure 7.1(b)]. In these cases, either the original formation processes involved few marked constraints or forces acting, or postdepositional factors had blurred an originally ordered pattern in such a way that a more 'random' pattern had been produced. There is very little that can be said about such unstructured data.

But in other cases the archaeological data gave better fits to the end results of relatively constrained processes. This was true, for example, of a sample of late Iron Age multivallate hillforts in the south of England [figure 7.1(a)]. What is known of the late Iron Age in Britain certainly supports the view that severe constraints might have been imposed on the hierarchy of settlement sizes. Literary evidence relates the presence of strict social and political classes, and the archaeological evidence includes parade armour and highly valuable items that seem to have symbolised and supported an elite. This markedly hierarchical society may indeed have been reflected in greater differences in the rank and importance of settlements. For these Iron Age hillfort data there is also relatively little blurring of the original pattern. It is fairly difficult for a large imposing multivallate hillfort to be destroyed without trace, the hillforts cover a relatively restricted period of time, and they have been fairly well surveyed and recorded. The structure of the rank-size graph is not obviously the result of any structuring in postdepositional factors, and there has been insufficient blurring to mask completely the original order. Similar results were found for other data sets. The Early Dynastic settlement pattern in Iraq (*circa* 3000 BC) is certainly the product of a fully-developed state system with strong centralised control and authority. It is precisely in this type of nonegalitarian system that constraints are imposed and deviations from the expectations of unconstrained stochastic processes are found.

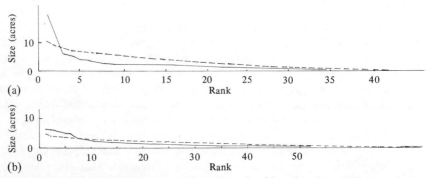

Figure 7.1. Relationship between rank and size for (a) multivallate hillforts and (b) univallate hillforts. The curves predicted as a result of the relatively unconstrained process of allocating sizes from the normal distribution are shown as dashed lines.

The simulation of rank-size relationships, then, used the 'Random Spatial Economy' idea (Curry, 1964) as a baseline for the sorting out of ordered data sets from those data sets with potentially less information. The type of order or structure could be indicated and fruitfully discussed.

Another example which uses a similar approach concerns the simulation of the dispersal of neolithic axes across Britain in the period between approximately 3500 BC and 2000 BC (Elliott et al, 1978). A number of axe factories existed in Britain at this time, for example in the Lake District, North Wales, and Cornwall, and their products are found spread widely across Britain. The polished stone axes found during fieldwork and excavation can be identified to source by petrological analysis. The archaeologist, of course, has absolutely no idea of how this dispersal of axes occurred; but he can take models derived from ethnographic studies to develop hypotheses about how the spread and exchange took place. Simulation can then be used to mimic these processes and the simulated output can be compared with the actual archaeological data.

But again these data are very poor, mainly because they accumulated over an extremely long period of time, 1500 years. It is not known which axes are precisely contemporary nor, with any reliability, which axe factories were producing at exactly the same time, nor how much they produced. One is, therefore, very much working in the dark, experimenting with different ideas to see which types of hypotheses fit the presently available data. Because the data are so poor, it is possible that any original structure in the data has been completely erased and blurred out, or changed into another structure. It is, therefore, useful to start again with the extreme null hypothesis that the axe distributions are the result of highly unconstrained stochastic processes. Constraints can then be added to the simulated process in order to see whether better fits to the data occur, and in order to identify the possible types and degrees of constraint that may have acted.

It was found that relatively good fits to the data could be achieved by simulating fairly unconstrained processes. Axes were allowed to spread over England by using a simple random-walk procedure. However, it was found that slightly better results occurred if some degree of competition between axes from different sources was introduced. Axes were given less chance of moving into areas where axes from other sources already existed. (For examples of results see figure 7.2.) The introduction of basic constraints of this form allowed the nature of the patterning in the data to be identified. The archaeologist still has to interpret what the patterning means in the context of neolithic society, but at least his discussion of the evidence is on a firmer basis than is possible when the traditional approaches are used.

Some advantages to the archaeologist of using the simulation of stochastic processes as a basis for the determination of process from spatial form have been discussed. But simulation has its own problems, not least of which is the impression sometimes obtained that the approach in archaeology perhaps involves rather a circular argument. On starting with a set of data, hypotheses are considered that could have produced the data. The hypotheses are simulated and the parameters of the model are tinkered with until the data can be adequately reproduced. Often little new information is learnt, and little is 'proved' because there is no independent check or test. One starts with a set of data and tests the simulated hypotheses against the same data. Fortunately archaeologists can often provide initial independent data derived, for example, from ethnographic information, and it is this type of simulation that is perhaps of greater value.

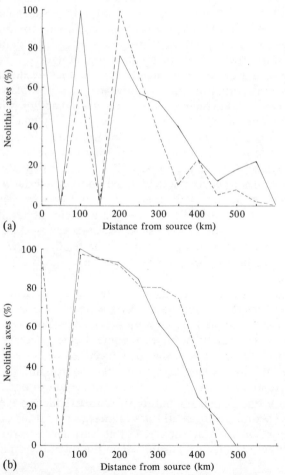

Figure 7.2. Neolithic axes: simulated (dashed lines) and actual (continuous lines) percentages of (a) Cornish, and (b) Lake District axes with distance away from the sources. Zero values near the sources occur because of lack of data.

A further special problem of simulating archaeological data concerns one of the supposed advantages. It has already been noted that the archaeologist knows very little about his data so that simulation can be used to run through and examine all the possible alternative hypotheses. Consider a situation in which a simulation procedure is controlled by just four parameters. With nothing to guide the archaeologist in his choice of parameter values, each parameter might be able to take on any one of, say, six states. By varying each parameter independently, the total number of parameter combinations is ninety-six. For each set of parameter values it might be thought necessary to carry out fifty separate simulations in order to identify the variability in the results owing to the stochastic component in the model. This necessitates a total of 4800 separate simulations. In fact, of course, many more than four parameters are usually needed to control a realistic simulation. Although search procedures might be employed to decrease the number of simulations needed, the problem remains of immense proportions. So the very advantage of simulation in archaeology, that it allows the large numbers of processes that could have produced the same spatial form to be examined rigorously, is also one of its disadvantages.

7.4 Archaeological data and the geographer

Up to this point some special problems posed by the spatial analysis of archaeological data have been identified, and some ways of tackling or bypassing those problems have been mentioned. But why might geographers and spatial analysts be interested in all this anyway? Might there be any value in the analysis of prehistoric as opposed to recent or modern data? The first and obvious point is the great time span covered by archaeological data. This may allow the geographer's models to be given a greater depth.

Archaeological data, bad as they may be, are unique in that they provide evidence of human activities over thousands and millions of years. How do geographical models stand up when applied over this great span of time? For example, the gravity model, already used in related studies such as linguistics (Trudgill, 1975), can also be shown to be of value in studying early and primitive societies (Hallam et al, 1976). Two recent applications of this model will be described. Jochim (1976) has used the gravity model to examine the location of settlements among hunter–gatherers. Ethnographic evidence suggests that the most important factor in the location of such settlements is nearness to resources, food, building materials, etc. Since all these resources occur at different places in the environment, the settlement must be placed between them such that the relative 'pulls' of each resource are reconciled and balanced to give the best location.

For example, some resources are at higher risk and are of greater value and prestige than others. The big game food source, for instance, is less secure but of higher prestige than gathered berries. By drawing analogies with industrial location, Jochim sees settlements locating near sure, low-value resources and worrying less about being near high-value, less sure resources. This is because people are willing to travel further to high-value resources.

Jochim considers the simple gravity model

$$I_{ij} = K \frac{P_i P_j}{d_{ij}^2} .$$ (7.1)

In this equation
I_{ij} represents interaction between settlements i and j,
P_i is the population of settlement i, and
d_{ij} is the distance between i and j.
Interaction with a food resource amongst hunter–gatherers is seen as being directly related to the dietary importance of that food resource. Jochim therefore reformulates the gravity model to

$$G = \frac{wn}{d^2} .$$ (7.2)

G here is the dietary proportion or importance of a resource. This is the value of a resource to an individual. G depends on the weight of the type of animal or plant, its density in a particular location, its size of aggregation, its mobility, its nonfood and fat content. In the equation, G is seen as being related to the weight of an animal (w) times the number of animals in a cluster (n), divided by the squared distance to the cluster. Since Jochim is interested in the relative distances between sites and resources he reorganises the equation to

$$d^2 = \frac{wn}{G} .$$ (7.3)

Thus the distance from a site to a resource is directly related to the weight and numbers of animals or plants in a location and inversely related to the dietary value of that resource. By using values for w, n, and G for different types of resources, Jochim shows that the gravity model predicts location of a site closest to immobile secure plant foods with fish and small game farther away and big game much farther away. Although there seems to be a certain amount of repetition within the terms of the formulated model, this result corresponds well with the observed behaviour of hunter–gatherer societies and suggests that the gravity model might be of value in the study of the location of palaeolithic and mesolithic archaeological sites.

The gravity model has also been recently applied to some anthropological data collected from the island of Tikopia. Kasakoff and Adams (1977) examined the pattern of marriages between the four clans on the island. In doing this they turned the gravity model into what they called a 'social gravity' model in which

$$I_{ij} = \text{marriage} = K \frac{P_i P_j}{d} ,$$ (7.4)

where P represents the population of the villages of the clans. Kasakoff and Adams then compared the numbers of marriages expected between the four clans according to this model with the tabulated data on marriages collected from the island. Traditional anthropological study of these data had suggested that there were clear clan preferences (some pairs of clans preferred to marry each other), while there was also a preference for marriages within one's own clan (endogamy). But application of the gravity model showed that this patterning and the apparent preferences could be fairly accurately predicted from the spatial distribution of the clans. Put simply, people were tending to marry nearby people whatever their clan. The *apparent* marriage preferences of the clans and the apparent endogamy had occurred because clans were spatially clustered, although it is also possible that the social customs influenced the original location of villages.

As seen by Kasakoff and Adams, the gravity model provided a spatial null hypothesis from which real marriage preferences could be more clearly seen. Anthropologists often fail to take into account the tendency of interaction to decline over distance. Many marriage 'rules' may simply be a function of distance constraints rather than of the complex ideological and structural explanations offered traditionally by anthropologists. For example, Levi-Strauss's analysis of elementary social structure assumes that the social units are 'isolates' in which access to preferred types of people is no problem even when they live very far apart (Kasakoff and Adams, 1977, page 52). So anthropologists would certainly benefit from the application of more geographical and spatial models like the gravity model. But geographers receive advantages in return. Information about the generality of a geographical model is gained from its working in a totally alien context.

7.5 Conclusion

So the application of geographical models to very early or primitive societies gives them greater depth. But this is only true if the models can be adequately tested against archaeological data. If all that has been said about the nature of those data is borne in mind, it remains true that the data sometimes do allow such testing if handled with care. In addition the data provide, at least potentially, a more direct reflection of intensities of interaction than could be provided by, for example, most of the historical texts available to historians and historical geographers. The archaeologist [for example, Hogg (1971)] can examine falloff curves of coins or pottery in much the same way as the human geographer studying interaction patterns in present-day societies.

The application of geographical models to archaeological data might result in the models being amended or being placed in a better perspective. For example, as already noted, anthropologists have paid little attention to the fact that intensity of interaction usually declines with distance. This is partly because archaeologists and anthropologists both know it is not true, except in certain social contexts. In spite of the growth of 'behavioural geography', most geographers would probably still assume that intensity of interaction was related to distance, size and attractiveness of centre, reason for journey, and so on. But this is only so in the special case of modern industrialised societies. In many types of societies interaction is by no means related to physical distance but is closely related to social distance—that is, the degree of kinship links between people. In such cases, patterns of interaction begin to look like those predicted by Soja (1971) for territorial societies, with plateaus and kinks instead of smooth patterns of falloff. To be universally valid, the geographer's models may need to be changed to take into account different situations which are not met with within the geographer's relatively limited field of experience.

However, even if archaeology is able to contribute towards geographical theory, the relationship between the two disciplines is interdependent. The archaeologist is unable to identify social constraints in patterns of interaction without being able to set up the null hypothesis of socially unconstrained interaction. He needs to know as his starting point what patterns of interaction should look like when unhindered by certain types of social structures. There has been sufficient analysis within geography to provide many of these basic expectations. It is only by using these geographical models that the archaeologist can identify and measure other, perhaps more interesting patterns in the past.

References

Ammerman A J, Cavalli-Sforza L L, 1973 "A population model for the diffusion of early farming in Europe" in *The Explanation of Culture Change* Ed. C Renfrew (Duckworth, London) pp 343–357

Athens J S, 1977 "Theory building and the study of evolutionary process in complex societies" in *For Theory Building in Archaeology* Ed. L Binford (Academic Press, New York) pp 353–384

Brainerd G W, 1951 "The place of chronological ordering in archaeological analysis" *American Antiquity* **16** 301–313

Chadwick A, 1977 "Computer simulation of settlement development in Bronze Age Messenia" in *Mycenaean Geography* Ed. J Bintliff (British Association for Mycenaean Studies, Cambridge) pp 88–93

Chapman G, 1970 "The application of information theory to the analysis of population distributions in space" *Economic Geography* **46** 317–331

Clarke D L, 1968 *Analytical Archaeology* (Methuen, London)

Clarke D L, 1972 *Models in Archaeology* (Methuen, London)

Clarke D L, 1977 *Spatial Archaeology* (Academic Press, London)

Cunliffe B W, 1972 "Saxon and medieval settlement patterns in the region of Chalton, Hampshire" *Medieval Archaeology* **16** 1–12

Curry L, 1964 "The random spatial economy: an exploration in settlement theory" *Annals of the Association of American Geographers* **54** 138–146

Doran J E, 1970 "Systems theory, computer simulation and archaeology" *World Archaeology* **1** 289–298

Doran J E, 1977 Review of Hodder and Orton (1976) in *Journal of Archaeological Science* **4** 96–98

Doran J E, Hodson F R, 1975 *Mathematics and Computers in Archaeology* (Edinburgh University Press, Edinburgh)

Elliott K, Ellman D, Hodder I, 1978 "The simulation of neolithic axe dispersal in Britain" in *Simulation Studies in Archaeology* Ed. I Hodder (Cambridge University Press, London) pp 79-87

Flannery K V, 1972 "The cultural evolution of civilizations" *Annual Review of Ecology and Systematics* 3 399-425

Hallam B R, Warren S E, Renfrew C, 1976 "Obsidian in the western Mediterranean: characterisation by neutron activation analysis and optical emission spectroscopy" *Proceedings of the Prehistoric Society* 42 85-110

Hamond F, 1978 "The contribution of simulation to the study of archaeological processes" in *Simulation Studies in Archaeology* Ed. I Hodder (Cambridge University Press, London) pp 1-9

Hawkins G S, 1973 *Beyond Stonehenge* (Harper and Row, New York)

Hodder I, 1978 "Simulating the growth of hierarchies" in *Mathematical Models and Culture Change* Eds K Cooke, C Renfrew (Academic Press, London)

Hodder I, Orton C, 1976 *Spatial Analysis in Archaeology* (Cambridge University Press, London)

Hodson F R, Sneath P H A, Doran J E, 1966 "Some experiments in the numerical analysis of archaeological data" *Biometrika* 53 311-324

Hogg A H A, 1971 "Some applications of surface fieldwork" in *The Iron Age and its Hillforts* Eds M Jesson, D Hill (Southampton University Press, Southampton) pp 105-125

Jochim M A, 1976 *Hunter-gatherer Subsistence and Settlement: a Predictive Model* (Academic Press, New York)

Kasakoff A B, Adams J W, 1977 "Spatial location and social organisation: an analysis of Tikopian patterns" *Man* 12 48-64

Mueller J W, 1975 *Sampling in Archaeology* (University of Arizona Press, Tucson)

Norton W, 1976 "Constructing abstract worlds of the past" *Geographical Analysis* 8 269-288

Redman C L, 1975 *Archaeological Sampling Strategies* Module in Anthropology 55 (Addison-Wesley, Reading, Mass)

Renfrew C, 1972 *The Emergence of Civilization* (Methuen, London)

Renfrew C, 1977 "Alternative models for exchange and spatial distribution" in *Exchange Systems in Prehistory* Eds T K Earle, J E Ericson (Academic Press, New York) pp 71-90

Robinson W S, 1951 "A method for chronologically ordering archaeological deposits" *American Antiquity* 16 293-301

Soja E W, 1971 "The political organisation of space" Resource Paper 8, Association of American Geographers, Commission on College Geography.

Sokal R, Sneath P H A, 1963 *Principles of Numerical Taxonomy* (W H Freeman, San Francisco)

Tainter J A, 1977 "Modeling change in prehistoric social systems" in *For Theory Building in Archaeology* Ed. L Binford (Academic Press, New York) pp 327-352

Taylor P J, 1971 "Distance transformations and distance decay functions" *Geographical Analysis* 3 221-238

Thom A, 1971 *Megalithic Lunar Observatories* (Clarendon Press, Oxford)

Thomas D H, 1973 "An empirical test for Steward's model of Great Basin settlement patterns" *American Antiquity* 38 155-176

Trudgill P, 1975 "Linguistic geography and geographical linguistics" *Progress in Geography* 7 227-252

Whallon R, 1973 "Spatial analysis of occupation floors: application of dimensional analysis of variance" *American Antiquity* 38 266-278

Wobst H M, 1974 "Boundary conditions for palaeolithic social systems: a simulation approach" *American Antiquity* **39** 147-180

Wrigley N, 1977a "Probability surface mapping: a new approach to trend surface mapping" *Transactions of the Institute of British Geographers* New Series **2** 129-140

Wrigley N, 1977b, review of Hodder and Orton (1976) in *Environment and Planning A* **9** 479-480

Zubrow E, 1975 *Prehistoric Carrying Capacity: a Model* (Cummings, Menlo Park, Calif.)

Part 2

Environmental science applications

Introduction
Progress in statistical analysis in physical geography has recently been summarised by Unwin (1977). He notes how many of the developments discussed in the introduction to the first section of this book have also characterised work in physical geography. Problems of spatial and temporal dependence have been confronted and the stochastic process perspective adopted. Concern with topics such as autocorrelation functions, forecasting models, model identification, series length and quality, and differencing has increased, though as yet most applications are confined to time or distance series rather than to the joint space–time series analysed by human geographers. The first two chapters in this section provide a perspective on these recent developments in statistical physical geography. Anderson and Richards review progress in the statistical modelling of channel form and process and illustrate the utility of the distance-series approach, and Ferguson adopts similar methods in his consideration of the regularity or randomness exhibited by river meanders. These are followed by a more general consideration of time-series analysis and forecasting provided by Bennett in the context of a critical appraisal of the problems encountered in forecasting long-term climate changes.

The section concludes with three chapters by nongeographers. In the first of these Marriott takes up the diffusion–spatial pattern theme, a topic of long-standing interest to human geographers. In this case, however, he examines it in the context of the uptake of nutrients by roots in soil, given that the uptake depends upon the spatial pattern of the roots. Woiwod then considers spatial analysis in the Rothamsted Insect Survey, with an outline of the sampling systems used to provide agricultural advisory services with information on aphids and moths, and the mapping techniques used in more fundamental studies of pest prediction, insect movement and diversity, and spatial population dynamics. Geographers may be somewhat disappointed to find that Rothamsted employs what Rhind has described as the 'granddaddy of all computer mapping programs', SYMAP, but they will find in entomology an area which is surely rich in potential for the application of standard spatial analysis techniques. Finally, Webster presents a discussion of current multivariate analysis in soil survey, and gives a description, with examples, of some of the uses of principal components analysis, canonical correlation and multiple discriminant analysis in soil science.

Reference
Unwin D J, 1977 "Statistical methods in physical geography" *Progress in Physical Geography* **1** 185–221

Statistical modelling of channel form and process

M G Anderson, K S Richards

8.1 Introduction

At the temporal and spatial scales that characterise geomorphological investigation, it is impossible to identify events and processes the outcome of which may be predicted deterministically. Consequently the theory of stochastic processes has been applied extensively to the study of geomorphological phenomena, particularly in the analysis of fluvial morphology. Certain types of stochastic process have proved useful as models of the organisation of channel networks; binary and random-walk processes, for example, are employed in the simulation of drainage networks (Langbein and Leopold, 1962; Kirkby, 1976), after appropriate methods of transforming binary strings and random walks into network 'trees' have been identified (Smart, 1970). The continuous variables describing various elements of channel form, such as width, depth, bed elevation, and direction changes, have been analysed by the series analysis techniques described in detail by Box and Jenkins (1970), and by spectral methods (Rayner, 1971). Usually this has implied discrete sampling of continuous processes, with attendant difficulties concerning sampling frequency. However, the parallelism between the discrete autoregressive/moving-average processes and the continuous equivalents described by differential equations is becoming clearer (Ferguson, 1976; Thornes and Brunsden, 1977, page 168).

Kendall (1973) and Chatfield (1975) both identify four primary objectives in time-series analysis: description, explanation, prediction, and control. The balance between these aims may be entirely different in the study of one-dimensional spatial series, and this may have some bearing on the strategy adopted in the analysis of such series. This is particularly germane to the relative weight attached to description and explanation on the one hand, and prediction and control on the other. Prediction is rarely important in distance-series analysis, since at the time of sampling the $(n+1)$th point is observable; thus the roles of description and explanation are paramount. For predictive purposes, the discovery that two equally parsimonious models fit the available data equally well may not present insuperable problems. However, if the aim is to describe important characteristics of the data and to explain its pattern of variability, this is likely to be completely frustrated by such an eventuality. Evidently, therefore, different types of data and varying problems in the study of channel form and process will require different modelling strategies; accordingly, it is worth reviewing the nature of the data involved.

Time-series data in fluvial geomorphology fall into two categories: observations of process variables and morphometric variables. The former include hydrological series, which have been analysed in a similar manner to economic time series (Carlson et al, 1970; Chow and Kareliotis, 1970; Matalas, 1963; Quimpo, 1968; Yevjevich, 1971). There have been insufficient regular observations of direct process variables such as suspended sediment or dissolved load concentrations for much progress to have been made in the analysis of such data, although it may prove possible to use bivariate series analysis as an alternative to rating curve estimation, with leads and lags between the hydrograph and sediment graph correlated to antecedent moisture conditions. This presupposes that the serious nonstationarity problems associated with short-term hydrographs may be overcome. As in economic time-series analysis, the objectives of time-series analysis of such hydrological parameters lean towards prediction. Direct time-series analysis of morphometric variables is rare, because the slow rate of change of most landforms mitigates against an extended sampling period capable of defining variations at geomorphologically relevant time scales. An exception to this is the movement of sand dunes past a river cross-section, which produces a time series described by a stochastic process similar to that underlying a series of bed elevations downstream at a given point in time (Nordin and Algert, 1966). This ergodic behaviour, and the close relation between dune movement and sediment transport (Willis, 1968), suggests that morphological measurements may be capable of providing information about the bed-load transport process (Nordin and Richardson, 1968).

Spatial-series analysis, largely restricted to one-dimensional series, has made considerable progress, with investigations employing space domain and frequency domain approaches. Studies of meander patterns have been numerous, with Thakur and Scheidegger (1970) and Chang and Toebes (1970) providing detailed investigations of the statistical characteristics of direction change series to build on the earlier work of Speight (1965). More recently, Ferguson (1975; 1976) has developed the approach considerably, reconciling the apparent distinction between regularity and randomness in meander forms. Channel bed forms have also been investigated extensively, with studies of sand bed phenomena such as ripples and dunes (Crickmore, 1970; Nordin and Algert, 1966; Squarer, 1970) and of larger-scale features in gravel bed streams (Melton, 1962; Church, 1972; Richards, 1976a). The ergodic characteristics of the former have already been noted.

Some attempts have been made to develop space–time models similar to those of Bennett (1975) and Martin and Oeppen (1975) that describe economic processes. These are in their infancy, and experience some problems. The best examples are the studies of ephemeral streams in Spain by Thornes (1976) and Lai (1977). The latter author has developed a transfer-function model predicting 1974 current channel width from

1974 total channel width upstream and 1973 current channel width at the same location and upstream. Current channel width is defined as the width of the most recently occupied channel, and total width refers to the potential width of flood channel. These channels are extremely unstable systems, reforming annually after torrential rains. Thus identification of the channel margins is difficult, although the models are fitted in spite of the random variability of measurement error. However, it is unlikely that equivalent sampling locations are defined in successive years; this is not a problem in the space–time modelling of economic data as long as boundary changes of the areal sampling units are negligible. In these distance series, a point located by distance downvalley may have a longer stream length above it in one year than in another, because storm and discharge contributing areas differ. Alternatively, sample points defined by channel distance may change location in the valley as the sinuosity varies, and so may experience different geological or valley gradient controls from year to year. Unfortunately, progress in space–time modelling demands the analysis of unstable systems where change is rapid, but these tend to introduce numerous practical difficulties.

Most of the progress to date therefore seems to have been in the analysis of spatial series of morphometric variables. Invariably this has initially been restricted to univariate processes, and the analysis of bivariate and multivariate series is a logical further development. This is particularly necessary in studies of large-scale bed forms (riffles and pools) and meander patterns, where an interrelationship between the two series is to be expected, but where analysis has proceeded independently. Cross-correlation of series of bed elevations and direction angle changes should be suggestive in descriptive and explanatory terms. An illustration of this is provided by the meander plan and profile of the Popo Agie River originally surveyed by Leopold and Wolman (1957), and illustrated in figure 8.1. In figure 8.2 the bed profile is quantified as residuals from a linear trend at equally spaced sampling locations, and the meander pattern as absolute direction changes. The absolute change is necessary because normally the direction of the angular deviation between successive reaches switches from positive to negative in alternate bends. Since there are thought to be two riffle–pool cycles to one meander bend, this means that the negative profile elevation residuals of the pool in the first bend would relate to negative direction changes, whereas the equivalent negative profile residuals of the pool in the next bend would relate to positive direction changes. Identification of cross-correlation would be extremely difficult. The two series are illustrated in figure 8.2, and the correlograms are shown in figure 8.3. The correlograms suggest a second-order auto-regressive process for the bed elevations, and a first-order autoregressive model for the angular deviations, conclusions which are reinforced by the form of the partial autocorrelation function. The cross-correlation function, also shown in figure 8.3, includes one significant term that

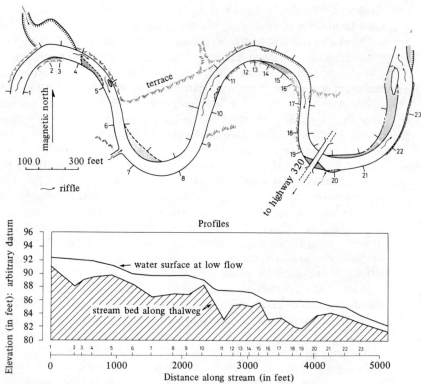

Figure 8.1. Plan and bed profile of the Popo Agie River, Wyoming. (After Leopold and Wolman, 1957.)

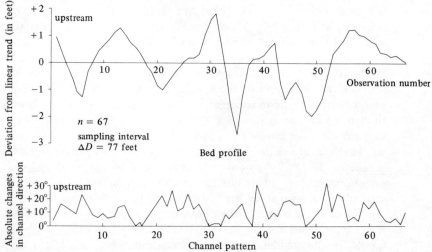

Figure 8.2. Series of bed elevation residuals and direction angle changes for the Popo Agie reach shown in figure 8.1.

indicates a negative correlation at a lag of +1. The series in this example begin at the upstream end, so that the correlation is between bed elevations at point j and direction changes at point $j+1$ downstream. The negative correlation, as expected, indicates maximum angular change in association with the negative profile residuals (that is, the pool is at the bend apex), and the downstream lag may be taken to indicate the tendency of change in channel direction to lag behind the downstream changes in the bed form. This lag is partly because of the primary nature of the bed form, and partly because the thalweg impinges on the outer bank just downstream from the bend apex, so that maximum bank erosion is concentrated there. Although on the one hand this model quantifies these common observations of meandering behaviour in the equation

$$|\nabla\theta|_j = -5\cdot97Z_{j-1} + e_j \qquad\qquad (8.1)$$

(where $|\nabla\theta|_j$ is the absolute direction change at point j, Z is the profile elevation residual, and the e_j are random errors), and offers a useful method for further investigation of bivariate aspects of channel pattern, one should note that in this case a perfect meander obeying the rules of the sine-generated curve model (Langbein and Leopold, 1966), only produces a cross-correlation of $-0\cdot26$ between profile and plan variables.

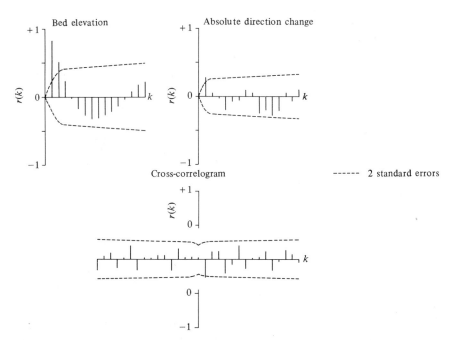

Figure 8.3. Individual autocorrelation functions for the series in figure 8.2, and cross-correlation function between the series after transformation by the Box–Jenkins approach.

This may be because the series are short ($n = 67$), and because of the lack of meander and riffle-pool cycles in the data, but does not inspire confidence in the approach. However, it may be possible to identify a spectrum of cross-correlation functions that describe the interrelationships of bed and pattern variations over a range of sinuosities [cf Ferguson's (1976) study of meander pattern alone]. Clearly, for comparative purposes some form of standardised measure of bed elevation residuals (for example, standard deviation units) and sampling frequency would be necessary to allow comparison of dimensionless coefficients.

8.2 Space domain analysis of distance series
The need for a reappraisal of the modelling strategy applied to physical distance series has been emphasised; the requirements of the descriptive and explanatory objectives being distinct from those of predictive modelling. This may be illustrated with reference to some of the statistical problems facing the research worker.

8.2.1 Stationarity in spatial data
Three separate approaches have been adopted in order to transform data to stationary series prior to analysis. Subtraction of least-squares harmonic curves is the least useful because it presupposes an initial trend removal and is therefore inefficient in terms of the number of parameters required, and because there is no *a priori* justification for the assumption of a regular repetition of wavelength and amplitude, as there is in many climatic series (Rayner, 1971). Subtraction of a regression trend alone has been criticised for creating autocorrelation where none exists, but is supported by Granger and Hatanaka (1964) who argue that it does successfully remove all low-frequency variation in the data. Differencing is strongly supported by analysts of economic data, whose interests in prediction and control are dominant. However, for purposes of description and explanation it seems preferable to remain in touch with the physical reality of the data throughout the analysis, and in spite of the occasional problems that may arise in using regression trends, this approach is preferable. Clearly in many physical distance series the trend itself is a physical phenomenon and as such is rather different in nature from the stochastic trends in economic data. Several other practical considerations support this view. First, early studies of meanders and bedforms aimed to identify gross descriptive parameters such as wavelength and amplitude, the subjective estimation of which was complicated by their natural irregularity. The pseudooscillatory second-order autoregressive model has proved a useful descriptor of such phenomena, and permits estimation of the wavelength and period of the process from the parameters of the model (Box and Jenkins, 1970, page 60; Richards, 1976a; Ferguson, 1976). If the bed profile trend is removed by differencing, however, the fitted model is often a first-order moving-average process from which no similar estimate

of wavelength is possible. Second, it is important to be able to correlate
the identified stochastic variation in form with some process model in
order to facilitate explanation. For example, Yalin (1971) derived a
theoretical model of large-scale turbulence in straight channels in which
velocity fluctuations along the channel follow a second-order autoregressive
scheme. This results in alternate erosion and accretion of the bed, and
links closely with the similar model fitted empirically to bed profiles
(Richards, 1976a). Thus the desire for physical interpretation demands
the minimum of transformation of the original data, although pragmatism
is necessary in deciding the strategy suitable in particular cases. An
exceptionally long series (several hundred metres) of profile elevations may
require a logarithmic trend rather than a linear or low-order polynomial
trend, and tributary entrances may create changes (Miller, 1958) that can
only be rendered stationary by treating them as stochastic changes in level
and adopting the differencing approach. Furthermore, both techniques
may be equally useful in the analysis of sand bed dune features where the
objective is to relate the stochastic process of bed elevation variability to
the flow resistance created by the bed forms (Crickmore, 1970; Squarer,
1970). Absolute wavelength and amplitude are critical controls of
roughness, so analysis of elevation residuals is useful. However, change of
elevation from point to point also influences roughness, so that analysis of
differenced data may also provide valuable insights. Although pragmatism
seems desirable, a final point in favour of analysis of regression residuals is
that an eventual objective may be to compare the models fitted to a
number of profiles in order to identify systematic variation in the model
parameters as possible independent variables change (for example, width,
valley gradient, bed material size). Clearly the approach must be consistent
before this objective may be realised; since the linear trend itself may be
one of the independent physical variables, it seems appropriate to use this
detrending method.

8.2.2 Bivariate series—the problem of transformation

The n observations of two variables X and Y may be regarded as a finite
realisation of a discrete spatial bivariate process $\{X_j, Y_j\}$. The cross-
correlation function measures the strength of association between the two
variables at various leads and lags, and the impulse-response function states
the nature of this relationship and may be estimated from the cross-
correlation function and the ratio of the variances of the two series.
However, successive estimates of the cross-correlations are themselves
autocorrelated, and their variances depend on the autocorrelation functions
of the two individual series. Thus straightforward cross-correlation is often
unsuitable because the resulting estimates of the weights in the impulse-
response function are inefficient and, perhaps more important, it is
impossible to test the terms in the cross-correlation function for significance
in order to determine the order of lead or lag between the two series.

This is particularly necessary since the general form of the relationship between Y_j and X_j is

$$Y_j = \sum_{k=-\infty}^{\infty} h_k X_{j-k} \, , \tag{8.2}$$

where h_k is the impulse-response function. Clearly it is necessary to identify which of the h_k weights are nonzero in order to express this in a parsimonious form.

The optimal procedure to adopt in studies of channel form is not clear. Jenkins and Watts (1968, page 340) propose that both series should be initially prewhitened by subtraction of the appropriate univariate stochastic process. This creates two white-noise processes, so that $\mathrm{var}[r_{xy}(k)] \approx 1/n$, and significance testing of the terms in the cross-correlation function is much simplified. This is an appropriate method for correlating two series that are considered on an equal footing. An alternative approach, more appropriate for identifying 'causal' models, is outlined by Box and Jenkins (1970, page 379) and involves prewhitening the independent 'input' series by its own model, and transforming the dependent 'output' series by subtraction of the input model. The relationship between the two variables provides the series analogue of conventional regression, so that causality is not implied by the relationship. The input series is assumed to be transformed into the output series by the operation of a linear system, which is represented by the transfer-function model defined by equation (8.2). Identification of this model by simple direct cross-correlation is difficult to achieve because spurious correlations may arise which make inference about the relationship impossible.

The choice between these approaches depends on the nature and objectives of the study. In geomorphological investigations, specification of the linear system may be less readily achieved than in control engineering problems, and the assumption that one variable is the independent system input may be questionable. As long as statistical inference is unnecessary, it may be satisfactory to opt for direct cross-correlation of the original series in order to demonstrate the form of interrelationship between them. The example in figure 8.4 illustrates this. After transformation by the Box–Jenkins approach, cross-correlation of width and elevation residuals for a reach of the River Fowey in Cornwall shows a single strongly significant correlation at zero lag, which indicates wider channels at riffle locations. There is a reasonable qualitative physical explanation for this correlation, and the low sinuosity streams of which this is an example commonly exhibit widening at riffle sections (Richards, 1976b). It might be expected that this correlation would exhibit a downstream lag of the width variation, but the cross correlogram for transformed data does not illustrate this (figure 8.5). However, simple direct cross-correlation illustrates a peak correlation at a lag of -1, which implies a downstream lag because the data are taken from the downstream end. This correlogram (figure 8.5) cannot readily be tested statistically, and cannot be used as the basis for estimating an impulse-response function. However, it illustrates four important characteristics that describe the system under investigation: the direct correlation of width and profile series, the possible downstream lag of changes in the width series, the pseudocyclic oscillation of the series, and the approximate period of this oscillation (here about five to seven sampling intervals, or 10 m to 14 m).

Further examples of cross-correlation between width and profile elevations are shown in figure 8.6, where all three analyses have employed the Box–Jenkins transformation. In these examples, the series are shorter and contain fewer riffle–pool cycles, so the correlations are weaker.

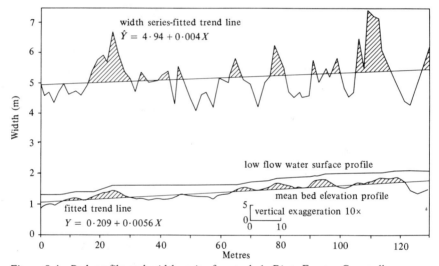

Figure 8.4. Bed profile and width series for reach 1, River Fowey, Cornwall.

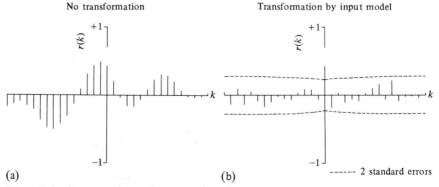

Figure 8.5. Cross-correlation function of profile elevation and width residuals for series in figure 8.4 (a) without prewhitening, and (b) with prewhitening.

Reach 3 on the River Fowey illustrates a downstream lag of the width
variations, with a two-tailed probability of $\alpha = 0 \cdot 06$ associated with the
correlation at lag -1. The Hoaroak Water, Exmoor, example has a similar
probability associated with the zero lag coefficient. An upstream lag is
just significant at $\alpha = 0 \cdot 05$ for the second reach on the River Fowey.
This is unusual, and seems physically improbable, but may be readily
explained in this case by the disruption of the general pattern of correlation
by a retaining wall on the right bank at the downstream riffle. This
perhaps suggests the possibility that the lack of expected correlation at
lags of 0 or -1 in these low sinuosity streams may reflect the existence of
disequilibrium conditions or at least some disruption of equilibrium.

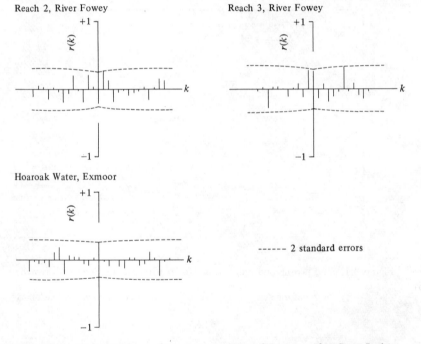

Figure 8.6. Three width–profile cross-correlation functions, after Box–Jenkins
prewhitening.

8.2.3 The problem of series length
As noted above, problems inevitably arise if series lengths are insufficient
to cover a number of significant oscillations: the three series in figure 8.6
are of forty-five to sixty observations at sampling intervals of 2 m, and
cover only two to four riffle–pool cycles. Inevitably, however, the
complex interaction between several independent and semi-independent
variables in river channel geometry increases the likelihood that as series
become longer, additional variables cease to justify the assumption of
invariance within the sample reach, and further complications arise in

model identification as well as in removal of trend. In particular, the most appropriate type of model becomes that of the linear system with added noise, which takes the general form

$$Y_j = \sum_{k=0}^{\infty} h_k X_{j-k} + N_j \, , \tag{8.3}$$

where the N_j series is a noise process which itself may be described by an autoregressive/moving-average model, and which subsumes the influence of the unmeasured independent variables that affect the output Y_j. The linear system (figure 8.7) in these examples is the impulse-response function. and in the examples discussed above, the evidence suggests that the relationships between channel morphometric variables are those with simple gain and delay of the form

$$Y_j = gX_{j-d} + N_j \, , \tag{8.4}$$

where g is the gain factor and d denotes the number of periods of delay (which in spatial series may be positive or negative according to the direction in which the series are analysed, and the tendency of the series defined as the output to lead or lag the input). In identifying this form of model, it is important that the corrupting noise process is independent of the input. In control engineering problems this can be ensured by generating the input by a random number generator, but in analysis of channel form the noise process may include the effects of unmeasured variables which are correlated with the input series. The residuals from the transfer-function model fitted to the width–profile elevation data for reach 1 on the River Fowey are fitted by a first-order autoregressive model. This may be because of an upstream 'memory' in the bank material type which also affects the channel width to 'contaminate' the variation in width generated by the profile–width transfer function. If, however, there is a correlation between the variability of bank material and the bed profile (for example, sandier banks at riffles), then the noise process and the estimation of the transfer function itself becomes impossible by the simple methods based on prewhitened and transformed input and output series (Chatfield, 1975, page 221).

It is possible that the pattern of residuals from the transfer-function model for reach 1 on the River Fowey is explained by a failure to remove satisfactorily the trend from the original data series, although the

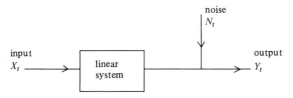

Figure 8.7. Linear system with added noise (for spatial series substitute the subscript j).

autocorrelation functions for the two series do not display the patterns expected when a trend remains. The reach in question is dominated by two large riffle features at the upstream and downstream ends, and these stand out as positive width residuals from the transfer function. The effects of other riffle features have been effectively removed by the fitted model (figure 8.8). Although the cross-correlation function of prewhitened profile residuals (input series) and the transfer-function residuals (noise series) contains a single significant term at a lag of +10, this alone does not point strongly to an invalidation of the essential requirement that the series should not be cross-correlated. Instead, it is suggested that in this case the linear regression trend satisfactorily removes trend in the mean bed elevation, but not in the variance, and that the large riffle features described create excessive widening that the linear model cannot handle. This problem is discussed in further detail in the section on frequency domain analysis.

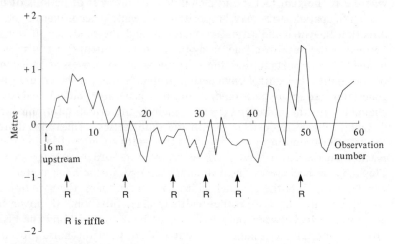

Figure 8.8. Residuals from linear system (transfer function) for width–profile correlation, reach 1, River Fowey.

8.2.4 Sampling frequency and spacing

One objective of series analysis of bed profile data which has been noted above is the possible identification of systematic variation in autoregressive parameters with associated physical controlling variables. However, this presupposes some initial knowledge of the probable variation in these parameters which results from variation in sample spacing. If the bed profile is adequately described by a second-order autoregressive model, the two parameters may be used to estimate the frequency of the process by the relationship (Box and Jenkins, 1970, page 60)

$$f_0 = \tfrac{1}{2}\cos^{-1}[\tfrac{1}{2}|\beta_1|(-\beta_2)^{\frac{1}{2}}] \ . \tag{8.5}$$

Generally the autoregressive process identified for bed profiles lies close to the boundary between the two types of second-order autoregressive [AR(2)] models with positive β_1 coefficients, and there is an approximate relationship between the β_1 and β_2 parameters fitted to the ten bed profiles analysed (figure 8.9). From this relationship it is possible to estimate the expected frequency of oscillation, as indicated in table 8.1. The wavelength of the process is simply the sampling interval divided by the frequency. The autoregressive parameters will vary either as the wavelength of the bed profile oscillation varies, or as the sampling interval varies. If, for a stream of a given width, the sampling interval is fixed, variation in β_1 and β_2 will result from variation in bed profile wavelength. However, it is generally the case that wavelength is correlated with stream width (Richards, 1976a; Keller and Melhorn, 1978). Thus if results are compared for streams of different width, the sampling interval must be defined in relation to width in each case. If this is not done, it will be difficult to interpret the variation in the parameters β_1 and β_2 in relation to changing wavelength to width ratios, because the sampling interval is inconsistent. For example, five reaches of varying mean width have all been surveyed at a sampling interval of 2 m. If in each case it is assumed that the expected wavelength of bed profile oscillation is $2\pi\overline{W}$, where $\overline{W}$ is mean width, the necessary sample spacing required to maintain a particular frequency is given in table 8.2. This indicates that much of the variation in the autoregressive parameters arises because of the difficulty in maintaining an appropriate relationship between channel width, bed form

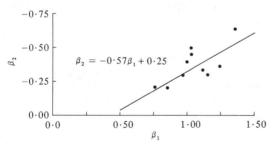

Figure 8.9. Typical β_1 and β_2 parameters for bed profile series.

Table 8.1. Expected frequency of oscillation of AR(2) process for various β_1 parameters.

β_1	β_2 [a]	f_0 [b]	β_1	β_2 [a]	f_0 [b]
0·70	−0·15	0·069	1·30	−0·49	0·061
0·90	−0·26	0·080	1·50	−0·61	0·043
1·10	−0·38	0·073			

[a] Estimated from linear relationship in figure 8.9.
[b] Calculated from equation (8.5).

wavelength, and sample spacing, which, of course, is compounded by the practical problems of estimating mean width in the field and measuring sample spacings of inconvenient dimensions.

Table 8.2. Sampling intervals required to maintain a constant frequency of $\frac{1}{15}$ (0·067).

Reach	Mean width (m)	Sampling intervals (m)
Reach 1, River Fowey	5·22	2·19
Reach 2, River Fowey	5·91	2·48
Reach 3, River Fowey	5·27	2·21
Hoaroak Water	2·79	1·17
Afon Elan	3·47	1·45

8.3 Frequency domain analysis of distance series

It has been suggested that the concept of frequency is not viable in spatial data series (Granger, 1969). Nevertheless, harmonic and spectral methods have been employed in the analysis of channel form, particularly again in relation to bed forms and meanders. At the riffle–pool scale, for example, simple harmonic analysis is capable of identifying the riffle–pool frequency, and the waveform at this frequency has been shown to explain 20% to 55% of the original variance of a bed profile series (Richards, 1976a). However, there is no *a priori* physical reasoning to support the assumption that a physical distance series of this type should exhibit a constant wavelength and amplitude, and the case of reach 1 on the River Fowey suggests this to be an unlikely occurrence. In spite of this scepticism, it is interesting to note that harmonic analysis of width and profile series for this reach and reach 3 on the River Fowey both indicate a phase difference between the width and profile cycles with a downstream shift of the former (which amounts to 4·5 m and 2·7 m for reach 1 and reach 3 respectively).

One of the constraining factors of harmonic analysis, as far as the analysis of spatial series is concerned—that of constant amplitude at a given frequency—is relaxed in the use of demodulation methods. Demodulation allows an estimate both of amplitude and phase of a harmonic of given frequency to be obtained for all j. The method consists of multiplying the series X_j by a function and then applying a low-pass filter F (that is, a filter that passes only low frequencies):

$$Z'_j = F(X_j \sin f_{0j}) , \qquad Z''_j = F(X_j \cos f_{0j}) . \qquad (8.6)$$

The simple moving-average filter seems the best one to employ (Granger and Hatanaka, 1964) and two are used of length m and p such that $n \geqslant m+p$ and $m > p$. Thus the estimate of the amplitude $\omega(j)$ and phase $\theta(j)$ is given by

$$\omega(j) = 2(Z'^2_j + Z''^2_j)^{\frac{1}{2}} , \qquad \theta(j) = \tan^{-1}\left(\frac{Z'_j}{Z''_j}\right) , \qquad (8.7)$$

for frequency f_0.

In relaxing the assumption of stability in time or space, it is possible to assess the occurrence of disturbances to a given periodic component, in order to explain their origin and identify the intervals in which they appear. There is growing evidence of application of this technique in hydrology, with specific reference to temporal changes in the annual amplitude of discharge series (Rodriguez-Iturbe et al, 1971; Anderson, 1975a; 1975b; Andel and Balek, 1975). Despite the unquestioned superficial attraction of the technique it must be recalled that in all empirical analyses the filter types used cannot be objectively determined, and trends in the amplitudes can only be qualitatively identified, because least-squares trend fitting and subsequent significance testing is invalidated by the use of the filters employed in the procedure. With these cautions borne in mind, figure 8.10 shows the results of demodulating the width and bed profile series for reach 1 of the River Fowey at the riffle–pool frequency $k = 6$. In accordance with the suggestion of Granger and Hatanaka (1964), a low-order trend is fitted to the estimates obtained. In this analysis the filter lengths applied were $m = 10$, $p = 6$. As with all statistical estimations, one is forced to compromise between variance and bias—use of too narrow a window (m and p large with respect to n) increases the correlation between adjacent estimates and increases the variance, thereby instilling a trend in the amplitudes. Granger and Hatanaka (1964) demonstrated this by applying progressively longer moving-average schemes to white noise, and showed that when $m = 0 \cdot 3n$, trends appeared. Figure 8.10 illustrates moderately well a further phenomenon encountered in the interpretation of demodulation results—the larger the amplitude, the larger the variance [Rice (1944) demonstrated this theoretically]. The demodulation results of the bed profile and width series for this reach of the River Fowey show the variation of amplitude to be relatively slow and smooth with distance upstream. The nature of the low-order trends fitted in this figure suggest that despite sampling fluctuations, there are true

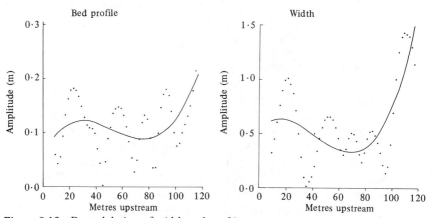

Figure 8.10. Demodulation of width and profile series, reach 1, River Fowey.

changes in the nature of the amplitudes both of width and bed profile along the reach. Particularly, the two large riffles which dominate the reach are identified, shown by the large zones of positive residuals at *circa* 25 m and 112 m upstream in the series in figure 8.4. Additionally, zones of minimum amplitude at the riffle–pool frequency are coincident at 70 m to 80 m upstream. Simple direct cross-correlation of the width and bed profile for this reach (figure 8.5) shows width to be lagging bed profile by *circa* 2 m in the downstream direction. Qualitative confirmation of this is apparent in figure 8.10, where the minima of the low-order polynomial trends occur at 79 m and 71 m upstream for the profile and width series respectively, and the downstream profile maximum leads that of the width by approximately 12 m.

In assessing this form of frequency domain analysis of distance series with particular reference to the example of the River Fowey reach, it must be remembered that the data length is short and that the frequency for demodulation is low (both factors that are rectified in the case study discussed below). Analysis at a higher frequency (a larger number of riffle–pool cycles) decreases the likelihood of interference with subharmonics. Nevertheless, the results of figure 8.10 suggest that there are 'disturbances' in the behaviour of width and bed profile amplitudes at the riffle–pool frequency. The disturbances in the bed profile series originate from two highly significant riffles in the reach, and these disturbances are propagated into the width amplitude series in a manner already described for the cross-correlation analysis.

Thus demodulation allows detailed analysis of the covariation of the amplitudes of spatial series at a particular frequency. Moreover, in a much longer reach, the riffle–pool frequency may be seen to vary with distance downstream, and demodulation at different frequencies would be necessary. In this manner, therefore, constant wavelength and amplitude assumptions of harmonic analysis can be relaxed, but only at the expense of increasing the difficulty of obtaining a large data set from a homogeneous reach, and of embodying a relatively qualitative analytical procedure as far as filter selection and trend approximation are concerned.

8.4 A case study—the Afon Elan

In order to test some of the points raised above more conclusively, a long reach of the upper part of the Afon Elan in mid-Wales was surveyed to provide 260 observations at 2 m spacing. A map of the reach in figure 8.11 shows the stream to be an underfit with large meanders cutting a valley floor between terrace bluffs. The number of identifiable riffle features over the whole reach is approximately nineteen, which gives a mean wavelength of 27·4 m, and a wavelength–width ratio of 7·89 (the mean width being 3·47 m). The majority of riffles in this stream take the form of asymmetric bars, rather than the medial bar form common in the low sinuosity River Fowey. The asymmetric bars include a lobate accumulation of gravel and pebbles on one bank, and a cross-channel shoal that diverts the thalweg towards the opposite bank, which experiences undercutting. At the time of survey, the low-flow water width was often narrower in the riffle sections, but the bank-to-bank width was noticeably wider than in adjacent pools. It is likely that the propensity for asymmetric shoals is caused by the high sinuosity of the stream; as the water passes round a right-hand bend, the successive riffle accumulations are on the right bank, as the water is forced against the left bank. The bed profile and bank-to-bank widths are plotted in figure 8.12. Trend removal was effected by

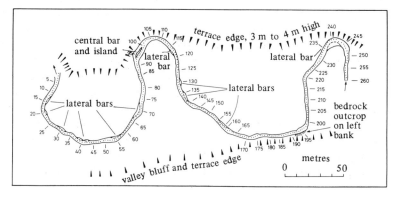

Figure 8.11. Surveyed reach of the Afon Elan, mid-Wales.

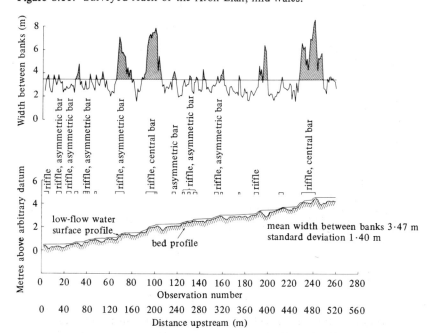

Figure 8.12. Bed profile and width series of the Afon Elan.

linear regression, although it is apparent that the residual variances may
not be constant, particularly in the width series.

Figure 8.13 illustrates the correlograms and partial autocorrelation
functions for the width and profile series. Both correlograms have a slow
decline, which suggests a possible residual trend. However, there is no
justification for higher-order polynomial and logarithmic trends in the
patterns evident in figure 8.12, and there are only five significant terms in
the profile correlogram, and six in the width correlogram. The partial
autocorrelation functions show very clear cutoffs after, respectively, the
second and first terms. Thus identification of AR(2) and AR(1) models
seems appropriate. Figure 8.14 illustrates the cross-correlation function,
which simply displays a single significant term at zero lag. A further
significant negative cross-correlation at a lag of −7 may be ignored because
this simply expresses the fact that the negative width residual of a pool is
correlated with a positive profile residual (a riffle) seven sample units
upstream (the data being analysed from the downstream end). From this
an alternative estimate of the riffle–pool wavelength is fourteen sample
units, or 28 m, which agrees well with the field estimate. However, a
cross-correlation analysis of untransformed series yields the strong peak at
zero lag, but a cycle whose wavelength is some twenty-four sample intervals,
or 48 m. This may be a reflection of the influence of the larger riffles,

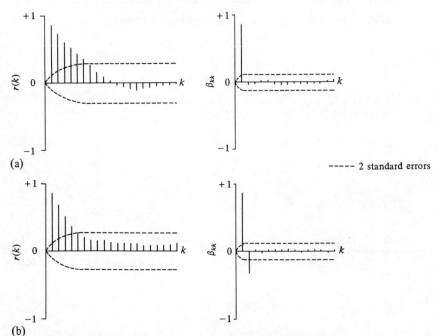

(a)

----- 2 standard errors

(b)

Figure 8.13. Correlograms and partial autocorrelation functions of (a) width and
(b) profile series of the Afon Elan.

which are more widely spaced, since the predicted wavelength–width ratio is 13·83, or approximately twice that noted in the field. Interestingly, this is the sort of wavelength–width ratio expected for the meander feature rather than for the riffle–pool cycle; however, no detailed inference can be drawn from this, since it is possible that there is some aliasing effect which causes this cycle.

A cross-correlation check between the noise series, defined as the residuals from the width–profile transfer-function model, and the input series of profile elevation residuals indicates that there is no such cross-correlation to hamper the efficiency of estimation of transfer-function parameters. The noise series (N_j) itself is adequately described by a first-order autoregressive scheme [AR(1)]

$$N_j = 0\cdot84N_{j-1} + e_j , \qquad (8.8)$$

which completes the progressive reduction of unexplained variance in the modelling of this reach. The bed profile series is described by an AR(2) process which explains 78·6% of the original variance. The width series (W_j) is described by a transfer-function model explaining 16% of the original series, and an AR(1) process which explains 70% of the remaining variance; jointly, these explain 75% of the variance of the width series. The relationships required [in addition to equation (8.8)] are

$$Z_j = 1\cdot16Z_{j-1} - 0\cdot33Z_{j-2} + e_j , \qquad (8.9)$$

(where the variance of the error series is 0·006) to describe the bed elevation residuals, and

$$W_j = 2\cdot27Z_j + N_j . \qquad (8.10)$$

The error series in equation (8.8), which models the noise process, has zero mean and a variance of 0·26. The correlogram for this error series is illustrated in figure 8.15, and clearly demonstrates that the errors are generated by a white-noise process. The frequency distribution of errors in figure 8.16 suggests a slight tendency for the errors to be drawn from a right-skewed distribution, although a Kolmogorov–Smirnov test does not

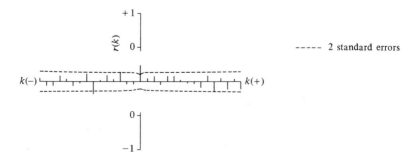

Figure 8.14. Cross-correlation function of width and profile series for the Afon Elan.

permit rejection of the null hypothesis that the distribution does not differ significantly from a normal one. The cross-correlation function of the prewhitened input series with the error series is also devoid of significant terms, so that these final checks on the error series lead to the conclusion that the model identification is adequate.

Figure 8.17 shows the results of demodulating the width and bed profile series at the riffle–pool frequency. Once again there is a smooth variation of amplitude with distance. However, there are pronounced maxima in both demodulated series. The bed profile results indicate high amplitudes

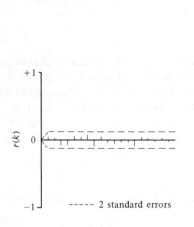

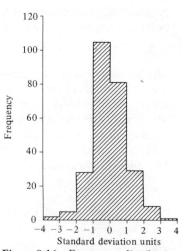

Figure 8.15. Correlogram of final error series of width variation, Afon Elan.

Figure 8.16. Frequency distribution of errors from fitted transfer-function–noise model for width variation, Afon Elan.

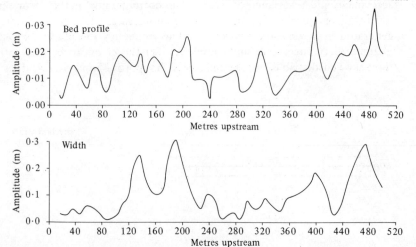

Figure 8.17. Demodulation of width and bed profile series, Afon Elan.

in the regions 100 m to 210 m and 390 m to 490 m upstream, and lower than average amplitudes at the riffle–pool frequency in the intervening section from 220 m to 390 m, where average amplitudes are *circa* 0·1 m compared to 0·15 m in the previously described sections. The demodulated width series in figure 8.17 illustrates a similar pattern, with a marked low amplitude section from 20 m to 100 m upstream in addition. On comparing these characteristics with those of the original width and profile data (figure 8.12) it can be seen, especially in the width series, that sections of the reach that exhibit high amplitudes reflect these zones of both width and bed profile extremes. The nature of the relative regularity of the riffle–pool sequence and channel width series means that demodulation results in figure 8.17 can be interpreted in terms of variance nonstationarity in the original data set. Thus in the segment 100 m to 210 m upstream, the demodulation results of the width series, for example, show a large amplitude (that is, a harmonic amplitude much larger than the average for the complete series) which indicates large oscillations (high variance) in the original series. The reverse is true for sections where the amplitude is low. The nature of the regularity of oscillation in the original data set therefore suggests that for bed profile and width series modelling, demodulation may be a particularly satisfactory way of rendering the initial series stationary.

8.5 Conclusions

Channel morphometry and related process variables such as sediment load can both be explored by spatial and temporal statistical models. This chapter has sought to examine models that relate channel width and bed profile series, and direction angle series. The important distinction between process and prediction as the desired output has been made, and process has been emphasised. This selected prerequisite has been seen to play a role in governing the initial method of elimination of stationarity in channel width and bed profile series, with regression detrending being preferred to the differencing technique for reasons of inference. In breaking away from the more classical approach to the study of channel characteristics (previously tackled by regression models) and in exploring spatial series, one major difference is encountered at the outset. This difference relates to the fact that in the latter approach it is more difficult to accommodate the situation of multiple input variables. Nevertheless, empirical studies both of the River Fowey and Afon Elan have been used to demonstrate the capacity of the cross-correlation function to show the form of interrelationships between the two variables of bed profile elevations and channel widths, and the example of the Popo Agie River illustrates the correlation between bed profile and channel direction changes. In both of these analyses, there is evidence of bed profiles leading the other series in the downstream direction, and in all cases the evidence suggests that very simple models satisfactorily relate the series.

These invariably are transfer functions with simple gain and zero or one period of delay between the input and output variables, and noise processes which summarise the effects of unmeasured input variables. These simple models provide a means of describing the essential form of the elements in the spectrum of channel types, which range from low sinuosity streams dominated by medial bars and gravel bed material to high sinuosity, true meandering streams with sandier bed material and point bar features. This corresponds to the spectrum of channel types defined with the use of sedimentological criteria by Bluck (1976), and illustrated in figure 8.18.

The previously aired constraints imposed by the concept of frequency analysis in spatial data sets have been relaxed by the use of demodulation techniques. Demodulation has shown that at the riffle-pool frequency there are significant disturbances in the amplitude of the bed profile series in both the reaches analysed, and similar disturbances are found in the width series. The existence of such marked disturbances in the demodulated series indicates nonstationarity in the original data, and it is suggested that demodulation could initially be applied to the series to render it stationary. Finally, series length has been emphasised as something of a restriction in the application of the models described, since the number of riffle-pool cycles along channel reaches which have near constant bed and bank material without significant tributary inputs is rarely sufficient to provide a statistically large sample. Despite these

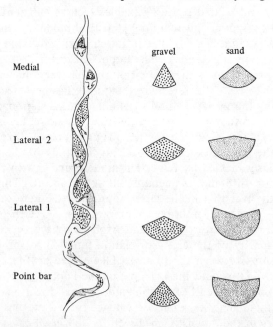

Figure 8.18. Typology of channel patterns in relation to bar features and directional properties of sedimentary structures (after Bluck, 1976).

cautions stemming from the empirical work described, spatial models have the resolution and potential to facilitate further examination of the interrelationships between distance based series of river channel variables.

References

Andel J, Balek J, 1975 "Model of the disturbances in hydrological sequences, based on the method of demodulation" International Association of Scientific Hydrology symposium, Bratislava, Czechoslovakia, September 1975 (mimeo)

Anderson M G, 1975a "Some statistical approaches towards physical hydrology in large catchments" in *Processes in Physical and Human Geography* Eds R F Peel, M Chisholm, P Haggett (Heinemann, London) pp 91-109

Anderson M G, 1975b "Demodulation of streamflow series" *Journal of Hydrology* **26** 115-121

Bennett R J, 1975 "The representation and identification of spatiotemporal systems: an example of population diffusion in North West England" *Transactions Institute of British Geographers* **66** 73-94

Bluck B J, 1976 "Sedimentation in some Scottish rivers of low sinuosity" *Royal Society Edinburgh Transactions* **69** 425-456

Box G E P, Jenkins G M, 1970 *Time Series Analysis—Forecasting and Control* (Holden-Day, San Francisco)

Carlson R F, MacCormick A J A, Watts D G, 1970 "Application of linear random models to four annual streamflow series" *Water Resources Research* **6** 1070-1078

Chang T P, Toebes G H, 1970 "A statistical comparison of meander planforms in the Wabash basin" *Water Resources Research* **6** 557-578

Chatfield C, 1975 *The Analysis of Time Series: Theory and Practice* (Chapman and Hall, London)

Chow V T, Kareliotis S J, 1970 "Analysis of stochastic hydrologic systems" *Water Resources Research* **6** 1569-1582

Church M A, 1972 "Baffin island sandar: a study of Arctic fluvial processes" *Geological Survey of Canada Bulletin* **216** 208

Crickmore M J, 1970 "Effect of flume width on bed-form characteristics" *Proceedings, American Society of Civil Engineers Journal, Hydraulics Division* **HY2** 473-496, paper 7077

Ferguson R I, 1975 "Meander irregularity and wavelength estimation" *Journal of Hydrology* **26** 315-333

Ferguson R I, 1976 "Disturbed periodic model for river meanders" *Earth Surface Processes* **1** 337-347

Granger C W J, 1969 "Spatial data and time series analysis" in *Studies in Regional Sciences* Ed. A J Scott (Pion, London) pp 1-24

Granger C W J, Hatanaka M, 1964 *Spectral Analysis of Economic Time Series* (Princeton University Press, Princeton, NJ)

Jenkins G M, Watts D G, 1968 *Spectral Analysis and its Applications* (Holden-Day, San Francisco)

Kendall M G, 1973 *Time Series* (Griffin, London)

Keller E A, Melhorn W N, 1978 "Rhythmic spacing and origin of pools and riffles" *Geological Society of America Bulletin* **89** America Bulletin **89** 723-730

Kirkby M J, 1976 "Tests of the random network model and its application to basin hydrology" *Earth Surface Processes* **1** 197-212

Lai P W, 1977 "Stochastic dynamic models for some environmental systems: transfer function approach" DP-61 16, Graduate School of Geography, London School of Economics, London

Langbein W B, Leopold L B, 1962 "The concept of entropy in landscape evolution" *United States Geological Survey Professional Paper* **500**-A 20

Langbein W B, Leopold L B, 1966 "River meanders—theory of minimum variance" *United States Geological Survey Professional Paper* **422-H** H1–H15

Leopold L B, Wolman M G, 1957 "River channel patterns-braided, meandering and straight" *United States Geological Survey Professional Paper* **282-B** 39–85

Martin R L, Oeppen J E, 1975 "The identification of regional forecasting models using space-time correlation functions" *Transactions Institute of British Geographers* **66** 95–118

Matalas N C, 1963 "Autocorrelation of rainfall and streamflow minimums" *United States Geological Survey Professional Paper* **434-B** B1–B10

Melton M A, 1962 "Methods of measuring the effect of environmental factors on channel properties" *Journal of Geophysical Research* **67** 1485–1490

Miller J P, 1958 "High mountain streams: effects of geology on channel characteristics and bed material" *New Mexico Bureau of Mines Memoir* **4**

Nordin C F, Algert J H, 1966 "Spectral analysis of sand waves" *Proceedings, American Society Civil Engineers Journal, Hydraulics Division* **HY5** 95–114

Nordin C F, Richardson E V, 1968 "Statistical description of sand waves from streambed profiles" *Bulletin International Association Scientific Hydrology* **13** 25–32

Quimpo R G, 1968 "Autocorrelation and spectral analyses in hydrology" *Proceedings, American Society Civil Engineers Journal, Hydraulics Division* **HY2** 363–373

Rayner J N, 1971 *An Introduction to Spectral Analysis* (Pion, London)

Rice S O, 1944 "Mathematical analysis of random noise" *Bell System Technical Journal* **23** 282–332

Richards K S, 1976a "The morphology of riffle-pool sequences" *Earth Surface Processes* **1** 71–88

Richards K S, 1976b "Channel width and the riffle-pool sequence" *Geological Society America Bulletin* **87** 883–890

Rodriguez-Iturbe J, Dawdy D R, Garcia L E, 1971 "Adequacy of Markovian models with cyclic components for stochastic streamflow representation" *Water Resources Research* **7** 1127–1143

Smart J S, 1970 "Use of topologic information in processing data for channel networks" *Water Resources Research* **6** 932–936

Speight J G, 1965 "Meander spectra of the Angabunga River" *Journal of Hydrology* **3** 1–15

Squarer D, 1970 "Friction factors and bedforms in fluvial channels" *Proceedings, American Society Civil Engineers Journal, Hydraulics Division* **HY4** 995–1017

Thakur T R, Scheidegger A E, 1970 "Chain model of river meanders" *Journal of Hydrology* **12** 25–47

Thornes J B, 1976 "Semi-arid erosional systems" *Geographical Papers* **7** London School of Economics, London

Thornes J B, Brunsden D, 1977 *Geomorphology and Time* (Methuen, London)

Willis J C, 1968 "A lag-deviation method for analyzing channel bed forms" *Water Resources Research* **4** 1329–1334

Yalin M S, 1971 "On the formation of dunes and meanders" *International Association of Hydraulic Research, 14th Congress, Paris, Proceedings* **3** paper **C13** 1–8

Yevjevich V, 1971 "Properties of river flows of significance to river mechanics" in *River Mechanics* Ed. H W Shen (Colorado State University, Fort Collins) pp 1.1–1.28

River meanders: regular or random?

R I Ferguson

9.1 Introduction

Streams and rivers flow downhill, but it is striking how seldom they follow the shortest, steepest, straight-line course. Instead most natural rivers are meandering, that is they point first to one side of the local downvalley direction and then the other. So widespread and familiar is this phenomenon that the verb to meander has come to be used as a metaphor meaning 'wander at random' (*Concise Oxford Dictionary*) in contexts far removed from geomorphology.

This popular usage stands in sharp contrast to the textbook view of river meanders. Definition sketches of meandering almost always depict symmetrical S-bends, and the term has often been specifically restricted to patterns that exhibit considerable regularity. There is a long history of physical explanations for meandering, and even though some of the proposed mechanisms seem quantitatively inadequate most earth scientists and engineers accept that some kind of deterministic meandering tendency is at work in rivers with erodible boundaries. Much effort has also been expended on the search for relationships between meander scale and possible environmental controls such as streamflow volume. This too presupposes that meandering is predictable and sufficiently regular for a single characteristic wavelength to be identified.

Not all geomorphologists agree with this regular and deterministic view, and two particularly original contributions (Speight, 1965; Langbein and Leopold, 1966) have led to a widespread impression that meanders are random after all. This is to some extent a distortion of ideas which can themselves be criticised. But the pioneering work of Speight, Langbein and Leopold in the statistical analysis and modelling of meander patterns contains the foundations of a more consistent and satisfactory theory.

9.2 Meanders as direction series

Meander patterns are difficult to describe in Cartesian coordinates because rivers often double back on themselves to form horseshoe bends and turn through more than 180 degrees. In such places the cross-valley coordinate V of the channel must be a triple-valued function of the downvalley distance U. The traditional description of meander bends as sinusoidal or parabolic does not allow for this common feature, and nor can any other single-valued Cartesian curve.

One simple alternative is to regard meander bends as circular arcs linked end to end, and subtending angles that may be greater or less than $180°$.

This geometry is implicitly or explicitly assumed in many discussions of meandering (for example, Chitale, 1970; Hey, 1976), but the corollary that curvature is constant round each bend is contradicted by the common observation that bends are tightest at the apex and straighten out progressively towards the intervening crossovers or inflexions.

The planimetric form both of individual bends and of longer stretches of river is best represented by measuring the direction θ of the river at various distances s along its course. A plot of θ against s is necessarily single valued, can readily be converted back into a map of the meander pattern, and has a local slope equal to the channel curvature $d\theta/ds$. It is convenient, but not essential, to measure either tangential directions at regular intervals along the channel or mean directions over steps of constant length. Measurements obtained in this way constitute a one-dimensional direction series $\{\theta\}$ and can be differenced to obtain a direction-change series $\{\nabla\theta\}$ that approximates the continuous downstream variation of curvature.

This direction-series representation of meander patterns was applied to semiarid gullies by Leighly (1936) but seems to have been forgotten for thirty years until its independent rediscovery by Speight (1965) and Langbein and Leopold (1966). These authors used it in different ways. Langbein and Leopold demonstrated graphically that sequences of two or three meander bends, whether from the Mississippi or a small laboratory stream, show an approximately sinusoidal variation of direction with distance downstream. This can be written as

$$\theta = \omega \sin ks , \tag{9.1}$$

where the channel direction θ is measured about its mean and oscillates between limits of ω and $-\omega$ with period $\lambda = 2\pi/k$. Differentiation of equation (9.1) shows that curvature oscillates with the same period as direction but out of phase.

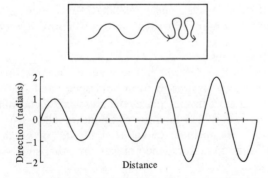

Figure 9.1. Map and direction–distance plot for sine-generated curve with $\omega = 1$ radian on the left and $\omega = 2$ radians on the right.

The spatial pattern corresponding to this regular oscillation of direction and curvature is not itself a sine wave, but what Langbein and Leopold termed a sine-generated curve. As figure 9.1 shows, the extremes of the $\theta-s$ curve, with locally zero curvature, occur at the inflexions of the meander pattern. Conversely at the apex of each bend θ is zero but curvature is maximum. If the amplitude ω of directional oscillation exceeds 90°, horseshoe bends are formed, as on the right of figure 9.1. The scale of meandering, measured by its wavelength λ along the channel, is determined by k. This wavelength must exceed the straight-line wavelength measured down the valley axis. The ratio between the two, called the sinuosity of the pattern, increases with ω and is typically between 1·5 and 2 for well-developed meanders.

The sine-generated curve has been widely accepted as a simple but satisfactory description of individual meander bends, but complete meander patterns seldom if ever consist of a string of identical bends. Direction and curvature series are plotted in figure 9.2 for a typical short meandering reach whose spatial pattern is also shown. Neither graph is much like a sine wave. Individual bends are not perfectly regular, and they differ in length, sinuosity, and mean direction. Only in laboratory conditions is something close to a repeating sine-generated curve ever observed, and even here some irregularity is apparent.

Irregularity in a meander pattern that embraces more than a few bends is best studied statistically. Two approaches are apparent: to examine

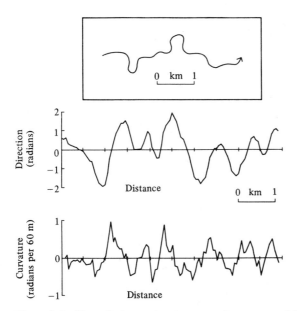

Figure 9.2. Map of ninety-nine-step meander pattern of River Trent, with plots of direction and curvature against distance to show irregularity.

the variability between bends in length, sinuosity, and other properties, or to treat the entire pattern as a single direction series and investigate its statistical structure by standard techniques of series analysis. The series approach has the advantages of taking into account any dependence between properties of adjacent bends, and of being far less affected by the spatial resolution at which the channel pattern is mapped and measured (Ferguson, 1975). It was first applied by Speight (1965) in a study of a river in Papua New Guinea. With the help of the statistician E J Hannan, Speight carried out spectral analysis of direction series measured by hand from aerial photographs of several reaches of the river in each of three years. A perfectly regular meander pattern with wavelength λ along the channel would have a spectrum with a single sharp peak at frequency $1/\lambda$. In contrast the computed spectra had multiple peaks, invariably strongest at lower frequencies (higher wavelengths) than the visually obvious scale of meandering.

Speight interpreted these results as evidence for a set of superimposed harmonics which he thought persisted both downstream and over the years. It can, however, be calculated from the graphs and figures he gives that none of the subsidiary spectral peaks is statistically significant at the 5% level. The evidence is thus more for an irregularly periodic pattern than for one with stable superimposed wavelengths, but either way Speight's study showed quantitatively that meanders are less regular than traditionally thought.

Despite this, it was Langbein and Leopold (1966), in the same paper in which they introduced the regular sine-generated curve, who were responsible for the notion that meander patterns may actually be random. The reasoning behind this paradoxical suggestion was that meandering is the outcome of deviations from a straight course in response to the superposition of many diverse physical causes, individually deterministic but random in their aggregate effect. Since the river has to maintain a certain average slope, its course between two points must be a random walk, constrained to have fixed length and therefore sinuosity. Langbein and Leopold proposed that the typical shape of a meander pattern is the most probable random walk of the required length. If the probability distribution of changes of direction per unit distance ($\nabla\theta/\nabla s$) is Gaussian, as seems reasonable for the aggregate of many independent effects, then the most probable path is the one with minimum variance of curvature, and its equation (von Schelling, 1951) is closely approximated by the sine-generated curve. The apparent regularity of meander patterns was thus held by Langbein and Leopold to be the expected outcome of a variance-minimising or entropy-maximising tendency which they had already invoked to explain other properties of river geometry (Leopold and Langbein, 1962; Langbein 1964).

Two rather different arguments are blended here. The general hypothesis that rivers minimise the variance of their adjustable properties has been supported and extended by some authors (for example, Dozier, 1976; Knighton, 1977) but criticised by others on the grounds that it involves arbitrary assumptions and neglects physical processes (Kennedy et al, 1964; Richards, 1973). The more specific claim that the most probable random walk of fixed length is a regular waveform has generally been accepted. But in fact the pattern with minimum variance of curvature is a single arc of von Schelling's curve, not a series of opposite bends of this or any other shape, since the curvature variance would then increase as the square of the number of bends. In any case the most probable path is not really typical, since the probability of occurrence of a single realisation of a continuous-state stochastic process is infinitely small.

Thus the view that meanders are random walks can neither explain the sine-generated curve nor be supported by it. The approximate normality of marginal distributions of direction in meandering rivers was advanced in support of the random-walk model by Thakur and Scheidegger (1968), but the same authors subsequently recognised that average or expected behaviour and serial statistics are what matter (Thakur and Scheidegger, 1970). Both they and Surkan and van Kan (1969) found that meander direction series have autocorrelation functions that tail off quite rapidly. They did not, however, note the contradiction between this and the formal random-walk model

$$\theta_j = \theta_{j-1} + \epsilon_j , \qquad \epsilon_j = N(0, \sigma^2) , \tag{9.2}$$

which is of course nonstationary in the mean and thus has infinite expected variance and undefined expected autocorrelations. Computer simulations of equation (9.2) in fact produce patterns that wander all over and off the paper, continually crossing over themselves (Ferguson, 1976, figure 2). Series variance increases indefinitely with length of realisation, and correlograms tail off only very slowly.

It is clear from all this that critical application of the statistical approach pioneered by Speight (1965) and Langbein and Leopold (1966) does not entirely bear out their views on meander patterns. Irregularity or randomness undoubtedly exists, but not in the ways originally proposed.

9.3 Disturbed periodic model

The fact that the sine-generated curve fits individual bends well, but requires different parameter values for different bends along the same river, suggests some kind of modulated oscillation of direction along the channel. This is consistent with the view that meandering is an essentially deterministic process but takes place in spatially variable conditions. Valley floors are not generally simple tilted planes underlain by uniform sediments, for the surface layer of materials is continually being redistributed and altered through lateral erosion and deposition by the meandering river itself. Langbein and Leopold's argument that individually deterministic effects may appear random in aggregate still applies, but this time to the passive conditions controlling the degree and direction of meander development rather than to the hydraulic processes involved.

A disturbed periodic model for the pattern produced by a regular meandering tendency in an arbitrarily varying environment was proposed by Ferguson (1976). For analytic convenience the meandering tendency is represented by the sine-generated curve in differential equation form,

$$\frac{d^2\theta}{ds^2} = -k^2\theta \,, \tag{9.3}$$

or

$$\theta + \frac{1}{k^2}\frac{d^2\theta}{ds^2} = 0 \,,$$

which is equivalent to equation (9.1) if $\theta = 0$ and $d\theta/ds = k\omega$ at $s = 0$. This wave equation describes a spatial version of simple harmonic motion. It can be disturbed and distorted by changing the centreline of oscillation from $\theta = 0$ to some randomly-varying function $\theta = \xi(s)$. With the addition of a first-order damping term to ensure stability, the final model is

$$\theta + \frac{2l}{k}\frac{d\theta}{ds} + \frac{1}{k^2}\frac{d^2\theta}{ds^2} = \xi \,, \tag{9.4}$$

with $0 < l < 1$.

A time domain version of this was suggested by Yule (1927) as a model for the sunspot cycle, based on the analogy of the motion of a frictionally-damped pendulum bombarded by small boys armed with peashooters. This transposes well to the meander problem if distance is substituted for time, and trees, clay-filled oxbow lakes, or other obstacles replace the juvenile disruption of the regular oscillation.

Expected serial statistics for this model are given in Ferguson (1976) on the assumption of a white-noise disturbance function. The direction correlogram should oscillate with period $2\pi/k(1-l^2)^{1/2}$ and be damped exponentially at a rate proportional to l. The expected spectrum is dominated by low frequencies. It has a peak if $0 < l < 1/2^{1/2}$, but drops continuously from its zero-frequency value for higher values of the damping or irregularity factor l. Expected statistics of curvature are broadly similar, but with more pronounced oscillatory behaviour in the correlogram and a spectral peak at the wavelength $2\pi/k$ whatever the value of l. This wavelength is of course that of the sine-generated curve from which the randomised model was derived, whereas the correlograms appear to oscillate at a higher wavelength.

The relationship between this model and the random-walk equation (9.2) is clearer if we switch from continuous to discrete distance. A second-order linear differential equation can be approximated by a second-order linear difference equation, which in turn is equivalent to a second-order auto-regressive [AR(2)] process of the form

$$\theta_j = \beta_1\theta_{j-1} + \beta_2\theta_{j-2} + \epsilon_j \,. \tag{9.5}$$

Strictly speaking the process described by equation (9.4) with step-function disturbances is matched at each step by equation (9.5) only if, in the case of the latter, disturbances are a first-order moving average of those of the former (Box and Jenkins, 1970, pages 355–367), but this modification has very little effect on the appearance of simulated meander patterns. Nor, with one exception noted later, does it alter appreciably the expected statistics of the process.

The parameters of the AR(2) approximation are related to those of the differential equation by

$$\left.\begin{aligned}\beta_1 &= 2\exp(-kl)\cos[k(1-l^2)^{\frac{1}{2}}]\\\beta_2 &= -\exp(-2kl).\end{aligned}\right\} \tag{9.6}$$

These relationships define the autoregression in terms of the scale and degree of regularity of the meander pattern. It is easily verified that for any $k > 0$ and $0 < l < 1$ the βs satisfy both the stationarity condition $-1 < \beta_2 < 1 - |\beta_1|$ and the oscillatory-behaviour condition $\beta_2 < -\frac{1}{4}\beta_1^2$. [See figure 9.3 and Box and Jenkins (1970, pages 58–60).]

If meander patterns are generated by a disturbed periodic process, then their direction series should have an approximately AR(2) structure with parameter estimates within this restricted part of the β_1, β_2 space, or just outside if we allow for sampling error. This provides a formal test of the model to supplement visual comparison of observed and expected correlograms and spectra.

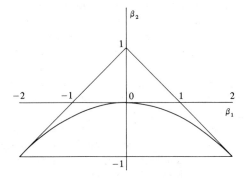

Figure 9.3. AR(2) parameter space showing stationary region (within triangle) and oscillatory subregion (below curved line).

9.4 Statistical confirmation
Both types of test have been applied to the meander patterns of nineteen reaches of English and Scottish rivers. These reaches are listed in Ferguson (1975) and were originally selected as having a flow gauging station, but lacking large tributaries, obvious signs of diversion by man, or other complications that might make the channel pattern unrepresentative or

internally nonhomogeneous. The rivers vary considerably in size (width 6·7 m to 67 m), sinuosity (1·11 to 2·33), and general appearance, and several would not normally be called meandering.

The direction series for each reach was obtained from a large-scale Ordnance Survey map by using a D-Mac digitiser to produce computer-compatible coordinates of points closely but unevenly spaced along the channel. A computer program then interpolated a smaller number of equally-spaced points along the zigzag line defined by the digitised coordinates, and computed the directions of the chords joining the interpolated points. Trials in which the same river was digitised several times showed that tracing reproducibility was good.

The interpolation procedure means that each direction is derived from four digitised points, two that were used for the previous direction and two which will be used for the next direction, so any error in digitisation will tend to introduce a lag-one correlation, but the same is true of errors in hand measurement. The automated method is far easier and faster and also allows various different values of the step length or sampling interval ∇s to be tried without redigitising the pattern. The choice of ∇s determines the number of direction values by which a given length of river is represented and consequently the resolution at which it is analysed. It involves a trade-off between loss of information (and degrees of freedom) as ∇s is increased, and greater likelihood of digitising error as ∇s is reduced. Speight (1965) chose a sampling interval equal to the average unvegetated width of his seasonally-flooding river, but subsequent workers have used arbitrary intervals that ranged from half to three-times channel width. Step lengths in this study are round numbers close to or a little above one-and-a-half times the channel width, and reanalysis with steps up to 50% longer or shorter produces almost identical correlograms and spectra once lags and wavelengths are converted from steps to absolute distances (Ferguson, 1975). Smoothing of the meander pattern at coarser resolutions does, however, lead to a slight fall in direction variance and a rather greater fall in curvature variance.
Series lengths at the chosen values of ∇s range from 99 to 370 terms, with a median of 182. This is long compared with many economic time series, but confidence intervals about serial statistics and parameter estimates are still quite wide. This is probably unavoidable, for longer stretches of river are increasingly likely to have nonstationary properties.

Correlograms and spectra were computed for each of the nineteen direction series, and for the curvature series obtained by differencing them. Direction statistics looked very varied, with some spectra possessing a peak at a finite wavelength and others not, and likewise some correlograms tailing off with, others without, pronounced oscillations. Curvature statistics varied less, all spectra having single significant peaks and all correlograms dropping to a significant negative value at a low lag before tailing off. These properties of course refute the random-walk model, in which successive changes of direction are independent.

This range of observed behaviour is in fact roughly as expected from the disturbed periodic model, and spectra and correlograms for several rivers are strikingly similar to those predicted when suitable values of the scale and irregularity parameters k and l are used. Two examples are shown in figures 9.4 and 9.5, one (River Trent) more regular than average and the other (River Aire) less. The parameter values used in computing the expected statistics are round-number values that give a reasonable

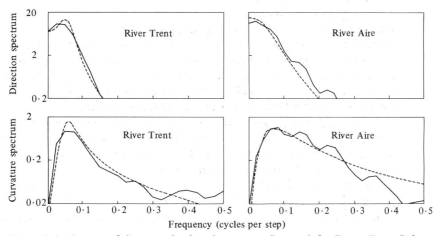

Figure 9.4. Spectra of direction (top) and curvature (bottom) for Rivers Trent (left, $\nabla s = 60$ m) and Aire (right, $\nabla s = 70$ m). Solid lines are observed spectra, broken lines are theoretical spectra for $\lambda = 1000$ m, $l = 0.4$ (Trent) and 0.8 (Aire), direction variance $= 0.85$ (Trent) and 1.0 (Aire).

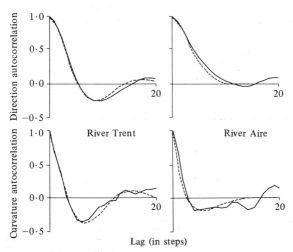

Figure 9.5. Observed (solid) and theoretical (broken) correlograms of direction (top) and curvature (bottom) for Rivers Trent (left) and Aire (right). Parameters as in figure 9.4.

visual match, not statistical estimates which would presumably give an even better fit.

Some of the other rivers can be matched just as well as the Trent and Aire, others less well. The biggest discrepancies appear to result from nonstationarity in mean direction, often attributable to valley bends round which the axis of meandering has to turn. This tends to raise the auto-correlations of direction and inflate the low-frequency variance in the direction spectrum. A good example is the River Dane (figure 9.6), whose correlogram oscillates without ever becoming negative and whose spectrum has a zero-frequency shoulder that largely conceals what might otherwise be a clear peak at a wavelength of about 400 m. Differencing to obtain the curvature series filters out low-frequency variance, and in all cases the curvature statistics can be matched closely by the disturbed periodic model. It may well be that nonstationarity of this type accounts for the strong low-frequency peaks in Speight's spectra.

The second way in which the disturbed periodic model can be confirmed is through the AR(2) approximation already discussed. This was fitted to the nineteen direction series by least squares, without prior detrending. The proportion of variance explained ranged from a minimum of 79% to a maximum as high as 96%, with a median of 92%. In every case both parameters are significantly different from zero, and define a point within or on the edge of the oscillatory region in the parameter space (figure 9.7).

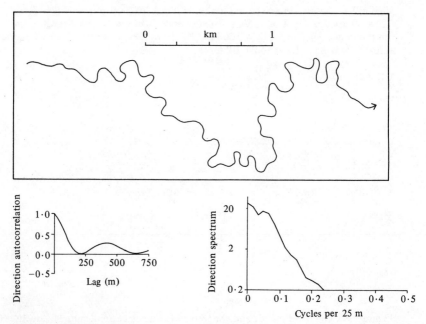

Figure 9.6. Map of River Dane showing valley bends responsible for the nonstationarity apparent in the correlogram and spectrum of direction ($\nabla s = 25$ m).

This reinforces the conclusion that the disturbed periodic model reproduces the main features of meander geometry.

A final statistical test is to inspect the partial autocorrelation function for each river to see if there is any evidence that a higher-order model would provide an even better fit. Since the discrete equivalent of the continuous model is not exactly AR(2) the partial correlogram should not be truncated at lag two but should tail off. For plausible values of k, the moving-average correlation in the AR(2) disturbances should introduce a small positive lag-three correlation. Twelve of the nineteen rivers have significant lag-three partial autocorrelations at the 5% level, positive in each case, so this prediction too is broadly confirmed.

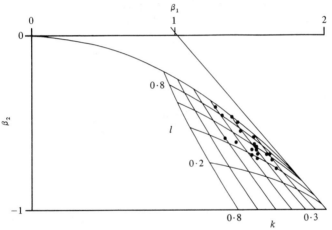

Figure 9.7. Contours of the disturbed-periodic parameters k and l in AR(2) parameter space of figure 9.3, with points for nineteen British rivers.

9.5 The regularity of meander patterns

It seems perverse to claim that meandering (or any other phenomenon) is both regular and random, but the two are not incompatible if one applies to the process and the other to the conditions in which it operates. What then is the status of the ensuing meander pattern?

From a systems engineering point of view the meandering process of equation (9.4) is a deterministic linear operator that transforms a spatially-varying input ξ to a spatially-varying output θ. If the valley conditions represented by the function $\xi(s)$ are treated as random, then $\theta(s)$ is also random, even though it is expected to behave in the quasi-periodic way outlined and tested above. But valley-floor irregularity can be treated as random without implying that it is indeterminate. The engineer may be able to predict the erodibility of river banks from their composition and material properties, and the sedimentologist may in turn be able to explain the spatial variability in these properties in terms of past river action. In principle, then, meander patterns are deterministic. The assumption of randomness is a mathematical convenience that permits the use of statistical methods, not a fundamental part of the disturbed periodic model.

Whether meanders are viewed as regular or random also has practical implications. Engineers and earth scientists have various reasons for wanting to know how to characterise meander patterns. For example, relationships between meander geometry and streamflow volume can be used for such diverse purposes as predicting how a meandering river will respond to a reservoir scheme and reconstructing the ancient climate responsible for meander traces in sedimentary rocks or valley patterns. In each case it is useful to be able to summarise meander form in a few indices.

The traditional parameters are wavelength, usually measured in a straight line for a single, supposedly typical, pair of bends, and sinuosity. In studies of individual meander bends, perhaps as they change through time, wavelength and other properties can be measured directly. But if the aim is to characterise the meander pattern at a broader scale, statistical methods are desirable because of the quasi-random differences between bends. Simple visual comparison of direction or curvature spectra (or both) can be very instructive, particularly when studying the changing pattern of a single river (Ferguson, 1977a). If a parametric description is wanted, the disturbed periodic model does not do away with the use of wavelength and sinuosity, for the relationship $\lambda = 2\pi/k$ allows wavelength estimation from the frequency of the curvature spectrum peak, the discrete–continuous parameter relations (9.6), or the direction–curvature variance ratio (Ferguson, 1975; 1976; 1977a); and sinuosity is a function of direction variance in meanders with a normal marginal distribution of direction (Ferguson, 1977b). The main departure from the traditional approach is the addition of a third parameter, the irregularity l, which like λ can be estimated by substituting least-squares $\hat{\beta}$s in equations (9.6).

Like many other geographical phenomena, the nature of meandering depends very much on the scale at which it is studied. Individual bends should certainly be treated from a regular, deterministic viewpoint. Broader-scale meander patterns are far less regular, but represent distorted waveforms rather than random walks. Statistical methods applied to direction series are of great value in describing, analysing, modelling, and summarising meander patterns, but it is regularity not randomness that emerges as the key feature.

References
Box G E P, Jenkins G M, 1970 *Time Series Analysis: Forecasting and Control* (Holden-Day, San Francisco)
Chitale S K, 1970 "River channel patterns" *Journal of the Hydraulics Division, American Society of Civil Engineers* 96 HY1, 201–221
Dozier J, 1976 "An examination of the variance minimisation tendencies of a supraglacial stream" *Journal of Hydrology* 31 359–380
Ferguson R I, 1975 "Meander irregularity and wavelength estimation" *Journal of Hydrology* 26 315–333

Ferguson R I, 1976 "Disturbed periodic model for river meanders" *Earth Surface Processes* **1** 337-347

Ferguson R I, 1977a "Meander migration: equilibrium and change" in *River Channel Changes* Ed. K J Gregory (John Wiley, Chichester, Sussex) pp 235-248

Ferguson R I, 1977b "Meander sinuosity and direction variance" *Geological Society of America Bulletin* **88** 212-214

Hey R D, 1976 "Geometry of river meanders" *Nature* **262** 482-484

Kennedy J F, Richardson P D, Sutera S P, 1964 "Discussion of 'Geometry of river channels' by W B Langbein" *Journal of the Hydraulics Division, American Society of Civil Engineers* **90** HY6, 332-341

Knighton A D, 1977 "Alternative derivation of the minimum variance hypothesis" *Geological Society of America Bulletin* **88** 364-366

Langbein W B, 1964 "Geometry of river channels" *Journal of the Hydraulics Division, American Society of Civil Engineers* **90** HY2, 301-312

Langbein W B, Leopold L B, 1966 "River meanders: theory of minimum variance" *US Geological Survey Professional Paper* **422-H** H1-H15

Leighly J B, 1936 "Meandering arroyos of the dry Southwest" *Geographical Review* **26** 270-282

Leopold L B, Langbein W B, 1962 "The concept of entropy in landscape evolution" *US Geological Survey Professional Paper* **500-A** A1-A20

Richards K S, 1973 "Hydraulic geometry and channel roughness—a nonlinear system" *American Journal of Science* **273** 877-896

Schelling H von, 1951 "Most frequent particle paths in a plane" *Transactions American Geophysical Union* **32** 222-226

Speight J G, 1965 "Meander spectra of the Angabunga River, Papua" *Journal of Hydrology* **3** 1-15

Surkan A J, Kan J van, 1969 "Constrained random walk meander generation" *Water Resources Research* **5** 1343-1352

Thakur T R, Scheidegger A E, 1968 "A test for the statistical theory of meander formation" *Water Resources Research* **4** 317-329

Thakur T R, Scheidegger A E, 1970 "Chain model of river meanders" *Journal of Hydrology* **12** 25-47

Yule G U, 1927 "On a method of investigating periodicities in disturbed series with special reference to Wolfer's sunspot numbers" *Philosophical Transactions of the Royal Society* Series A **226** 267-298

Statistical problems in forecasting long-term climatic change

R J Bennett

10.1 Introduction

The problem of forecasting long-term changes in climate is an interesting one from a statistical point of view for four main reasons. First, it highlights the difficulties in making long-term projections from short data sets and only partially observed trends. Second, the problem of forecasting climate is typical of an important class of environmental systems in which the chief inputs are not easily measured. Such systems which lack data on crucial inputs can be approached only with noncausal models. A third reason for a statistical interest in climatic forecasting is that the physical mechanisms which generate patterns of climatic change, as opposed to weather patterns, are not known with any degree of certainty, nor is any totally accepted deductive theory of climatic change available. This again drives the forecaster to recourse to statistical models which are usually noncausal in structure. A fourth motive for interest in climatic forecasting is that, like many other forecasting problems both in environmental and socioeconomic systems, forecasts are stimuli to action and have been urged as such. Hence policy based on statistical forecasts also have a statistical content which it is interesting to explore. On the basis of statistical analysis, changes in climate have been forecast over various periods, but are normally considered to be restricted to changes in average weather over several decades. Some analysts, however, have considered climatic change over periods of thousands and even hundreds of thousands of years. Our response to the long-term forecasts in particular might be similar to that of Beachcomber who, reporting the claim of a group of researchers that, at the rate the sea was encroaching upon the land in East Anglia "Ely Cathedral will be completely surrounded by water in 30000 years time", he remarked only, "I have sent word to the Bishop". But the reason for analysing such forecasts here is that they have been urged with considerable authority and have been claimed as the basis of action especially by such writers as Lamb, Bryson, and Winstanley. There is certainly considerable evidence of past changes in climate and it is the intention of this study not to question the fact that past changes in climate have occurred. Instead it is sought to question the evidence and statistical assumptions upon which forecasts of future changes in climate have been based. Kopec (1976) identifies three strands to the debate over climatic change. First those workers who feel that the causes of climatic change are extraterrestial. Second those who hold that it is terrestially caused, especially by man-induced changes. Finally there are those workers who accept the inevitability of climatic changes, but are extremely sceptical of the possibilities of forecasting changes either from terrestial or extraterrestial causes. This study falls into the last category, and seeks to expose the areas of considerable statistical weakness of the forecasts that have been urged most forcefully to date. It is hoped that such a critical examination of the statistical basis of the forecasts will reveal those areas in which advance can be made by statistical techniques, and those other areas where instead it is the development of deductive theory that is required.

A major strand in the argument that underlies forecasts of climatic change has been the assurance with which some workers have adduced long-term changes from short-term fluctuations in climate. Thus Lamb (1966; 1974) has argued that many recent changes in a wide variety of different weather patterns, such as the five- to six-year drought in the Sahel, the low frequency of westerly winds over the British Isles, the failure of the monsoon in India, and reduced harvests of coffee, groundnuts, sorghum, and rice in the tropics, represent short-term extremes "superimposed on the underlying trend, which nevertheless *is* in the direction of this extreme" (Lamb, 1974, page 22). Hence Lamb has sought to adduce the interrelation of short-term trends and periodicities with the long-term trend towards a new ice age. Although the policy implications that can be drawn from such forecasts are very unclear and may be not imperative, policy advice over the medium- and shorter-term forecasting horizon has been proffered with considerable urgency to action and hence deserves critical appraisal. The imperative to action is nowhere more evident than in the discussion of the causes and consequences of desertification where, especially in respect of the Sahel region of central West Africa for example, a variety of policies have been urged with important human and economic consequences. For example, policies that have been suggested include changes to agricultural practice which include marked reduction in cattle stocks, enforced migration of population, settlement of nomads, and even food stockpiling.

Such policies, although feasible in the physical sense, have severe implications for the basis of existing societies and cultures, and require strong and certain imperatives to implement them. Hence in assessing forecasts of climatic change, not only must we bear in mind the validity of the models used and the certainty of their forecasts from a statistical point of view, but also we must encompass explanations which can be understood and made credible to those affected by the consequences of any policy decisions. In this chapter various statistical approaches to climatic forecasting are discussed in section 10.2. The main difficulties in applying statistical methods to such problems are discussed in section 10.3, and in section 10.4 the implications of these difficulties are drawn. It is concluded that although long-term climatic forecasts may deserve considerable credence, on human time scales their certainty is not sufficient to form a sound basis for economic policy. Instead economic policy

should be oriented towards constructing robust economies in the face of
uncertainty deriving from climatic variability.

10.2 Approaches to statistical forecasting of climatic change

There have been a variety of approaches to statistical forecasting of long-
term changes in climate, all of which rely, to a greater or lesser extent, on
the historical record of data available, which is generally fairly sparse. The
most straightforward approach, and that which requires a minimum of
assumptions, is that of direct eyeballing of the historical sequence. From
visual inspection it is then sought to adduce long-term trends and changes.
This approach has been applied to patterns of global temperature variation
(Mitchell, 1972), ocean temperature changes (Rodewold, quoted in
Lamb, 1974), global rainfall changes including latitudinal shifts (Lamb,
1966), anomalies in global atmospheric circulation [for example, mean
atmospheric pressure at sea level (Lamb, 1972; 1974)], and changes in
westerly winds over the British Isles in 1861–1973 (Lamb, 1972). Visual
inspection of historical records of these data is usually combined with
statistical, or more usually subjective, extrapolation of past trends into the
future. For example, Lamb (1966; 1972; 1974) and Bryson (1974) have
hypothesised, on the basis of eyeball fitting of present climatic statistics to
past curves, that the present period is an analogue of the transition from
the Pleistocene to the Holocene, that is, into a new ice age. In each case
such extrapolations are heavily constrained by the length of the historical
record, which in no case is greater than about 200 years.

To overcome the shortness of the available record, a second approach to
climatic forecasting has been to extrapolate the data sequences into the
past with the use of statistics of other *surrogate* variables as indicators.
Thus Lamb (1972) has used the famous time series on the opening date of
the Port of Riga to extend the temperature record back to 1535, and
there are other similar historical sequences. The ice record locked up in
ice caps has provided another source of surrogate information. For
example, Bergthorsson (1962, quoted in Bryson, 1974) has used recent
climatic data for the past 150 years to regress on the ice records. This
regression equation is then used to extrapolate temperature backwards for
1000 years before the present. A more ambitious study by Bryson (1974)
has been used to reconstruct statistics for temperature, sunshine, growing
season, snowfall, and total precipitation for Minnesota by using pollen and
isotope data for Greenland. By making use of 'canonical transfer functions',
regression equations for the present interrelation of these climatic variables
with pollen and isotope data are used to extrapolate to 30000 years BP.
This study and others by Lamb (1972; 1974) and Winstanley (1973) have
used a further assumption that it is possible to extend the statistical record
by combining data for several spatial locations. For example, on the basis
of the work of Lamb (1966), Winstanley identifies a general downward
trend in rainfall in the Sahel, based on regression tests by Lamb on a

reconstructed fifty-year average rainfall and temperature series for England since 1100. This evidences a 100–200-year periodicity superimposed on a 700–800-year cycle. On the basis of these statistical results Winstanley (1973, page 194) concludes that for the Sahara and for Rajasthan (India): "the strength of the general circulation of monsoon rainfall have now been decreasing for about 50 years. Assuming a symmetrical 200-year cycle, this downward trend will continue for a further 50 years ... There will, presumably, still be a regular fluctuation superimposed on this downward trend. At a low point on one of these fluctuations around the year 2030, the isohyets and the desert climate will be some 150–180 kilometres further south than in 1930. If 1930 marked the peak of a 700-year 'cycle', and we are now in a long downward trend again, then these figures for the southward shift of the desert climate could well be underestimates". Such spatial aggregation contains a large number of difficulties and assumptions, since spatial stationarity and ergodicity are implied. Such methods also assume that constant regression relationships hold between surrogate and derived climatic information and that spatial aggregation is indeed valid. Similar difficulties are present in the CLIMAP (1976) work. This ambitious study is involved in reconstructing climates for various periods up to 18000 BP and simulating the effects of climatic changes inferred from shifts in oxygen isotope ratios in marine shells. Even here, however, it is admitted that the accuracy of the dating of the results is not greater than ±2000 years, and depends a great deal on the rapidity of climatic change over the period for which reconstruction is desired: the accuracy of dating being the greater the slower the change.

A third approach has been via more 'black-box' techniques: to search for trends and periodicities by statistical methods. For example, in important attempts to determine long-term trends in the Sahel climate record, Bunting et al (1976) have applied linear and quadratic regression equations, and also harmonic analysis with an assumed period (of 180 years). By using only between fifty and eighty-one years of data for eleven stations they were unable to detect any statistical trend. A further study of spectral components by the same authors does reveal some evidence of cycles of three years and six to eight years, but with such short data sequences these estimates have wide variance limits. Other analysts (for example Benroit, 1974; Jenkinson, 1973) have used moving-average and band-pass filters, but again the data sequences are usually too short to determine if the observed pattern is a cyclic phenomenon or merely a chance configuration.

Because of the limitations of eyeballing, backward extrapolation, and statistical trend detection, many analysts have sought to develop theoretical approaches to forecasting based on plausible mechanisms which might generate climatic change. The problem with this approach is that the generating mechanisms are only imperfectly known and are extremely difficult to measure. The most plausible theory postulates changes in climate arising from long- and short-term variation in the seasonal and

latitudinal distribution of solar radiation owing to variation of the Earth's
orbit around the sun. Milankovitch (1920) hypothesises three components
of this variation: eccentricity (stretch of the size of the Earth's orbit),
wobble (of the Earth's axis), and roll of the perihelion (which shifts the
longitude). Each component has different cyclic periods (Vernekar, 1971;
Calder, 1974):

1 stretch: 90000–100000 years,
2 wobble: 41000 years,
3 roll: 20000 years.

The superimposition of these components provides a hypothesised series of
cyclic harmonics which can be correlated with available fossil evidence:
botany and oxygen isotope changes of marine fossils. Comparisons of ice
volume predicted by the Vernekar–Milankovitch theory with oxygen
isotope changes from 860000 BP are used by Calder (1974) to argue that
a new ice age, which is just beginning, corresponds to the coincidence of
small wobble and large stretch. Calder forecasts that this ice age will
intensify before ending fairly abruptly in 120000 years time. The Vernekar–
Milankovitch hypothesis is now fairly widely accepted although its effects
on climate are very unclear. But whether or not it is correct, the general
process of climatic change is clearly related to shifts in atmospheric
'centres of action' (the semipermanent pressure patterns). In addition, a
number of analysts, for example Charney (1975), have argued that other
more local factors are involved. Charney postulates that changes in albedo
that result from reduction of vegetation cover owing to past drought and
overgrazing set up a process of intensifying aridity: high albedo gives
radiative heat loss relative to the surroundings, which induces horizontal
circulation shifts with subsiding air to give a feedback stimulus to further
desertification. The problem in validating this hypothesis and the Vernekar–
Milankovitch theory, or any other hypothesis, is similar to that of purely
black-box models of trend extrapolation, that without longer statistical
records and more detailed physical theory it is very difficult to accept
unequivocally any one explanation. Despite these difficulties, however, it
is probably only by more detailed search for theoretical explanation that
credible forecasts of climatic change can be made.

10.3 Problems in applying a statistical approach to forecasting climatic change

Each of the approaches to forecasting climatic change discussed above is
underlain by statistical objections that arise from two fundamental and
interrelated impediments: first the present inadequacy of theoretical
knowledge of climatic change; second the inability with presently available
data to make clear inference. The theoretical impediment arises from the
fact that the causes of climatic change are not at all clear. It is well-known
that this change must arise from variations in the general atmospheric
circulation which result from adjustments in the distribution of heating

and cooling. The fundamental driving force of this mechanism, which acts as the independent variable, is solar radiation. There is only inconclusive evidence of patterns of variation in solar input with time, and most theorists have relied instead on explaining the changes in receipt of solar radiation as being due to the varying distance of the Earth from the sun, as discussed above. Independent variables other than the receipt of solar radiation are even more difficult to analyse. WMO-ICSU (1975), for example, identified changes in the composition of dry air, and changes in the physiography of the Earth, including the configuration of ocean basins, as two other exogenous factors. Other less truly independent variables can also be introduced to account for patterns of climatic change, for example adjustments to the atmospheric transmittance of solar radiation, the albedo of the Earth's surface, the greenhouse effect of the infrared flux from the earth as modified by changing levels of CO_2, dust, and other pollutants, etc (Bryson, 1974; WMO-ICSU, 1975). In addition, some dependent variables act as independent variables at shorter time scales. These explanations can now be accepted with some confidence, but it is difficult to be at all confident in the detailed results that follow from any deductive calculation. Thus the arithmetic of Vernekar (1971) or Calder (1974) used to infer the onset of the supposed present period of atmospheric cooling can be expected to be seriously in error in timing, even if the overall theory is correct. In fact it is not even necessary to adduce serious errors to undermine these forecasts, since even slight shifts in periodic time constants and equation parameters lead to differences of up to 1000 years in cyclic timing. The Vernekar–Milankovitch model is very sensitive to changes in its assumptions which, although unimportant to the overall theoretical pattern of forecast, are highly significant on human time scales. In essence, then, we just do not know with sufficient certainty in the human time domain, the effect or magnitude of likely climatic changes.

The sensitivity of the theoretical conclusions to miniscule changes in assumptions which cannot be evaluated one way or the other brings us to the second inferential impediment. If we are not clear on the theoretical causes of climatic change, it is probable that almost all reasonable theories can be validated with the data presently available on monitored climatic variables. Directly measured records of rainfall, temperature, wind, or pressure are short and no more than 200 years in length, whereas surrogate records are subject to degrees of uncertainty both in their measurement and in the implications for climatic variables. Given the time constants and the trends that it is sought to identify, it is not usually possible to differentiate between reasonable rival hypotheses. Connected with this difficulty is the historicist's dilemma: but we cannot be sure of the degree to which future events can be predicted from past patterns of occurrences and periodicities. Most accepted theories have been unable to cope adequately with changes in variance or relationship in the process of climatic change hypothesised. Yet it is likely that current trends in

industrial growth, pollution, and energy consumption have significant
effects on the general atmospheric circulation: especially on atmospheric
transmittance and the greenhouse effect. If these changes are likely to
affect the variance and, more importantly, the relationship between
processes of climatic change, it is problematic how far the patterns of past
events allow future events to be knowable in advance.

From these two major impediments flow a series of difficulties which
affect any application of statistical methods to forecasting long-term
climatic changes. These can be summarised as follows.

(1) *Choice of variables.* Since the theory of climatic change is not at all
clear, it is not obvious which variables should be examined let alone
forecast. In addition, most of the variables that it might be important to
use as leading indicators are either not directly measurable or cannot yet
be measured. This applies to variations in solar output, albedo, atmospheric
transmittance, and global greenhouse effect. The lack of ability to measure
applies to almost all of the inputs or independent stimuli to the climatic
system, although satellites will offer a great deal in the future, after
sufficient data have been collected. Hence the forecaster is left with
measuring only outputs or dependent variables such as changes in
temperature, rainfall, wind, and pressure. This leads inevitably to what
Kendall (1973) terms autoprojective models, that is, autoregressive and
moving-average structures rather than those based on bivariate or transfer-
function causal structures. This in turn has severe consequences for the
legitimation of policy decisions, as discussed in the next section.

(2) *Range of variation.* In almost all cases the possible range of variation
of all climatic variables, inputs and outputs is unknown. Moreover, in
most marginal areas, where changes in climate are most important, the
degree of variability is very large. Hence it is difficult to make strong
inferences based on statistical tests with specified distribution assumptions,
and we are left with the historicist's dilemma of assuming that observed
past patterns of variation will continue into the future.

(3) *Equation structures.* Lack of sufficiently well-defined *a priori* theory
necessitates the definition of equation structures on the basis of observed
statistical relationships alone, or tempered by the degree of theoretical
knowledge available. This leads either (a) to forecasting models that are
essentially extrapolation devices, or at most black- or grey-box models, the
internal dynamics of which are not known, or (b) to simulation models
with highly certain theoretical structure, but in which it is not known
whether the total simulation pattern adequately represents the climatic
system. Both approaches lead to difficulties in legitimising policy decisions.

(4) *Sources of nonlinearity and nonstationarity.* Sufficient deductive and
empirical knowledge is available of the existence of nonlinear and nonstationary trends and also of sudden 'flips' from one climatic regime to
another. But theory is again often inadequate to predict the structure of
these particular relationships. Hence many analysts have been reduced to

making extrapolations based on linear and stationary models, for which the certainty of the forecast on human time scales can be expected to be low.

(5) *Uniqueness of observed record.* As in many inference problems in environmental systems, the historical record of statistics available is unique. Hence recourse to the classical theory of statistical hypothesis testing is dubious to say the least. Alternatives grounded in Bayesian inference suffer from the difficulty that they are unlikely to be of much utility for extrapolation of long-term changes. The ensemble approach, of using a sample of spatial locations both to extend limited statistical records and to overcome objections to classical statistical inference, has been utilised by some analysts. But there are severe limitations to this approach since spatial stationarity (ergodicity) cannot usually be assumed. Again the margin of error introduced undermines the statistical forecasts that result in terms of human time scales.

(6) *Feedback loops.* There are considerable difficulties of forecasting and inference in systems that involve substantial degrees of feedback. Such feedback may be engendered in climatic systems by autonomous mechanisms, or by conscious feedback controlled by man. This latter, although desired by many, is of less importance in climatic systems than in many other environmental systems. However caused, the presence of feedbacks between subsystems requires that we approach the climatic forecasting problem as a simultaneous equation block. This will reduce the confidence of our estimates by raising the estimated confidence limits which will undermine the certainty of the resulting statistical estimates.

(7) *Confidence intervals.* Related to the previous point, it is widely known that as the period of forecasting into the future is increased, the statistical confidence levels which can be associated with the forecast tend to increase exponentially (see for example Haggett, 1973). Thus the forecasting error expressed as a function of lead time follows a trajectory as shown in figure 10.1 which is asymptotic to the value σ_Y^2, the variance of the forecast variable $\{Y_t\}$. Thus forecasts for lead times of one-thousand years from the present, or even only ten to twenty years into the future,

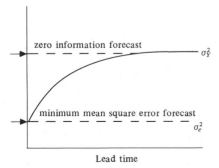

Figure 10.1. The forecasting error expressed as a function of lead time (after Wold, 1965).

can be made with only a very low degree of statistical confidence. Such statistical forecasts are based only on the information available in historical records, so that the only way to reduce the width of the confidence intervals is to add significantly to our *a priori* understanding of the mechanisms that generate climatic change.

(8) *Series length.* Confidence levels of forecasts and levels of statistical significance of trends are fundamentally constrained by the limited length of available historical records. Moreover, it is difficult to see how this impediment can be removed within human time scales. Not only are unrecorded data lost (except for reconstruction exercises that use surrogates), but also new data that become available as time passes add little to our range of statistical confidence. Confidence levels and sample estimates are usually proportional in some sense to the inverse of the square root of the data length, that is $f(1/T^{1/2})$. Hence to halve standard errors of estimates it is necessary to increase the data length four times; to reduce standard errors to a third it is necessary to increase the data length by nine times, and so on. Hence new data will give little increase in the confidence of existing statistical estimates.

The total effect of these various statistical difficulties is to undermine greatly the credibility of any forecast of long-term climatic change based on statistical procedures alone. Indeed we may well ask if it is feasible statistically to forecast such changes at all. The difficulty is that it is sought to extrapolate from severely limited historical records and inadequate theoretical understanding for extremely long time periods. Shortage of data or theory in short-term forecasting may be quite acceptable, and there are many studies which evidence the utility of such forecasts, but for long-term forecasting the statistical approach has much less utility. It is impossible to determine if past changes are part of a long-term trend, a long-term periodicity, or are merely random. This conclusion should not be surprising since there are various rules of thumb which suggest that no confidence can be placed in estimates of lag or frequency components greater than two tenths or three tenths of the available data length (Bennett, 1978). Although this range can be extended to perhaps a third or a half of data length if confirmatory information is available, for example, by extrapolation across ensembles, when we have at most about 200 years of data for only a limited range of stations it is not possible to extrapolate with any confidence for the periods of hundreds or thousands of years that have been suggested. This problem is recognised by Lamb (1977, page 19), though he is optimistic about its consequences: that we should attempt "To establish statistical rules based on the frequency with which various evolutions over periods from a few months to some years follow various initial conditions, which shall be identified in terms of physical circumstances and processes recognized in the past record of climate. A very long past record is required, particularly in the case of processes with characteristic time scales of a century or more, so that many past cases may be included in the statistical survey and the sequels mapped (in some cases globally) in terms of frequency contours and statistical significance". The statistical difficulties detailed above show that these hopes are almost certainly overoptimistic. In addition Lamb's (1974) attempt to infer long-term trends from short-term fluctuations on the basis of statistical evidence alone also contains great difficulties. As can be seen from figure 10.2, although it is possible to infer trends in systems with low output variance, when the system outputs are subject to a very high degree of variation, as is the case in climatic change, it becomes impossible to determine any trend elements unless the data sequences are very long. This condition can be loosely stated as that the bandwidth of the process must be less than the bandwidth of the trend in order to allow trend elements to be identified (Bennett, 1978). This condition is satisfied in figure 10.2(a) but not in figure 10.2(b). As a simple example it can be noted that it is not possible to distinguish from statistical records alone whether an observed trend results from true periodicities or from a model with integrators:

$$(1 + \delta z^L)Y_t = X_t ,$$

where L is a long-term lag, $|\delta| > 1$, and z is the unit shift operator, that is, $zY_t = Y_{t-1}$. Thus for black-box systems, integrators of system dependent variables, and trends in system independent variables, or a number of other hypotheses, are indistinguishable.

As a result of this conclusion, many analysts have chosen to invoke statistical or mathematical criteria in order to choose between models. Thus particular system equations are chosen for their canonical structure, properties of minimality (based on realisation theory), or parsimony of parameters. Such criteria must always be invoked when there is insufficient *a priori* evidence or deductive theory to allow a choice based on physical generating mechanisms. Use of such criteria leads to considerable problems if a decision is made to act on the results of forecasts of climatic changes. This and other important implications are discussed in the next section.

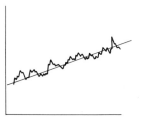

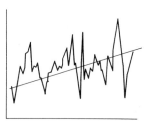

bandwidth trend > bandwidth process   bandwidth trend < bandwidth process
(a)   (b)

Figure 10.2. Relation of trend and bandwidth of time series.

10.4 Implications of statistical forecasts of climatic change

From the preceding discussion we are drawn to accept that, as yet, it is not possible to produce reliable statistical forecasts of climatic change. Despite this inadequacy, forecasts have nevertheless been produced and policy implications drawn. The response of analysts grappling with the problems of forecasting climatic change seems to have been first to try and improve the existing technique and explanation, and second to accept some inadequacy in explanation, provided some insight into the future can be obtained. Certainly as yet there are no widely accepted mechanisms of climatic change, and forecasts rely heavily either on extrapolating past trends or on building models of past relationships. In both cases, however, the analysis that can be brought to bear is limited to the development of models that are at present essentially black box in that the internal mechanisms remain largely unknown. This is justified by Winstanley (1973, pages 193–194) in that "until we do have a generally accepted theory for general circulation and until we can predict the variables that cause climatic change, our best hope for climatological forecasting lies in the extrapolation of trends, and in determining what climatic changes accompany and follow the establishment of different circulation anomalies". Hence Winstanley advocates a black-box approach to climatic forecasting in the absence of sound physical models. Indeed the preceding discussion has shown that the acceptance of black-box models is almost inevitable in statistical forecasting of long-term changes. The fundamental inputs into the climatic system cannot be measured, their range of variation and sources of nonlinearity and nonstationarity are unknown, and the forecasting equation structures and feedbacks are very uncertain. Moreover the short and unique historical records available limit the degree to which long-term changes can be confidently forecast.

The consequence of using black-box models is that we are forced to make choices between equations based on statistical and mathematical criteria rather than on the basis of symmetry of the models with the real mechanisms that generate climatic change. The difficulties inherent in this approach are well-illustrated by considering the use of mathematical approximants for the generation of forecasting models. Most forecasting schemes reduce to the form of a mathematical approximant, for example, exponential smoothing, Fourier harmonics, moving averages, distributed lags, and autoregressive moving-average models. The most general form of mathematical approximant is that of the Padé series, since other mathematical approximants, for example Legendre, Laguerre, Chebyschev, Jacobi, Lagrange, and Hermite schemes, reduce to the Padé form. A Padé series of order (p, q) is given by the expression

$$H_{p,q}(z) = \sum_{r=0}^{\infty} h_r z^r ,$$

where z is again the backward shift operator, $zY_t = Y_{t-1}$. This infinite series is represented by a Padé approximant given as the ratio of two polynomials

$$f_{p,q}(z) = \frac{\sum\limits_{i=0}^{q} b_j z^j}{\sum\limits_{i=0}^{p} a_i z^i} \, .$$

This is equivalent to a Box–Jenkins transfer function or autoregressive moving-average model and is an extremely general structure. Its importance is that it represents an infinite series, such as a time series, by a finite number of terms, and can be used for almost all series whether convergent or divergent, or even for sums of two or more independent series. Most important, the approximants of extremely complex time series can be defined with very low order. This is possible because a finite approximant is chosen which contains no linear dependency or parameter degeneracy, or is parsimonious, that is, the Padé approximant is a special form of minimal realisation. The problem of interpreting the resulting forecasting model arises because the approximant is in one-to-one correspondence with the underlying climatic processes only in a mathematical sense, since without prior knowledge of the generating process we have no reason to prefer one approximant to another except in terms of its mathematical structure. Hence any representation of an infinite series of parameters by a finite one involves committing a misspecification error generally of unknown proportions.

Of course it has been accepted that the black-box approach to climatic forecasting is only a first approximation pending the development of better theoretical models. But the approach does lead to severe difficulties when policy decisions are involved. The main problem with the use of such an approach to policy is one of legitimation. We are forced to accept methods of foreseeing the future which are based on criteria other than physical meaning or mechanisms that can be generally understood. It then becomes difficult both to justify decisions and actions in response to these forecasts, and to obtain popular assent for these actions based on an understanding of the reasons for them. Instead we are forced to accept new criteria of acceptability of policy related to those criteria by which we judge the efficacy of a model. Such criteria are based not on how well the model or forecast reproduces the mechanisms of the real world, but instead on criteria of accuracy based on mathematical and statistical limits. As an example consider the frequently used criterion of parsimony, popularised by Box and Jenkins (1970), by which a model is considered optimal in some sense if forecasts are produced within the smallest possible parameter set. It is argued that such models are accurate, and also that being simpler they are often more comprehensible than larger more realistic models. In addition statistical confidence in parameter estimates of parsimonious models may be higher for a given sample size, and the production of a small model often leads more easily to understanding of the interactions between the model and its environment. But any mathematical or statistical benefits of parsimonious models are usually far outweighed in policy applications by their lack of physical interpretability. It was G J A Stern, in discussion of the paper by Newbold and Granger (1974, page 151) before the Royal Statistical Society, who adapted St Jerome in saying ' "The whole world groaned to find itself saying 'parsimonious parameterization' ' ". Box–Jenkins could be described as a method of stating in a short but opaque equation much of what is to be said about a time series: other methods could be characterised as less complex in expression and possibly less accurate but gaining in comprehensibility'.

A more disturbing conclusion drawn by many workers is that models can be accepted as legitimate provided only that they yield good forecasts; as stated by Haggett (1973, page 234) "models are to use, not to believe". This point of view can be supported for very short-term forecasting in which the effects of transient elements and lead–lag patterns are preeminent; but it creates severe difficulties when long-term forecasts are involved or when policy actions are required. Long-term forecasts based on purely black-box and statistical methodology are highly sensitive to statistical and other assumptions. Although slight inaccuracies in assumptions or minor shifts in behavioural structure may have little effect on the fit over past data or short-term forecasting accuracy, the effect becomes greater the longer the time period over which forecasts are made. Adaptive forecasting methods, such as those of Kalman filtering or the Harrison and Stevens Bayesian predictor, may extend the period of forecasting possible but do not extend the range to that required for the climatic problems discussed above. For policy applications it becomes important that we disagree with the view that we need not believe in our models. If, in response to the climatic exigencies of today, we are to impose stockpiling of food, destocking, resettlement, or changed agricultural practices in the Sahel or in Rajasthan, then we shall probably require to limit the standard of living of the present in favour of a longer-run lower equilibrium level either of population or of social well-being, and certainly we shall affect present cultural and social structures most deeply, perhaps irreversibly. Such policies are necessarily political in content. Indeed Haggett (1973) recognises that when there is a 'generation effect' and policy decisions today affect the social well-being of future generations, then there are severe ethical difficulties in using unreliable forecasts. If policies must be implemented on the basis of forecasts which cannot be understood by the population who must suffer these possible hardships, and cannot be justified by belief of those who produce the forecasts, then it is highly unlikely that the decisions will be given popular support. It is, after all, difficult enough to gain acceptance of oil conservation in response to the more obvious and immediately credible ultimate exhaustion of oil resources.

If the adjustments required of society in response to climatic changes are unquestionable 'stone tablets' based on acts of faith derived from black-box statistical models, then it is not possible to open the forecasts to general debates since only their statistical validity can be questioned. Hence policies based on models in which we cannot believe are unlikely to give the governments implementing them credibility or support.

10.5 Conclusion
Statistical forecasting of long-term climatic change is of considerable scientific and human interest. It is hoped that this chapter has shown that, for the present, climatic forecasts based on statistical procedures alone are subject to a number of inadequacies. Most important of these is the fact that *on human time scales* the uncertainty inherent in the specification and estimation of statistical forecasting models is too great to allow these models to be made the basis of reliable economic policy. Moreover, attempts to use statistical justification for economic policy to ameliorate the exigencies of climate run into severe difficulties of legitimation. This is not to suggest that as a scientific exercise statistical analysis of climatic change is not useful or that climatic changes have not or will not occur. The difficulties emerge only when it is sought to advise on policies using presently available forecasts.

What then of the future? Should we disregard the very real human and economic problems created by climatic variability in many marginal, and even nonmarginal areas. A well-balanced view is given by Hare (1977, pages 28–29) that although observed climatic variations of the 1970s have evidenced the importance of studies of climate and its role in affecting the economic system, there is little evidence that current variations represent a slide into a new ice age. Although accurate prediction of climatic variations is not yet possible, Hare concludes that in the future more credible forecasts may be available. These may derive from identifying periodic phenomena and trends, but they will mainly emerge from the use of general circulation and related models of the atmosphere which incorporate the underlying causes. For the present, it must certainly be accepted that climatic variability must be taken into account in economic planning. In particular, communities that live in marginal areas, especially deserts, will continue to be at risk and the economies of these areas should be adjusted to cope with renewed climatic variations. The message, for the present at least, is that if we cannot predict with accuracy, then we must plan more robust economies in the face of a climatic variability, that is, plan rather than forecast for a high degree of uncertainty.

Acknowledgements. I would like to thank Paul Richards for originally drawing my attention to this problem and to John Thornes for reading an earlier copy of the paper. The figures were drawn by Ric Davidson.

References

Bennett R J, 1978 *Spatial Time Series: Analysis, Forecasting, and Control* (Pion, London)

Benroit P, 1974 "Rainfall trends in the Sahel and Sudan zones of West Africa" unpublished paper, Centre of West African Studies, University of Birmingham, England

Box G E P, Jenkins G M, 1970 *Time Series Analysis, Forecasting and Control* (Holden-Day, San Francisco)

Bryson R A, 1974 "A perspective on climatic change" *Science* **184** 753–760

Bunting A H, Dennett M, Elston J, Milford J R, 1976 "Rainfall trends in the West African Sahel" *Quarterly Journal of the Royal Meteorological Society* **102** 59–64

Calder N, 1974 "Arithmetic of ice ages" *Nature* **252** 216–218

Charney J G, 1975 "Dynamics of deserts and drought in the Sahel" *Quarterly Journal of the Royal Meteorological Society* **101** 193–202

CLIMAP, 1976 "The surface of the ice-age Earth" *Science* **191** 1131–1137

Haggett P, 1973 "Forecasting alternative spatial, ecological and regional futures: problems and possibilities" in *Directions in Geography* Ed. R J Chorley (Methuen, London) pp 219–235

Hare F K, 1977 "Is the climate changing? The story so far" *Mazingira* **1** 19–29

Jenkinson A F, 1973 "A note on variation in May to September rainfall in West African marginal rainfall areas" in *Drought in Africa, Report of 1973 Symposium* Eds D Dalby, R J Harrison Church (Centre for African Studies, School of Oriental and African Studies, University of London, England)

Kendall D G, 1973 *Time Series* (Charles Griffin, London)

Kopec R J, 1976 *Atmospheric Quality and Climatic Change* Papers of the Second Carolina Geographical Symposium, Studies in Geography number 9 (University of North Carolina, Chapel Hill, NC)

Lamb H H, 1966 *The Changing Climate: Selected Papers* (Methuen, London)

Lamb H H, 1972 *Climate: Present, Past and Future. 1. Fundamentals of Climate Now* (Methuen, London)

Lamb H H, 1974 "The current trend of world climate—a report on the early 1970's and a perspective" Research Report 3, Climatic Research Unit, University of East Anglia, Norwich, England

Lamb H H, 1977 "Understanding climatic change and its relevance to the world food problem" Research Report 5, Climatic Research Unit, University of East Anglia, Norwich, England

Milankovitch V, 1920 *Théorie Mathématique des Phénomènes Théoriques Produits par la Radiation Solaire* (Gauthier-Villars, Paris)

Mitchell J M, 1972 "The natural breakdown of the present interglacial and its possible intervention by human activities" *Quaternary Research* **2** 436–445

Newbold P, Granger C W J, 1974 "Experience with forecasting time series and the combination of forecasts" *Journal of the Royal Statistical Society* **A137** 131–164

Vernekar A D, 1971 "Long-period global variations of incoming solar radiation" *Meteorological Monographs* **12** 1–20

Winstanley D, 1973 "Rainfall patterns and general atmospheric circulation" *Nature* **245** 190–194

WMO-ICSU, 1975 *The Physical Basis of Climate and Climate Modelling* Global Atmospheric Research Project, GARP Publication Series 16 (World Meteorological Organisation and International Council of Scientific Unions, Geneva)

Wold H O (Ed.), 1965 *Bibliography on Time Series and Stochastic Processes* (Oliver and Boyd, Edinburgh)

Diffusion and spatial pattern: a model for competition

F H C Marriott

"And since even in Paradise itself, the tree of knowledge was placed in the middle of the Garden, whatever was the ambient figure; there wanted not a centre and rule of decussation. Whether the groves and sacred Plantations of Antiquity were not thus orderly placed, either by *quarternio's*, or quintuple ordinations, may favourably be doubted"

Sir Thomas Browne, *The Garden of Cyrus*, 1658.

11.1 Patterns of points and descriptive statistics

Early work on geometrical probability was concerned chiefly with the properties of points falling at random in space of two or three dimensions. The comparison of these theoretical properties with actual patterns of points—usually regarded as random samples from an infinite plane or three-dimensional space—led naturally to tests of randomness. These include, for example, the Pearson χ^2 test applied to a Poisson distribution, the dispersion test based on the variance/mean ratio, tests such as the T square test (Besag and Gleaves, 1973) that depend on the distributions of nearest-neighbour distances between pairs of sample points and between random points and sample points, and tests for spatial or space–time clustering that are used in medical statistics and usually based on techniques of randomisation.

Later, interest developed in nonrandom distributions and in the stochastic processes that produced them. The theory of the negative binomial distribution and various clustering processes were studied, as well as processes with restrictions on nearest-neighbour distances. Most of the processes considered included the Poisson process as a special case, and sometimes statistics used for tests of randomness could be used for parameter estimation in the stochastic model. The next step was to test the goodness of fit of these models, and recent work has been mainly concerned with the spectrum of spatial stochastic processes and its estimation, or with the covariance function.

This is probably as far as it is possible to go in studying naturally occurring spatial patterns. Certainly different stochastic processes and different patterns may have the same spectrum, but in view of the complexity of the processes involved and the heterogeneity that is almost inevitably associated with very large samples, it seems unlikely that processes with the same second-order properties can ever be distinguished in samples. In this sense the second-order structure may be regarded as a complete description of the pattern.

The study of spatial patterns has been concerned mainly with the
processes that generate them. The problem discussed in this chapter is not
of that sort, and techniques for the study of stochastic processes in space
are not directly relevant. They have, however, generated a large number
of statistics used for parameter estimation and significance tests. These
can be regarded, if we choose, merely as descriptive statistics that relate to
a particular spatial pattern—which may or may not be a sample from
some population—and it is in this way that they will be used here.

11.2 Spatial pattern and competition

If a given number of plants are to be placed in a given area it is obviously
desirable, from the point of view of competition for nutrients and moisture,
that they should be spaced out as well as possible. Many plants have
elaborate mechanisms for seed dispersal, which may produce an effectively
random distribution of seeds and, by restricting nearest-neighbour distances
when they germinate, an even better dispersed pattern of plants. Farmers
and gardeners try to space plants as evenly as possible. Sometimes they
are placed at the vertices of a regular lattice, sometimes in evenly spaced
rows (but irregularly in the rows), and sometimes seed may be broadcast
'at random', and then thinned to give a minimum nearest-neighbour distance.

Obviously the ideal two-dimensional pattern from this point of view is
the triangular lattice, in which each plant can be regarded as the centre of
a regular hexagon, these hexagons forming a honeycomb that covers the
space available. This was discussed by Sir Thomas Browne (1658), who
attributes to this 'quincunctial' arrangement very great antiquity. Plants
so arranged will clearly exploit the available nutrients better than plants
randomly scattered, which in turn will do better than plants clumped in
groups. The problem is to quantify these differences, to decide how much
advantage one arrangement has over another, and to compare them with
other arrangements, such as regularly arranged clusters.

11.3 The Oxford analogue

A device to investigate this problem was produced in Oxford some time
ago in the form of an electrical analogue of a simple diffusion process.
This instrument consists of a more or less circular plate covered by a square
lattice (figure 11.1). Plugs can be inserted in holes at the intersections of
the lattice, and any charge on the plate leaks away through these plugs.
The procedure, then, is to insert a number of plugs in chosen positions,
charge up the plate, then switch on and allow the charge to leak away for
a given time and measure the residual charge. Each plug represents a small
circular sink into which the charge diffuses, and the rate of discharge
depends on the number of plugs inserted and their positions. By switching
on for different times, an 'uptake curve' can be measured, and this is the
uptake curve that would be obtained by planting cylindrical parallel roots

in corresponding positions in a cylindrical flowerpot. A full description of the Oxford analogue is given by Sanders et al (1971).

As a model of plant competition, the analogue has some limitations. Obviously it refers only to competition for nutrients or moisture and does not reflect the effects of shading and competition for light. The effect of mass flow (Nye and Marriott, 1969) is ignored. A cylindrical root does not behave like a perfect sink in the sense that any particle whose random movements would cause it to cross the root boundary is absorbed, though the model of cylindrical sinks, possibly of smaller diameter than the actual roots, may give a good approximation to the real distribution of particles. These practical problems, however, need not concern us. The mathematical or statistical question is how to relate the observed uptake curves to the corresponding patterns of points in the analogue, and these curves and patterns may be regarded as the basic data for analysis.

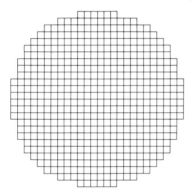

Figure 11.1. The electrical analogue. Plugs may be inserted at the nodes of the lattice to give sinks into which the charge diffuses. Each sink has radius about 0·23 units of the lattice spacing.

11.4 The uptake curve
All these curves have two points in common.
(1) The initial rate of uptake is proportional to sink density. This situation lasts only as long as there is no competition—that is, as long as particles from any small area have had time to reach only the nearest sink. This period will be very brief when sinks are close together.
(2) The rate of uptake steadily decreases. The concentration at every point decreases and in particular so does the concentration around the sinks.

Consider now some typical uptake curves (figure 11.2). Suppose A is the curve for a regular triangular lattice. If a sink is inserted between each pair of sinks in curve A, thus quadrupling the density, we get curve B. Suppose now that each sink of curve A is replaced by a compact group of four sinks (curve C). The initial rate of uptake will be as in curve B, but

soon the curve will flatten and when the area around the sinks has been
depleted each group will be little more effective than a single sink, so that
the uptake curve will have a slope similar to that of curve A.

This brings out one important point. Curve C *crosses* all the curves for
triangular lattices with density between that of curve A and curve B. We
could define a measure of efficiency of uptake in terms, say, of equivalent
density on a triangular lattice, but such a definition would have to be tied
to a particular uptake level or a particular time. We might, for example,
define efficiency in terms of time to 50% uptake, but this would be purely
arbitrary and would be quite misleading if the final uptake level (at harvest,
say) were 10% or 90%. Instead of defining an arbitrary measure of efficiency
of this sort, it would be much more useful, though more ambitious, to try
to predict the uptake curve from the density and pattern of sinks.

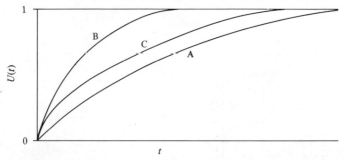

Figure 11.2. Typical uptake curves $U(t)$. Curve A represents the uptake for sinks on
a triangular lattice. Curve B corresponds to a similar arrangement with four times the
density. Curve C is obtained by replacing each sink of the lattice corresponding to
curve A by a compact group of four sinks; the initial rate of uptake is the same as
for curve B, but when the area round the group has been depleted the group behaves
rather like a single sink and the rate of uptake is similar to curve A.

11.5 Statistics of pattern
A bewildering variety of statistics is available for describing the pattern of
sinks, and the question of which to use for predicting uptake is difficult
to resolve. Four suggestions may be considered.

(1) The simplest approach is to define some measure of dispersion and, by
simulation, find an average uptake curve for the appropriate density and
value of the dispersion measure. This was the approach used by Baldwin
et al (1972). Accuracy is necessarily limited, but for a long series of
patterns they were able to predict uptake to within about 20%.

(2) It has been suggested that the uptake curve could be related to the
size distribution of the set of polygons that define the areas nearer to each
sink than to any other sink. This is the 'Dirichlet tessellation', sometimes
also called Delthiel, Theissen, or Voronoi polygons. The underlying idea is
that each sink exploits a certain area defined by the associated polygon

and that, for a given density, efficiency will be highest when the polygons are all of equal size.

Unfortunately this idea does not stand up to scrutiny. Figure 11.3 shows an example of a pattern with equal polygons and a strongly clumped distribution. Clearly the uptake depends not only on the size of the polygons but also on their shapes and on the positions of the sinks within them.

(3) To any stochastic model there corresponds a second-order structure which can be described in terms of the spectrum, the covariance function, or the $K(r)$ function (Ripley, 1976). These functions can be estimated from samples, and, although for point processes the second-order structure does not define the model, it is arguable that the estimates contain all the · usable information that can be derived from the sample.

This suggests that we might estimate a function from the pattern of points as if it were a sample, and use this function to predict the uptake curve. Unfortunately there are various difficulties.

(a) Since we are dealing with processes on a lattice, all these functions are discontinuous. Further, since we want to discuss regular arrangements of sinks, it would be inappropriate to consider smoothed spectra in the usual way. The prediction of a smooth curve from these discontinuous functions may make the mathematics more difficult.

(b) There is no obvious intuitive connexion between, say, the spectrum and the uptake curve. In trying to relate the two we should be working in the dark, and if we succeeded in finding a relationship that gave a satisfactory prediction, it might not throw much light on the way in which pattern determines uptake.

(c) It is necessary to take account of edge effects, and it is not at all clear how to do so. Corrections customarily used (for example, Ripley, 1976) are aimed at obtaining unbiased estimates from random samples. In the present context, bias is a meaningless concept. The analogue is not in any sense a random sample from a population, and its border is strictly impermeable[1].

In view of these difficulties, I have not attempted to relate the second-order properties of the pattern with the uptake curve. I think, however, that it is an approach worth further consideration.

(4) One way of describing uptake by a sink is to associate with it a 'circle of depletion', with area at time t proportional to the uptake from time 0 to t. This circle grows with time until when the nutrient is nearly exhausted it almost fills the whole space. If two or more sinks are present, the circles

[1] The difficulty might have been avoided if the analogue had been designed specifically for a statistical analysis of this sort. A square rather than circular patch might be regarded as a sample from a square lattice in which each edge reflected the points on one side to the other. This system would have zero net flow across any boundary element and a well-defined spectrum. In fact, however, the analogue was designed to model a different problem.

begin to overlap and consequently the total uptake is less than double that from a single sink (figure 11.4). The area is proportional to r^2 until the circles begin to overlap. As r increases, the proportion overlapping increases until as θ approaches $\frac{1}{2}\pi$ the area and the rate of increase are nearly the same as for a single sink.

This representation must not be regarded as an accurate quantitative representation of competition effects; in particular, the rate of growth of the depletion circle for a single sink in the middle of a circular field obviously depends on the size of field. All the same, it provides a reasonable demonstration of what happens when sinks compete, and may even give some quantitative idea of the effects of competition when uptake is fairly low.

There is one immediate consequence of the model; the uptake curve is determined by the distribution of distances of points in space from the nearest sink. Again this suggestion must not be pushed too far. It is certainly not true, for example, that the density at a given point depends only on the distance to the nearest sink—depletion at a point will be much more rapid if it is at the centre of a ring of sinks at equal distances. The most that can be said is that the distribution of distances of points in space from the nearest sink may provide a suitable summary statistic from which to predict the uptake curve.

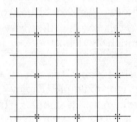

Figure 11.3. The square lattice may correspond to the Dirichlet tessellation of points on a regular square lattice, but it also arises from the pattern shown.

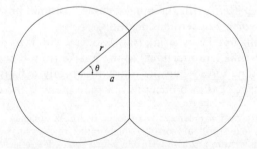

Figure 11.4. 'Depletion circles' for two points a distance $2a$ apart. The total area $A = 2r^2(\pi - \theta + \cos\theta \sin\theta)$, where $\cos\theta = a/r$ (if $a > r$, $\theta = 0$). The rate of growth $\mathrm{d}A/\mathrm{d}r = 4r(\pi - \theta)$.

11.6 The model

Consider the cumulative distribution F(r) plotted against r^2. The initial slope (when r is less than the distance between any pair of sinks) is proportional to the density of sinks. Thereafter it decreases; in particular, as r increases, the function F(r) for a close group of sinks resembles more and more that for a single sink of larger size.

There is, in fact, a remarkable resemblance between $F(r^2)$ and the uptake curve U(t), with one important difference. The uptake curve approaches unity asymptotically whereas the distribution curve reaches it at a definite value of r, the maximum distance between a point of the circumference of the analogue and the nearest sink.

This suggests an integral transform model of the form

$$U(t) = \int \phi(r, t) f(r) dr ,$$

where $f(r)dr = dF(r)$ and the integral is taken over all values of r. The function ϕ represents the average uptake at points distance r from the nearest root at time t. Notice that when $t \to 0$, $\partial \phi/\partial t \to 0$ except at $r = a$, where a is the sink diameter. Hence $[\partial U/\partial t]_{t=0} \propto f(a)$, that is, the total sink circumference or equivalently the sink density.

Now there is one special case in which the model is exact and the uptake, the distribution of r, and the function $\phi(r, t)$ are all known. If a sink of radius a is placed at the centre of a circle of radius b, the distribution of r^2 is uniform between a^2 and b^2, and the diffusion equation has been solved. Unfortunately the solution is complicated. U(t) is in terms of Bessel functions—as one might expect—and $\phi(r, t)$ can be expressed as an infinite series of Bessel functions involving the roots of equations in Bessel functions (Crank, 1975, page 86). Further, a and b appear not only as the limits of integration but also in the expression for $\phi(r, t)$; clearly the uptake at any value of r depends on the radius of the impermeable boundary.

The basic equation $U(t) = \int \phi(r, t) f(r) dr$ is only approximate. The exact solution must be a very good approximation to the case of a single sink at the centre of the analogue, or to the case of a regular triangular lattice in which the bounding circle $r = b$ is replaced by a regular hexagon. We might try to fit the model by putting in the known function $\phi(r, t)$, but whereas in the analogue the effective sink radius a is a known constant (approximately $0 \cdot 23$ [2], in terms of the side of the square lattice), b is not. We might simplify the problem in several ways.

(1) Set $b = \infty$. This considerably simplifies the expression for ϕ, and it would be fairly easy to insert the value of ϕ and evaluate U(t) for a known f(r). It amounts to removing the impermeable barrier and considering the depletion of a circle of known radius in an infinite plane. This might give

[2] Derived from the electrical properties of the analogue; see Sanders et al (1971).

a reasonable approximation for low uptakes and even for the start of competition, but could not predict $U(t)$ for high values of t and U.
(2) Set $b = r_{max}$. At the cost of extra complexity, this should give a better approximation, but again would break down for high t and U.
(3) Consider a distribution of values of b. This should give a still better approximation but at the cost of prohibitive complexity.

   I have not followed up any of these possibilities. The computations would be extremely heavy and the simplicity of the original model would be destroyed if b were varied. I have therefore preferred to try to fit the model directly from the analogue results.

11.7 Fitting the model
Instead of trying to find acceptable approximations to $\phi(r, t)$ mathematically, by using the known function for a regular array, an alternative approach is to look at the patterns and uptake curves actually obtained on the analogue. It is fairly easy to get a reasonable approximation to $f(r)$ by finding the values of r for a regular lattice spread over the analogue. Having done so, it is easy to calculate $\int \psi(r, t) f(r) dr$ for any function $\phi(r, t)$ by summation, for a range of values of t.

   To find a reasonable guess for $\phi(r, t)$, I examined the uptake curve for a single sink in the middle of the analogue (corresponding roughly to $a = 0 \cdot 23$, $b = 13 \cdot 5$) and for other regular arrangements of different density. There are, of course, many functions of t that could be fitted, and it would be tedious to describe all the trial functions investigated. The curve starts from zero and approaches unity asymptotically, and the general form suggests $\tanh k\rho t$, where ρ is the sink density.

   For these regular arrangements, a plot of $\log \tanh^{-1} U(t)$ against $\log t$ gives more or less a straight line, but the slope is rather less than unity. A better approximation to the uptake curve is given by

$$U(\tau) = \tanh \pi \rho \tau ,$$

where $\tau = 1 \cdot 12 t^{0 \cdot 84}$. The appearance of $\rho \tau$ on the right-hand side suggests that $\phi(r, \tau)$ may be written $\phi(\tau/r^2)$, and so

$$\int_a^{b = 1/(\pi\rho)^{1/2}} \phi\left(\frac{\tau}{r^2}\right) 2\pi\rho r dr = \tanh \pi \rho \tau ,$$

or on writing $\zeta = 1/r^2$

$$-\int_{a^{-2}}^{\pi\rho} \phi(\tau\zeta) \frac{\pi\rho}{\zeta^2} d\zeta = \tanh \pi \rho \tau .$$

Differentiation with respect to ρ gives

$$-\frac{1}{\rho} \phi(\pi\rho\tau) + \frac{1}{\rho} \tanh \pi \rho \tau = \pi \tau \operatorname{sech}^2 \pi \rho \tau ,$$

or

$$\phi(r, \tau) = \tanh\left(\frac{\tau}{r^2}\right) - \frac{\tau}{r^2} \operatorname{sech}^2\left(\frac{\tau}{r^2}\right).$$

11.7 Calculations

With f(r) known, or having obtained a numerical estimate of it, it is now simple to calculate the predicted $U(t)$. This has been done for four cases, chosen to illustrate extremes of well-dispersed and clumped distributions. The results are shown in table 11.1; the uptake figure obtained on the analogue is to be compared with the prediction given in brackets.

Case A. A single sink in the centre of the analogue [figure 11.5(a)]. The fit is excellent, as of course it must be since the model was derived from this case.

Case B. Forty sinks on a rectangular lattice [figure 11.5(b)]. Again the predicted results are extremely accurate.

Case C. Twenty sinks clustered in the centre [figure 11.5(c)]. The predictions are good although a little low for higher uptakes.

Case D. Forty sinks clustered in the centre [figure 11.5(d)]. The predictions for low uptake are good but the tendency to underestimate at higher uptakes is more marked.

In an attempt to improve the predictions, I have tried more complicated formulae for $\phi(r, t)$ and obtained slightly better fits. All the same, the effect of underestimation at high uptake for clumped distributions is always present, and I believe that it is a necessary feature of the model,

Table 11.1. Observed and (predicted) $U(t)$ for four sink patterns.

Time	A	B	C	D
2	0·010 (0·011)	0·39 (0·40)	—	0·16 (0·15)
3	0·012 (0·015)	0·48 (0·51)	—	0·18 (0·17)
4	0·018 (0·019)	0·56 (0·58)	—	0·20 (0·19)
5	0·020 (0·023)	0·62 (0·63)	0·10 (0·10)	0·22 (0·20)
10	0·039 (0·041)	0·78 (0·76)	0·15 (0·13)	0·27 (0·25)
20	0·063 (0·074)	0·91 (0·86)	0·22 (0·19)	0·39 (0·33)
30	—	0·96 (0·92)	0·28 (0·23)	0·49 (0·39)
40	0·12 (0·13)	0·98 (0·95)	0·34 (0·27)	0·56 (0·44)
50	—	0·99 (0·97)	0·39 (0·31)	0·62 (0·49)
60	0·17 (0·18)			
80	0·22 (0·23)			
100	0·26 (0·28)			
150	0·36 (0·38)			
200	0·44 (0·47)			
300	0·59 (0·62)			
400	0·69 (0·72)			
500	0·77 (0·80)			

Note: Dashes denote that values have not been recorded.

imposed by the variation in the size and shape of the associated polygons.
The slight improvements possible do not justify the extra complexity.

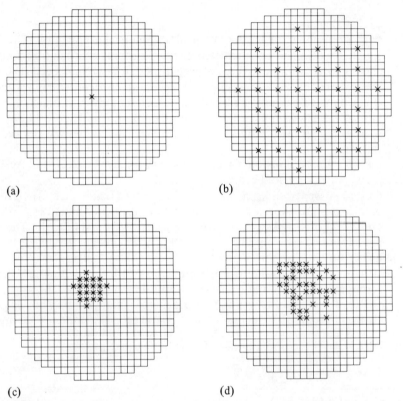

(a)   (b)

(c)   (d)

Figure 11.5. Four patterns examined on the analogue. (a) A single sink in the centre,
(b) forty points on a rectangular lattice, (c) twenty points clustered at the centre, and
(d) forty points clustered at the centre.

11.8 Discussion

The analogue makes it possible to obtain the uptake curve when a number
of small zero sinks are placed in a uniform area and deplete it by simple
diffusion. The theoretical results for a circular sink in a circular field must
correspond closely with those for arrangements on a lattice, particularly a
triangular lattice. The effects of competition in random or clumped
arrangements are necessarily determined by the spatial pattern of the sinks.

It must then be possible to predict the uptake curve from a mathematical
description of the pattern. The most convenient description for this
purpose is the distribution of distances from random points to the nearest
sink. This is suggested by the convenient, though imprecise, concept of
overlapping depletion areas around each sink.

On theoretical grounds, the model cannot work well when there is great variation in the distance from sinks to points on the perimeters of their associated polygons. In practice, remarkably good fits are obtained for regular patterns and for the onset of competition in clumped patterns, but the model begins to break down for high uptake in clumped patterns. In these cases the uptake is underestimated. The accuracy is within about 20%, which is comparable with that of Baldwin's model, but it is possible to see exactly where and why the predictions are inaccurate.

It is possible that some other statistical description of the pattern would give a better prediction of the uptake curve. The spectrum, or some equivalent function, is a natural candidate but there are difficulties in associating the two curves.

Acknowledgements. I am grateful to past and present members of the Department of Agricultural Science, and especially to Dr J P Baldwin, who supplied the data from experiments with the electrical analogue, and Mr P H Nye, who made valuable suggestions for revising the first draft of this paper.

References

Baldwin J P, Tinker P B, Nye P H, 1972 "Uptake of solutes by multiple root systems from soil. II. The theoretical effects of rooting density and pattern on uptake of nutrients from soil" *Plant and Soil* **36** 693-708

Besag J E, Gleaves J T, 1973 "On the detection of spatial pattern in plant communities" *Bulletin de l'Institut International de Statistique* **45** 153-158

Crank J, 1975 *The Mathematics of Diffusion* second edition (Clarendon Press, Oxford)

Nye P H, Marriott F H C, 1969 "A theoretical study of the distribution of substances around roots resulting from simultaneous diffusion and mass flow" *Plant and Soil* **39** 459-472

Ripley B D, 1976 "The second-order analysis of stationary point processes" *Journal of Applied Probability* **13** 255-266

Sanders F E, Tinker P B, Nye P H, 1971 "Uptake of solutes by multiple root systems from soil. 1. An electrical analog of diffusion to root systems" *Plant and Soil* **34** 453-466

The role of spatial analysis in the Rothamsted Insect Survey

I P Woiwod

12.1 Introduction

Both in the total number of individuals and in their diversity of form, insects are one of the most successful groups of organisms. One particularly important characteristic of most species is the ability of adults to fly. Indeed, this feature has been suggested as the reason for the 'near domination of the world by insects' (Wigglesworth, 1976). Flight probably evolved originally as a means of dispersal (*loc cit*) and many species can travel long distances, sometimes under the direct control of the individual but often through passive transport by the wind (Johnson, 1969).

Such movement, whether long- or short-range, is an important component in population dynamics, as is human migration in demography (for example, Hägerstrand, 1962). It is also of immediate practical relevance in the distribution of pest species, especially in well-known migrants such as aphids (greenfly) and locusts.

To investigate the problems associated with such movement requires a spatial framework of quantitative information that has not been available because of the practical difficulties of monitoring insect populations simultaneously over large areas. By sampling aerial populations of different groups of insects, the Rothamsted Insect Survey is now providing some of this information at least for Great Britain.

After a brief description of the way the Survey functions, examples are given of how the spatial nature of the data are being investigated. The examples chosen do not cover all aspects of the work of the Survey, but only those most relevant to workers with an interest in similar spatial data in different fields of research. Other research interests of the Insect Survey not so directly concerned with spatial aspects are covered in several of the references cited (for example Taylor, 1974; Taylor et al, 1978a).

12.2 The Survey

Two networks of traps operated by the Survey sample aerial insect populations throughout Great Britain. One network uses suction traps which sample at forty feet (12·2 m) and the other uses light traps which sample at four feet (1·2 m). As a routine, aphids are identified from the suction trap catches and macrolepidoptera (the larger moths) from the light traps, although other insect groups are identified in a less systematic way. These two types of traps serve slightly different purposes and hence have different sampling characteristics.

The suction traps were designed for the Insect Survey and have been described and illustrated elsewhere (Taylor and Palmer, 1972). Air is sucked down a ten-inch (25 cm) diameter tube from a height of forty feet by means of a powerful centrifugal fan and passes through a conical gauze net to remove and deposit the insects in a liquid-filled collecting bottle. The collecting bottle is emptied daily and the catch sent to Rothamsted for sorting and identification.

Suction trapping is very efficient and the number of insects caught is directly related to the volume of air passing through the trap. An absolute measure of aerial insect density can therefore be calculated if required (Taylor, 1962). The sampling height (12·2 m) was chosen after analysis of vertical profiles of insect density showed that air drawn from this height gave an adequate sample of migrating insects whilst excluding most individuals only involved in local low-level flight. Figure 12.1 shows why aerial samples from such traps are so effective in comparison with direct crop samples. The log variance x log mean relationship for insects caught in aerial samples is very close to the random line (dotted), where variance is equal to the mean at all densities and can be estimated from a single sample. In contrast, the crop sample has a much steeper regression, that is, at most densities the insects are highly aggregated and sample variances,

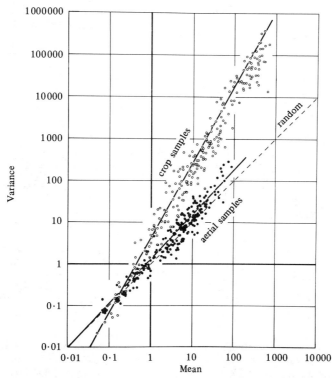

Figure 12.1. Log variance x log mean relationship from nearly random aerial insect fauna (●) and highly aggregated aphids on crops (○). After Taylor (1974).

being very much greater than the mean, require many samples to give a statistically adequate estimate, whereas an aerial sample is less expensive, being taken from a population randomised by mixing in the air currents.

Aphids, the main group of insects identified from the suction trap catches, are often serious pests, causing damage to crops by direct feeding and also as vectors of virus diseases. The primary purposes of the suction trapping system is to provide immediate information on aphid aerial distribution and abundance and to produce long-term data for developing aphid forecasting systems (Taylor, 1977b). There are at present twenty-two such traps operating in Great Britain (figure 12.2), mainly run by agricultural research stations or on experimental farms, and the immediate product of the system is a weekly aphid bulletin which lists the number of aphids caught during the previous week in the thirty-two most important species (from about 300 species which are regularly identified). These bulletin figures are used by agricultural advisers and research workers as background information or are incorporated into local warning schemes.

The network of survey light traps uses the standard Rothamsted trap (Williams, 1948) with a 200 watt tungsten filament light. Night-flying insects are attracted by the light and are collected in a killing jar which is emptied daily. There are about 160 light traps now operating throughout Great Britain (figure 12.3), run on a voluntary basis, with the identification

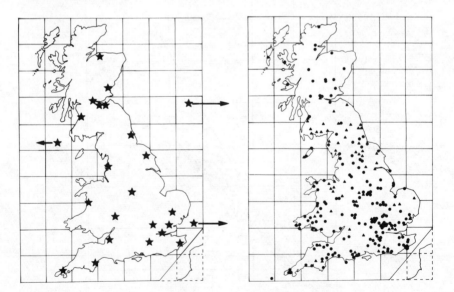

Figure 12.2. Network of suction traps at forty feet in Great Britain (★). Traps are also in operation in Denmark, Holland, and Northern Ireland (direction of traps from indicated position given by arrows).

Figure 12.3. Network of light traps in Great Britain. (●) Sites with one or more year's complete sample; (▲) recently started sites.

of the larger moths often being made by knowledgeable amateurs who also help to operate the traps. Light traps are less efficient than suction traps and their catches more difficult to interpret but work is continuing to standardise this type of trap (Taylor and French, 1974). However, for a group of insects that are relatively rare, like the moths, a trap which concentrates the catch is necessary to obtain adequate samples for statistical analysis, and only by working on a group that can be identified by amateurs and by relying on voluntary labour has it been possible to obtain such a good coverage of traps thus giving greater detail than would be possible in any other way. The data from the light traps is of particular interest in providing insight into spatial population dynamics and multispecies population problems, although some of the moth species are also of economic importance and an annual summary of pest species is published (for example, Taylor and French, 1976).

12.3 Data handling
About 300 species of aphids and 600 species of moths are regularly recorded in the Survey's traps. The catches are recorded daily and the data bank already has records of over five million individual insects. Initially the data are recorded on sheets for each site giving daily catches of individual species. These records are usually accumulated for a complete year and transferred to punched cards and then to magnetic tape. Programs have been written for most routine extractions.

12.4 Analysis
12.4.1 Mapping
From the inception of the survey it was apparent that some type of mapping would be required to condense and display the data being accumulated. By 1968 there were sufficient sites operating to make mapping worthwhile. The irregular nature and the changing number and position of sites, particularly for the light trap network, required an interpolating program and the SYMAP IV program was chosen mainly because it became available at the time and provided most of the facilities required.

SYMAP is a large flexible contour and choropleth mapping program written in Fortran IV and developed by the Laboratory for Computer Graphics Harvard, initially for the mapping of geographical census data. The interpolation function in the program produces a uniform grid of values by using a distance-weighted mean algorithm which is described in detail in Shepard (1968). Each grid value is basically calculated from the mean value of the surrounding four to ten data points weighted by the inverse square of their distance from that grid point. The flexibility of the program means that a wide choice of input and output is possible including the number, symbolism, and values of the contour levels. This flexibility is very useful in the initial stages, where a wide range of options can be tried to produce a suitable size and form of output for routine use, but once a satisfactory output has been achieved very few of the facilities of the program are used again and a tailor-made program with these facilities would probably be more efficient.

An important facility in SYMAP is that the data matrix of interpolated values used as a basis for the maps can be stored for further manipulation. This data matrix is used by the sister program SYMVU which produces three-dimensional figures useful for display rather than for analysis. Manipulation of the data matrix has also been used in extensions of the program to produce mean and difference maps and examples of how these are used will be given later. With these programs the data matrix from individual maps is manipulated and the resulting data is fed back into the SYMAP program to display the secondary type of map. In effect, SYMAP is being used as an interpolation program for the irregularly spaced samples as well as to display the results of the manipulation of these interpolated values.

Maps produced by SYMAP are very similar to ones drawn from the same data by hand. However, the computer-produced maps must always be interpreted with reserve since the patterns of contours are influenced by the number of samples, the distribution of the sample points, and the quality of the samples themselves. The two sets of Insect Survey samples are different in all three respects and this is apparent in the maps presented.

The maps of aphids from the suction traps have a maximum of twenty-two points which, from a mapping point of view, is very few for an area the size of Great Britain. In this instance the samples are very efficient and the nature of the movements of the aphids on wind currents means that densities change only gradually over large distances and trap catches can show high correlation over tens of kilometers (Taylor, 1974). Also the trap sites have been chosen in exposed situations where the samples are influenced as little as possible by local conditions. The simple nature of the patterns produced is then to some extent due to the small number of sites but more to the actual distribution of the aphids themselves and recent increases in the trap coverage in England has not appreciably altered the resulting maps. However, in Scotland, where there are no traps in the Northwest, the extrapolation of the catches from the transect of traps on the east of the country causes horizontal bands to be formed by the attempts of the program at extrapolation [figure 12.4(a)].

The maps from the light-trap system are very different because they provide a much better coverage; at the moment over one-hundred sites are operating in a particular year (figure 12.3). However, the sampling system is less efficient because more local and less mobile populations are being sampled, thereby producing greater between-site variability, and a larger number of traps are required to characterise the moth population of an area. An additional complication is that these traps are run on a voluntary basis so tend not to operate as consistently as the suction traps, and the coverage changes more from year to year. This also has to be taken into account when analysing yearly pattern changes in the contour maps.

With these considerations borne in mind, mapping has proved invaluable in gaining an understanding of the spatial population changes in insects, and two examples where a mapping approach has proved useful are now given in more detail.

Figure 12.4. Mean density distribution maps 1970–1975 of the peach-potato aphid. (a) Summer migration; (b) autumn migration; (c) spring migration; and (d) difference map of spring/autumn migration. Logarithmic contour intervals, approximately ×3 for each level, darker areas indicate higher densities.

12.4.1.1 *Mapping an aphid migration*

Aphids have a complicated life history the details of which vary from species to species but in many there are three seasonal cycles of population growth and redistribution. In a typical crop pest, the first redistribution (migration) in the spring is from an overwintering host plant into the crop, the second migration in the summer is from the crop onto secondary host plants or other suitable crops, and the final autumn migration is a return from these secondary hosts to an overwintering one. The relationship between these three migrations has an important bearing on aphid forecasting (Taylor, 1977b) but the overlap between them can cause problems of separation which have to be solved before analysis can begin. Figure 12.5 shows the typical migrations in the bean aphid (*Aphis fabae*) detected by one suction trap in the south of England and one in Scotland. In this example the migrations have been separated by a program that fits Gaussian curves to the log of the suction trap catches. In the southern trap the typical three migrations are evident. However, the overwintering host plant, for this species spindle (*Euonymus europaeus*), is rare in Scotland and the first migration from the overwintering host plant is consequently missing there. The crop infestation in Scotland probably originates from England's summer crop migration, but the size of the migration from the crop can be just as large although it takes place later. The return autumn migration is similar in timing in both traps because it is synchronised by the length of day, although it is unlikely that any of this migration in Scotland will succeed in locating a suitable host plant.

Another aphid species, the peach–potato aphid (*Myzus persicae*), gives a good example of how a mapping approach has been applied to these migrations with useful results. *Myzus persicae* is an important pest of potatoes and sugar beet, not only because of direct feeding but also because it is an efficient vector of virus diseases in these crops. The problems caused by this species have become even more serious in recent years owing to the development of resistance to organophosphorus insecticides by the aphid.

In sugar beet (*Beta vulgaris*), *Myzus persicae* is the main vector of virus yellows, named after the characteristic yellowing of the leaves caused by this group of virus diseases. The economic importance of virus yellows has promoted much research into the relation between the incidence of the disease and meteorological factors. This work has met with some success (Hurst, 1965; Watson, 1966) but produced the unexpected result that the weather at Rothamsted, which is outside the main sugar-beet growing area of eastern England, gives better prediction of the incidence of yellows than that in the beet-growing areas where the disease occurs. In a recent analysis (Watson et al, 1975), regressions of virus incidence against the number of 'frost days' in January, February, and March and the mean temperature in April accounted for 66% of the variance when the Rothamsted weather was used but only 35% when using meteorological records from individual beet-growing areas, even though some of these are up to 250 km away from Rothamsted. The explanation for this result became clear when migrations of *Myzus persicae* were mapped from the Insect Survey records (Taylor, 1977a).

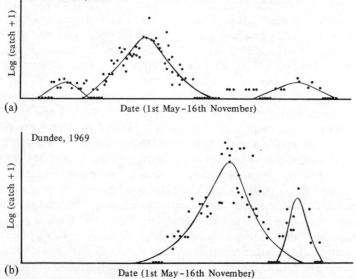

Figure 12.5. (a) The typical three migrations of the black bean aphid in the south of England; (b) only two of the migrations are present in Scotland (see text). Each point represents one days catch.

Figures 12.4(a), 12.4(b), and 12.4(c) are geometric mean maps of the three migrations of *Myzus persicae*, summarising all the information from the suction traps from 1970 to 1975. The summer migrations [figure 12.4(a)] show a strong concentration which reflects the distribution of crop hosts in eastern England (as already pointed out the area in Scotland is over-emphasised in the west owing to lack of trap coverage). The autumn migration [figure 12.4(b)] has lost this strong regional pattern which reflects the wide distribution of the secondary host that this migration comes from. The spring migration [figure 12.4(c)] is concentrated in the southeast. Figure 12.4(d) is a log difference map produced by dividing the spring by the autumn migration maps. This gives the differential change in population between these migrations which is a measure of survival during winter. From this map the explanation for the predictive relationship between Rothamsted weather and virus yellows is clear. The main overwintering area is in the southeast of England including Rothamsted (40 km northwest of London) but does not include the main crop areas of eastern England. The overwintering area is thus the source of the aphid vectors of the virus infection and it is largely the winter weather conditions in this area that will determine the size of the population available for migration into the crop-growing areas in spring. The overwintering survival map enables us to go one step further and actually suggest a possible explanation for the pattern that has been found. Figure 12.6 shows likely limiting factors for overwintering survival, winter temperature for the northern limit and winter rainfall for the western limit, although at the moment these limits are only speculative (Taylor, 1977a).

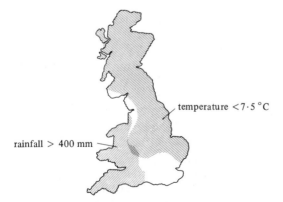

Figure 12.6. Possible limits for successful overwintering of *Myzus persicae*. Mean temperature >7·5 °C, total rainfall <400 mm, October to May. Expected survival zone unshaded. After Taylor (1977a).

12.4.1.2 *Mapping diversity*

For multispecies populations, such as the moths sampled by the Insect
Survey light traps, a measure of community structure is required for
comparisons of differences between, and changes within, sites (Taylor et al,
1978a) as well as for more fundamental ecological analyses into community
structure itself (Taylor, 1978). One such measure commonly employed is
species diversity, a combination of number of species and individuals in a
sample, and many statistics have been suggested for this property, which
has long been considered important in ecology. One of the early statistics
proposed was based on the observed frequency distribution of individuals
in species of moths caught in a Rothamsted light trap. The form of this
distribution is commonly found in biological samples, a hollow curve in
which many species are represented by a single individual, a number of
species are moderately common and one or two are abundant. This
distribution was fitted by the logarithmic series characterised by two
parameters, one of which, α, appeared to reflect closely the ecological
concept of diversity (Fisher et al, 1943).

Since this original suggestion of the use of a parameter from the log series
there has been a proliferation of alternative indices suggested. Recently
those in common use have been carefully compared by using the replication
of sites and years provided by the Survey's light trap records (Kempton and
Taylor, 1974; Taylor et al, 1976; Kempton and Taylor, 1976), and despite
the popularity of more recently developed nonparametric diversity
statistics based on information theory, α from the log-series was found
to excel in nearly all the characteristics most desirable in such a statistic:
independence of sample size, consistency of values at sites of known
stability, change in statistic reflecting population change at sites known
to be under environmental stress, robustness to the fluctuations in
abundance of the commonest species, ability to discriminate between
different sites, as well as reflecting the subjective assessment of diversity
at an individual site made by experienced ecologists.

Having decided on a suitable statistic to measure ecological diversity,
the next step is to try to understand its determining factors and at this
stage the mapping approach is invaluable. Figure 12.7 shows a map of α
diversity for moths from the Survey's light trap records. Since α is so
consistent from year to year at a site, and independent of sample size, the
measurements do not need to be made simultaneously and the map has
been produced by accumulating records from 1968 to 1974 to give
maximum detail.

The map shows immediately two important aspects of diversity. Firstly,
the latitude effect, with greater diversity occurring in the South. Secondly,
urban areas such as London and Birmingham stand out as areas of
particularly low diversity (see Taylor et al, 1978a). Also by comparing
maps of other variables such as botanical diversity and soil or geological

formations, likely explanations have been found for some of the other patterns apparent from inspection of the map.

Once possible factors have been identified, these can be quantified and a more formal statistical approach adopted to explain as much of the variance of α as possible. Such analysis has already begun, although a considerable amount of variability still remains to be explained (Taylor, 1978).

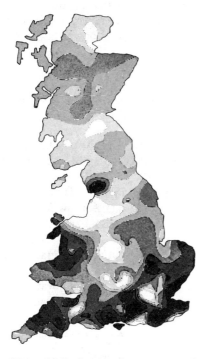

Figure 12.7. Diversity contour map for moths (1968-1974). Contour intervals <15, 15-20, 20-25, 25-30, 30-35, 35-40, 40-45, 45-50, >50, darker areas denote higher diversity. After Taylor (1978).

12.4.2 The variance/mean relationship

The way in which organisms arrange themselves in space is often highly characteristic of a species and hence of interest to ecologists who want to know how organisms are so distributed and why. The frequency distribution of counts of individuals per unit area have usually been used in attempts to analyse and understand this spatial characteristic, the negative binomial distribution being one of the most commonly applied for this purpose. However, this approach to spatial analysis of fitting distributions to sample counts has limitations because different distributions can fit at different densities (McGuire et al, 1957; Taylor, 1965) and very large samples are usually required to distinguish between different distributions.

An alternative approach has been to look for a relationship between two fundamental characteristics of such sample counts, the variance (s^2) and the mean (m). Taylor (1961) found that a power law of the form

$$s^2 = am^b \tag{12.1}$$

held for sets of samples from a wide range of organisms and suggested that the parameter b was a measure of aggregation characteristic for a given species (Taylor, 1965; Taylor, 1971). In a more recent analysis, over 150 sets of samples were used to compare Taylor's power law with two alternative models which have been proposed:

$$s^2 = m + b'm^2 \, , \tag{12.2}$$

from the negative binomial with a common k, where $k = 1/b'$, and

$$s^2 = a''m + b''m^2 \, , \tag{12.3}$$

proposed, although in a different form, by Iwao (1968). The detailed analysis showed Taylor's model (12.1) to fit considerably better than either of the alternatives (Taylor et al, 1978b).

Figure 12.8 illustrates a selection of the data from this analysis. These examples illustrate the wide applicability of the power law and it now seems unlikely from this analysis of the available data that there will be any major group of organisms to which it does not apply. Some of the samples are taken simultaneously at different spatial coordinates, others have been taken over the same area at successive time intervals; the range of sampling scale is very large, from the surface of a flatworm (Protozoa) to the whole of the United States of America (man); the sampling

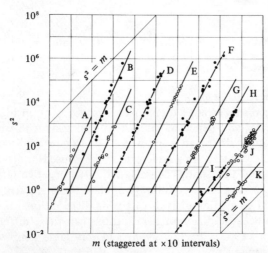

A garden dart moth of Great Britain from Insect Survey
B Haddock from trawls in the North Atlantic
C sea urchins on bed of North Sea
D cattle ticks from pasture samples in Australia
E human population of the United States of America
F bean aphid in Great Britain from Insect Survey
G great tit in Great Britain
H buff ermine moth in Great Britain from Insect Survey
I collared dove population in Great Britain
J a ciliate protozoon population on flatworms in North Wales
K a bee orchid population in south England

Figure 12.8. The power function relating population variance (s^2) to mean density (m) for a selection of organisms. Modified after Taylor et al (1978b).

methods are equally varied and the organisms range from orchids to birds and from moths to man.

The variance of the number of organisms in a set of sample units is a measure of the range of spacing of individuals within those units at that mean density and hence an estimate of this characteristic in the section of the population from which the samples are taken. It is the rate of change of this spacing with mean density that the power law describes, and the fact that the variance equals the mean [that is, $a = b = 1$ in equation (12.1)] if the distribution of organisms is random is a useful starting point for any comparisons.

From the 156 sets of data analysed, only two were not significantly different from random and only two were significantly more regular than random over a range of densities ($b < 1$), which demonstrates how unusual these two conditions are in the distribution of living organisms. Unless $b = a = 1$ in equation (12.1) the variance increases disproportionally with the mean which indicates that the spacing of nearly all organisms has a density-dependent component.

Before the b from equation (12.1) can be used effectively as a characteristic of density-dependent aggregation behaviour, we need some understanding of how it relates to known behaviour, either by observation or experiment, although modelling may be an alternative (see later). The sets of data used in the analysis of Taylor et al (1978b) were so varied in method, scale, and quality of sampling that they are not suitable for this purpose. The records from the Insect Survey appear to be more promising. Here we have usable records of over one-hundred species of aphids and two-hundred species of moths collected consistently by their respective sampling methods. Figure 12.9 shows variance/mean plots for individual

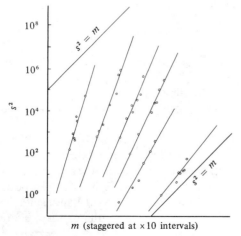

m (staggered at ×10 intervals)

Figure 12.9. The power law relationship for a selection of aphid species from the Insect Survey.

species of aphids where each point represents a year's sample from all the traps. The species selected show a range of values of b obtained and a similar figure can be derived from the moth records. It is hoped that comparison of these records with similar data from bird censuses will enable this relationship to be related to particular aspects of known behaviour and so be applied more widely in the analysis of spatial behaviour.

12.4.3 Modelling of spatial behaviour

All organisms have some element of mobility even if this is less pronounced than in aphids and those moths which are compulsively mobile. Unfortunately until recently very few mathematical models in population biology have been devised to investigate this spatial characteristic, migration being included as a form of unaccountable mortality, if not ignored altogether.

Taylor and Taylor (1977) proposed a model based on density-dependent movement which suggested that population regulation could be produced by such movement. They also showed that the model successfully reproduced known spatial properties. The general principle of the model is the interaction of two opposing biological forces. First, there is a tendency to congregate wherever resources are most abundant and, second, a tendency to reduce competition by moving away from any competitors and if possible to find an underutilised resource somewhere else. These two opposing forces of attraction and repulsion are density dependent and Taylor and Taylor proposed that they might each be proportional to a fractional power of the mean. Thus the net displacement (Δ) of an individual would be proportional to the difference between these fractional powers giving

$$\Delta = E\left[\left(\frac{\rho}{\rho_0}\right)^p - \left(\frac{\rho}{\rho_0}\right)^q\right], \tag{12.4}$$

where

ρ is population density,

ρ_0 is the biologically identifiable parameter at which migration balances congregation, and

E is a scale factor containing the dimensional units.

p and q are rate constants for density-dependent migration and congregation respectively.

Computer simulations have been made under the assumption that individuals move according to the Δ function on a heterogeneous environmental grid on which there are some areas where survival and reproduction can occur, some where only survival but not reproduction is possible, and others which will kill any immigrants. A more detailed background to the assumptions, methods, and results of these simulations is given elsewhere (Taylor and Taylor, 1978).

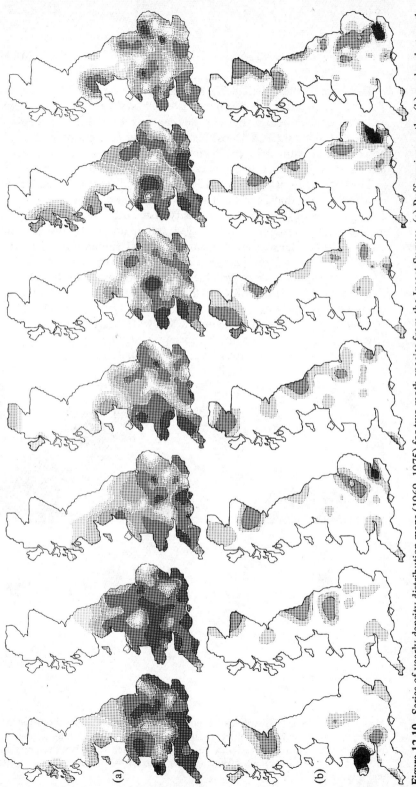

Figure 12.10. Series of yearly density distribution maps (1969–1975) for two moth species from the Insect Survey. (a) Buff ermine moth; (b) garden dart moth. After Taylor and Taylor (1978)

There are several particularly interesting outcomes of these simulations. The total population can reach a quasi-stable state after very few generations and this level is below the carrying capacity of the environment. Thus the model demonstrates a possible mechanism for the internal regulation of population which, in the form of Wynne-Edwards's social control mechanisms, has recently aroused so much controversy.

The changes in patterns of spatial distribution from the model correspond well with those that have been found occurring naturally. Figure 12.10 shows a series of yearly SYMAP density distribution maps of two contrasting moth species from the Insect Survey and figure 12.11 shows a similar set of SYMAP density maps from the model simulation. The buff ermine *Spilosoma lutea* [figure 12.10(a)] is a common species that persists throughout its geographical range, and areas of very low density (represented by blank holes on the map) appear, move around, and eventually disappear over several generations in a manner similar to that of the simulated population in figure 12.11(a). The garden dart *Euxoa nigricans* [figure 12.10(b)] is much more sparsely distributed although it can cause considerable damage to crops locally and usually unpredictably. In this

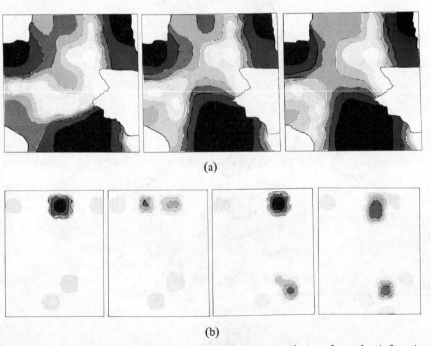

(a)

(b)

Figure 12.11. Density distribution maps of computer simulations from the Δ function. (a) A population mapped at three generation intervals after reaching a quasi-stable condition. Migration is in response to central mean density. (b) A population producing periodic outbreaks after migration in response to local density. After Taylor and Taylor (1978).

species, areas of high density appear as islands which grow, move, and decline in a similar way to the simulated population shown in figure 12.11(b). The general features of the two sets of maps and the way they change are therefore remarkably similar.

Censuses from simulations of the model show one other feature which is of interest, namely, they all obey the power law relationship, one of the few spatial analyses which has been shown to apply to populations of all types of organisms. Figure 12.12 shows log variance/log mean plots of seven different Δ simulations. As can be seen such simulations are indistinguishable from samples of actual organisms shown in figures 12.8 and 12.9.

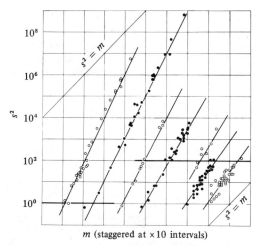

m (staggered at $\times 10$ intervals)

Figure 12.12. The power law relationship for populations produced by seven simulations from the Δ function. Parameter values are given in Taylor and Taylor (1978).

Although it is possible that other models based on different functional mechanisms could produce similar results, the biological assumptions behind this particular one are simple and realistic and, as well as being successful in reproducing known spatial situations, open up new possibilities in the theories of population dynamics and evolution (Taylor and Taylor, 1978). If further work on the model confirms its general applicability and usefulness, simulations could improve our understanding of the power law relationship and also help in analysing types of spatial patterns and how they change. To be useful such simulations should be compared with actual spatial data of the type now being produced by the Rothamsted Insect Survey.

References

Fisher R A, Corbet A S, Williams C B, 1943 "The relation between the number of species and the number of individuals in a random sample of an animal population" *Journal of Animal Ecology* **12** 42–58

Hägerstrand T, 1962 "Geographic measurement of migration" in *Les Déplacements Humains* Ed. J Sutter (Union Européenne d'Éditions, Monaco) pp 61–83

Hurst G W, 1965 "Forecasting the severity of sugar beet yellows" *Journal of the Royal Agricultural Society* **122** 101–112

Iwao S, 1968 "A new regression method for analyzing the aggregation pattern of animal populations" *Researches on Population Ecology* **10** 1–20

Johnson C G, 1969 *Migration and Dispersal of Insects by Flight* (Methuen, London)

Kempton R A, Taylor L R, 1974 "Log-series and log-normal parameters as diversity discriminants for the Lepidoptera" *Journal of Animal Ecology* **43** 381–399

Kempton R A, Taylor L R, 1976 "Models and statistics for species diversity" *Nature (London)* **262** 818–820

McGuire J U, Brindley T A, Bancroft T A, 1957 "The distribution of European corn borer larvae *Pyrausta nubilalis* (Hbn.) in field corn" *Biometrics* **13** 65–78

Shepard D, 1968 "A two-dimensional interpolation function for irregularly-spaced data" *Proceedings of the 23rd National Conference of the Association for Computing Machinery, ACM publication P-68* (Brandon/System Press, Princeton, NJ) pp 317–323

Taylor L R, 1961 "Aggregation, variance and the mean" *Nature (London)* **189** 732–735

Taylor L R, 1962 "The absolute efficiency of insect suction traps" *Annals of Applied Biology* **50** 405–421

Taylor L R, 1965 "A natural law for the spatial disposition of insects" *Proceedings of the 12th International Congress of Entomology, London, 1964* (The Royal Entomological Society of London, 41, Queensgate, London SW7) pp 396–397

Taylor L R, 1971 "Aggregation as a species characteristic" in *Statistical Ecology, Volume 1* Eds G P Patil, E C Pielou, W E Waters (Pennsylvania University Press, University Park, Pa) pp 357–377

Taylor L R, 1974 "Monitoring change in the distribution and abundance of insects" *Report Rothamsted Experimental Station for 1973, Part 2* pp 202–239

Taylor L R, 1977a "Migration and the spatial dynamics of an aphid, *Myzus persicae*" *Journal of Animal Ecology* **46** 411–423

Taylor L R, 1977b "Aphid forecasting and the Rothamsted Insect Survey" *Journal of the Royal Agricultural Society of England* **138** 75–97

Taylor L R, 1978 "Bates, Williams, Hutchinson—a variety of diversities" in *The Diversity of Insect Faunas* Eds L A Mound, N Waloff (Royal Entomological Society of London, Blackwell Scientific Publications, Oxford) pp 1–18

Taylor L R, French R A, 1974 "Effects of light trap design and illumination on samples of moths in an English woodland" *Bulletin of Entomological Research* **63** 583–594

Taylor L R, French R A, 1976 "Rothamsted Insect Survey. Seventh annual summary" *Report Rothamsted Experimental Station for 1975, Part 2* pp 97–128

Taylor L R, French R A, Woiwod I P, 1978a "The Rothamsted Insect Survey and the urbanization of land in Great Britain" in *Perspectives in Urban Entomology* Eds G W Frankie, C S Koehler (Academic Press, New York) pp 31–65

Taylor L R, Kempton R A, Woiwod I P, 1976 "Diversity statistics and the log-series model" *Journal of Animal Ecology* **45** 255–272

Taylor L R, Palmer J M P, 1972 "Aerial sampling" in *Aphid Technology* Ed. H F van Emden (Academic Press, London) pp 189–234

Taylor L R, Taylor R A J, 1977 "Aggregation, migration and population mechanics" *Nature (London)* **265** 415–421

Taylor L R, Taylor R A J, 1978 "The dynamics of spatial behaviour" in *Population Control by Social Behaviour* Eds F J Ebling, D M Stoddard (Institute of Biology, London, Symposium 1977) pp 181-212

Taylor L R, Woiwod I P, Perry J N, 1978b "The density-dependence of spatial behaviour and the rarity of randomness" *Journal of Animal Ecology* **47** 383-406

Watson M A, 1966 "The relation of annual incidence of beet yellowing viruses in sugar beet to variations in weather" *Plant Pathology* **15** 145-149

Watson M A, Heathcote G D, Lauckner F B, Soway P A, 1975 "The use of weather data and counts of aphids in the field to predict the incidence of yellowing viruses of sugar-beet crops in England in relation to the use of insecticides" *Annals of Applied Biology* **81** 181-198

Wigglesworth V B, 1976 "The evolution of insect flight" in *Insect Flight* Ed. R C Rainey (Royal Entomological Society of London, Blackwell Scientific Publications, Oxford) pp 255-269

Williams C B, 1948 "The Rothamsted light trap" *Proceedings of the Royal Entomological Society of London A* **23** 80-85

13

Exploratory and descriptive uses of multivariate analysis in soil survey

R Webster

13.1 Introduction

Soil survey is a general procedure whereby information is obtained about the soil of areas. The information is needed for a variety of purposes. Some of it is easy and cheap to come by; some is difficult, expensive, or time-consuming, and often all three. Such information necessarily derives from samples, and those who gather the information try to achieve the best compromise between good representation of the soil of the areas concerned and the cost or effort of obtaining that information. Soil varies substantially from place to place within national territories, and even within small administrative districts and farms. Separate information is therefore needed for different parts of an area. As in many other kinds of survey, if the individual measurements are costly then it pays to stratify a survey area by a cheap means, and then sample each stratum independently and relatively sparsely for the information that is dear.

Although it has not been planned in statistical terms, this has been the standard approach to soil survey for approximately fifty years. The soil has been classified and information on its qualities for use assembled for each class. Initially there were big differences to be placed on record. Usually they were already known to people on the land, and soil scientists could simply adopt the layman's classes and nomenclature for their own purposes. However, when scientists turned their attention to finer differences and tried to predict the behaviour of soil with greater precision and to communicate with one another throughout the world, they found that their early classifications were too crude or only locally applicable. They therefore tried to refine their classifications and make them consistently applicable at least over national territories if not worldwide. They have found it extraordinarily difficult. No classification of soil has been singularly successful or agreed. Attempts to understand why this is so and to improve general purpose classifications of soil have been the main driving force behind multivariate analyses in soil science. The aims have been to discover relationships among sampling sites and the general structure of soil populations. This chapter describes how these aims have been pursued by reference to several case histories in which I have been involved.

13.2 Soil populations and the nature of data
13.2.1 Source

The information available for analysis comes from sampling sites. That concerning the soil itself may be from the face of a pit, 30 cm to 50 cm across and extending back from the pit face just a few centimetres. Alternatively the soil examined may be a core, commonly from 3 cm to 15 cm diameter, extracted by using an auger. The soil varies not only laterally but also vertically. A vertical section through the soil is known as its profile, the nature of which can have important bearings on plant growth and land management, and can provide clues to the depositional history or weathering regime of the land surface.

The soil profile may be from a few centimetres to several metres deep. There are distinct layers—horizons—in some profiles, and these are usually described separately. A profile description thus consists of several horizon descriptions, which join top to bottom. Similarly, laboratory measurements are usually made horizon by horizon on material from the whole thickness of each horizon. It is often possible to recognise the same kinds of horizons in profiles at different sites, and so to match measurements in them even though they are not at the same depths [figure 13.1(a)]. In other profiles change is gradual. Nevertheless, conventional practice is to divide the profiles into horizons for description and measurement. Some workers, recognising the arbitrary nature of such horizons, consider this susceptible to inconsistency and prefer to describe soil profiles of this sort between the same fixed intervals. Any match between measurements in different profiles is then based strictly on equal depths [figure 13.1(b)].

Normal practice is to record several characters in each horizon or interval of each profile. The result is that survey data consist of a number of measurements or qualitative descriptions for several contiguous

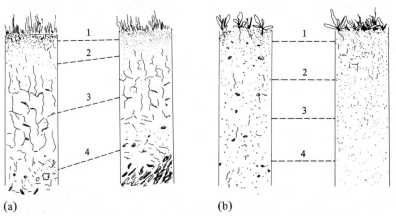

(a)   (b)

Figure 13.1. Soil profiles showing (a) distinct horizons, and (b) gradual change with increasing depth. Pecked lines join comparable parts of the profiles.

layers at isolated sampling sites. The data are to some extent inter-dependent. There is rarely strong correlation between characters recorded within any one horizon because surveyors avoid recording two characters that carry the same information. But there may be substantial correlation among values of the same property in different horizons.

13.2.2 Recording scales

Normal practice in soil survey was to record field observations qualitatively and laboratory determinations as measured values on continuous scales—that is, fully quantitatively. However, in recent years there has been a steady trend towards more quantitative field recording. Hodgson (1974) gives current procedures and recording scales used in England and Wales. A few examples illustrate these. Soil colour is matched in the field with standard colour panels and recorded by using the near-equal interval scales of the Munsell system. Estimates of several other characters such as proportions of mottles, clay, and stones are often made by eye or feel or with the use of simple field equipment. Other characters are ranked, but the classes have been carefully chosen so that the rank steps are of equal significance. The abundance of roots may be classified as none, few, common, many, or abundant, and recorded as 0, 1, 2, 3, or 4. The classes actually represent estimates of the number of roots per 100 cm^2 cross section of soil as <1, 1–10, 11–25, 26–200, and >200. Such scales, though coarse stepped, may be regarded as linear and of equal interval. As a result a modern soil survey records only few properties entirely qualitatively, either as binary or unordered multistate characters.

It will be evident that the intervals on the scale of root abundance form an approximately logarithmic progression that is effectively transformed by ranking. Root distribution is usually strongly skewed, and recording it as above makes it more nearly symmetrical. The distributions of measurements on fully quantitative scales are often skew too, and for these transformation either to logarithms or square roots will usually stabilise variances and, for homogeneous populations, approximately normalise them.

13.3 Principal component and coordinate analysis

Rayner (1966) first used Gower's (1966) principal coordinate analysis (PCO) to examine relationships among soil profiles in South Wales. Since then PCO and principal component analysis (PCA) have both been applied increasingly to study soil populations, especially in Britain. In most instances the aim has been to display and understand population structure, and in a few cases to reduce the number of variates for subsequent analysis.

13.3.1 West Oxfordshire

An early study by Cuanalo and Webster (1970) well represents the situation facing soil surveyors and systematists. The data have been reanalysed with minor changes for present purposes.

In a survey of West Oxfordshire by Beckett and Webster (1965a) that covered about 1000 km^2, the land had been classified into seventeen physiographically distinct types and these then mapped. The area was thus stratified, and each stratum was sampled in as near random a way as was practicable (Beckett and Webster, 1965b). At each site several morphological and physical properties of the soil were measured at two fixed-depth intervals centred at 13 cm and 38 cm. Here fifteen characters at each of the two depths for eighty-five sites are analysed. They are listed in table 13.1. The variates were standardised to zero mean and unit variance, and a component analysis carried out on the correlation matrix. Table 13.2 lists the six largest latent roots and the proportion of sample variance for which they account.

The first latent root is much the largest, and the first two account for nearly half the variance. There was evidently substantial correlation

Table 13.1. Variates analysed for West Oxfordshire soil and their code numbers used in figures 13.4 and 13.5.

Property	Codes for two depths	
	13 cm	38 cm
Colour hue	1	16
Colour value	2	17
Colour chroma	3	18
Chroma of mottles	4	19
Mottle abundance	5	20
Clay and silt	6	21
Fine sand	7	22
Stone content	8	23
Plastic limit	9	24
Liquid limit	10	25
pH	11	26
Mean strength in summer	12	27
Mean strength in winter	13	28
Mean matric suction in summer	14	29
Mean matric suction in winter	15	30

Table 13.2. Latent roots of correlation matrix for thirty variates in West Oxfordshire.

Order	Root	Percentage of sample variance	Cumulative percentage
1	9·695	32·32	32·32
2	4·833	16·11	48·43
3	2·342	7·81	56·24
4	1·776	5·92	62·16
5	1·668	5·56	67·72
6	1·304	4·35	72·06

among the variates; pairs of variates representing the same soil property
at the two depths were especially strongly correlated. Projection of the
population scatter onto the plane of the first two principal axes therefore
gives quite the most informative single display of the sample distribution,
and figure 13.2 is the result (see also figure 13.3). The symbols indicate
the broad classes to which the sampling sites would be assigned in a fairly
conventional classification.

The relationships between sites apparent in figure 13.2 are generally
speaking what would be expected. In fact Cuanalo and Webster (1970)
prepared a diagram of what they expected from the analysis before

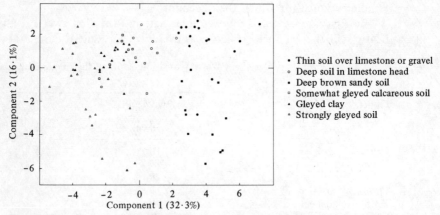

Figure 13.2. Scatter of West Oxfordshire sampling sites in plane of first two principal
axes. Six broad classes are distinguished by symbols.

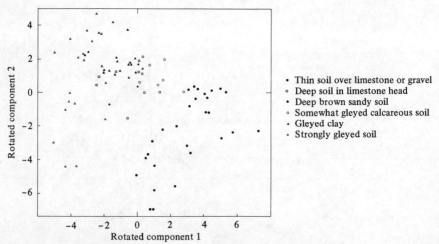

Figure 13.3. The same scatter of sites as in figure 13.2 but rotated through
approximately 40° by Varimax method.

carrying it out. In some ways the agreement between the result and expectation was remarkably close. However, there was also a prominent difference between the two. We had thought that each stratum in our sample would be represented by a cluster of points widely separated from those of other strata. The analysis proved us wrong. This was the first ordination of a probability sample of soil profiles. From subsequent work we know that clustering is the exception rather than the rule.

As in other fields of study, principal components can be interpreted either from the dispositions of individuals whose characters are well-known or from the elements of the latent vectors, or a combination of the two. In this study the first principal axis was originally thought to represent gleying—the result of waterlogging—which should be characterised by small values of matric suction, at least in winter, colour chroma, and abundant mottles. The second characteristic was thought to represent particle size. More careful examination of the latent vectors, shown in figure 13.4 suggests that interpretation is less simple, and that further rotation would help.

Accordingly the first two components have been rotated by making use of Kaiser's (1958) Varimax method, and the results are shown in figures 13.3 and 13.5. By restricting the rotation to the first two components the relationships both of sample points and vectors to one another are exactly retained, and figures 13.3 and 13.5 are strictly comparable to figures 13.2 and 13.4. Clearly there has been a clockwise rotation through approximately 40°, and since the vectors are clustered around the new axes we can give the latter meaning. The new horizontal axis now

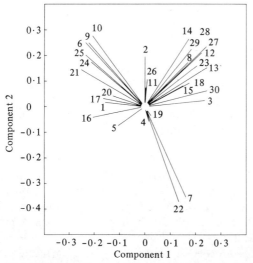

Figure 13.4. Latent vectors projected into the plane of the first two principal components for West Oxfordshire. Table 13.1 lists the variates.

strongly represents colour chroma, matric suction, and strength—properties that are associated with the water regime of the soil. The driest, strongest, and most brightly-coloured soil profiles are on the right; the wettest, weakest, and dullest are on the left. The vertical axis has strong contributions from particle size and plastic and liquid limits—properties embodied in the texture of the soil. The heaviest textured (most clayey) profiles are at the top, the lightest (sandiest) at the bottom.

Although the population cannot be divided readily into distinct clusters there are two portions of the projection that deserve mention. In the upper third of figure 13.3 the sampling sites are concentrated on the left, leaving the right-hand side empty: all the clay profiles are wet. There are also very few profiles falling immediately to the left of centre with negative values of the second rotated component: there are evidently very few moderately wet, light-textured soil profiles in the sample. If the soil in the area surveyed is light-textured then it is likely either to be dry or very wet. Medium-textured soil shows almost the full range of wetness.

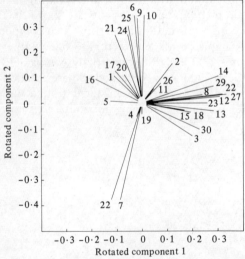

Figure 13.5. Varimax rotation of latent vectors for West Oxfordshire.

13.3.2 Australian Capital Territory

The analysis of the West Oxfordshire data could be said to do no more than describe and display quantitatively what an experienced soil surveyor would discover intuitively given sufficient time. There were few surprises. A similar analysis of data from a soil survey in the Australian Capital Territory (ACT) was different.

Soil survey in Australia proceeded as elsewhere by delineating parcels of land that were reasonably homogeneous with respect to profile, grouping parcels with similar profiles, and then sampling to characterise each class and evaluate its potential for use. However, in the Southern Tablelands of New South Wales and the Capital Territory the propriety of the method was in doubt. Was it practicable to classify soil in this way, and if so were the classes recognised appreciably more homogeneous than the landscape at large? A study was made of an area at Ginninderra in the ACT to answer these questions (Webster and Butler, 1976).

The area, covering some 400 ha of rolling terrain, was sampled at 111 intersections of a rectangular grid at intervals of approximately 180 m. Ten-centimetre diameter cores were taken at each point on the grid, and the soil was described comprehensively in the field at fixed depth intervals, chosen as a result of previous experience, to represent the main horizons. Several chemical properties of the soil were also measured on the core material in the laboratory.

Several ordinations were made of the sample with different combinations of variates by using PCO on a distance matrix calculated with the use of Gower's (1971) general similarity coefficient. PCO was preferred to PCA, mainly because there were a number of nonapplicable comparisons. All the ordinations were similar in general, and we shall consider just one computed on morphological data alone from three depths, namely 0 cm– 10 cm (representing the A1 horizon), 15 cm–25 cm (representing the A2 horizon where present), and 41 cm–51 cm (representing the B, usually Bt horizon). Also included were depth to top of B horizon where present, abruptness of transition between A and B horizons, and degree of bleaching in the lower part of the A horizon. There were thirty-three variates. The first five latent roots in this analysis are listed in table 13.3, and figure 13.6 shows the scatter in the plane of the first two principal axes. The axes are labelled with a rough interpretation made by comparing the properties of the sampling points that lie near the ends of the range.

We should notice two features of this analysis immediately. First, the largest latent roots account for a smaller proportion of the total variance than in the previous example. The sample distribution is much more nearly hyperspherical. Second, the scatter is very even. There is no evidence of clustering at all.

Figure 13.6 shows a further feature. The mapping units (classes of parcel listed in table 13.4) in which the sample points lay are shown by

Table 13.3. Latent roots of dissimilarity matrix for sampling sites at Ginninderra calculated from morphological data.

Order	Latent roots	Percentage of trace	Cumulative percentage
1	3·63	14·6	14·6
2	2·84	11·4	26·0
3	2·41	9·8	35·7
4	1·38	5·6	41·4
5	1·24	5·0	46·4

symbols, one for each class. There is little evident relation between the mapped classes and this projection of the data. Nor was there any with projections on other low-order axes, either in this or in other principal coordinate analyses. To some extent this arises from the fact that each mapping class contains a mixture of profile types (table 13.4). However, the systematic classification of the profiles fared little better. Some relation was evident, but it was far from strong [see figures 2 to 4 of Webster and Butler (1976)].

Despite the doubts expressed earlier, this seemed an unlikely result, and further analyses were carried out to ensure that the PCO was not misleading. Correlation coefficients were calculated for all possible pairs of variates, and within- and between-group variances were compared for a number of classifications including a numerically optimised one. All results pointed

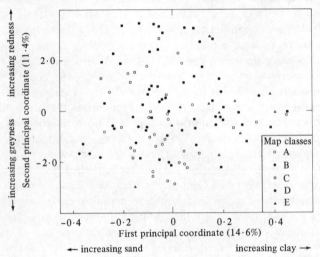

Figure 13.6. Scatter of Ginninderra sampling sites in plane of first two principal axes. Symbols show the mapping classes in which sites lay (table 13.4).

Table 13.4. Mapping classes for the Ginninderra survey, summarised from Webster and Butler (1976).

Group	Description
A	Undulating to rolling land dominated by red earth and skeletal soil with minor occurrence of yellow podzolic and bare rock
B	Steep hilly land dominated by skeletal soil with red earth and bare rock
C	Gently undulating land dominated by red and yellow podzolic and red earth with minor rock outcrop
D	Very gently undulating land dominated by red and yellow podzolic and red earth
E	Broad depressions dominated by yellow podzolic with minor red podzolic

to the same conclusion. Correlations are weak, the distribution is well inflated in many dimensions, and classification based on many variates did not result in within-class variances of others that were appreciably smaller than that of the total sample.

In these circumstances PCA or PCO show fairly quickly the true situation faced by the soil surveyor. General-purpose classification is unlikely to be profitable, and the surveyor should instead concentrate his attention on those properties, which will usually be few, that most affect the use envisaged for the soil.

As it happens this is only a part of the picture, and I shall consider in the next section the relation between soil and the landscape in which it occurs.

13.4 Canonical correlation

Whereas PCA and PCO have found increasing use in soil science in the last dozen years, there have been very few applications of canonical correlation. The recent studies by Hernandez (1976) and myself (Webster, 1977) are the only ones that I know to have been reported in detail, despite widespread interest for many years in relations between soil, its environment, and its agricultural potential. I believe this situation derives partly from the difficulties of appreciating the technique and partly from a lack of good examples illustrating it in the statistical textbooks. Further, several natural scientists in other fields have been disappointed that such a potentially powerful technique should tell them so little that they did not know already.

I first applied canonical correlation to relate the soil in the Dee Catchment of North Wales to its environment (Webster, 1977). Fairly large and significant correlations were found, but as above, they seemed only to express quantitatively what was common knowledge among British soil scientists.

13.4.1 Ginninderra

A second application to the soil at Ginninderra in the ACT was more revealing, and we consider this now.

In the soil survey of Ginninderra not only was the soil described at the sampling sites but also the local physiography. The broad environment of this small area—temperature, rainfall, country rock, and to a large extent native vegetation—is constant. But in rolling terrain with several small valleys and about 55 m of relief, local physiographic variation is appreciable. Characters of sites measured were slope gradient, distance to nearest ridge crest upslope, and absolute altitude. A water dispersal area was defined for each site. Its upper side was a contour line 30 m long with the sample point at its centre. Two sides extended downslope from the ends of this line normal to the contour until they met a drainage channel, or until their gradient was sensibly zero, or until they met one another. Except in the last case the dispersal area had a fourth side, which was the drainage channel if present or a line at the foot of the slope otherwise. Four characters of this dispersal area were then measured, namely its size, its mean gradient, its mean length, and the vertical difference between its mean height and the height of the sample point.

These seven physiographic characters were taken as right-side variates in several analyses with various subsets of soil characters on the left side. Use of all the recorded soil variates produced ill-conditioned matrices with this size of sample. The analysis illustrated here included the seventeen soil morphological variates listed in table 13.6, and is fairly typical. Variates were standardised to unit variance so that their contributions to the correlations could be compared and the vectors thereby interpreted. The canonical correlations are listed in table 13.5. The first two are significant at $\alpha < 0\cdot001$ as judged by the usual test for the kth and higher-order canonical correlations:

$$\chi_k^2 = -[n-1-\tfrac{1}{2}(p+q+1)]\ln \prod_{i=k}^{q} (1-\lambda_i) \,,$$

with $(p-k+1)(q-k+1)$ degrees of freedom, where

n is the number of sampling points, equal to ninety-eight in these analyses,

p and q are the numbers of left- and right-side variates respectively, and

λ_i is the ith canonical root.

Table 13.6 lists the first two pairs of canonical vectors. Sand content at 0 cm–10 cm is the largest element in the first left-side (soil) vector, and there are substantial contributions from colour hue at both depths and from porosity at the lower depth. In the second vector, colour value, porosity, hardness, and coalescence at 41 cm–51 cm are the largest elements. On the right side (physiography) the vector elements both for absolute height and height above mean height of dispersal area are large.

Table 13.5. Canonical roots and correlations between soil (seventeen variates) and physiography (seven variates) at Ginninderra (ACT).

Order	Root	Percentage of trace	Cumulative percentage	Canonical correlation
1	0·5945	29·6	29·6	0·77*
2	0·5134	25·6	55·2	0·72*
3	0·2908	14·5	69·7	0·54
4	0·2080	10·4	80·1	0·46
5	0·1658	8·3	88·3	0·41
6	0·1337	6·7	95·0	0·37
7	0·1008	5·0	100·0	0·32

* The first two roots are significant at $\alpha < 0\cdot001$.

The dispersal area itself carries appreciable weight and is the largest element of the second vector.

The first pair of vectors can be interpreted to mean that high land with large area for dispersal of water has soil that is redder (hue was scaled from $1 = 7 \cdot 5R$ on the Munsell scale, red, to $8 = 5Y$ yellow) and more sandy at the surface than that elsewhere. Vector 2 suggests that sites with large dispersal areas but close to their local base level have pale, hard, and relatively strongly coalescent subsoil.

A relation between soil colour and height is observed in many parts of the world, and is usually attributed to variation in water regime. The relation between height and sand content probably arises from differential movement of particles during denudation. Although dependence of soil colour and texture on the height of the land above its local base level is widespread it is only obvious where there is a simple catenary sequence.

Table 13.6. First two canonical vectors for soil (left side) and physiography (right side) calculated from standardised data for Ginninderra.

		Vector 1	Vector 2
Soil (left side)			
	colour hue	−0·370	−0·134
	colour value	−0·130	−0·077
	colour chroma	0·096	−0·044
0 cm–10 cm	structure grade	−0·103	−0·269
	structure size	−0·217	−0·150
	sand	0·537	−0·263
	silt	0·100	−0·081
	colour hue	−0·473	0·320
	colour value	0·217	0·561
	colour chroma	−0·040	−0·140
	structure grade	−0·096	0·072
41 cm–51 cm	structure size	0·014	0·066
	porosity	0·494	0·611
	hardness	0·345	0·437
	coalescence	0·304	0·508
	sand	0·042	0·208
	silt	0·122	0·065
Physiography (right side)			
absolute height		0·732	0·273
dispersal area (log)		0·404	0·665
slope gradient		−0·085	−0·376
height above drainage channel[a]		−0·129	−0·555
horizontal distance to drainage channel[a]		0·053	0·036
height above mean of dispersal area		0·330	−0·298
horizontal distance to crest (square root)		0·082	0·343

[a] Or zero gradient (see text).

At Ginninderra the landscape pattern is complex, and so the relationships were not at all easy to see. The analysis revealed local relationships that might otherwise have remained hidden.

Hernandez (1976) analysed the physiographic variates with other subsets both of morphological and chemical soil data, and also analysed chemical data with morphological data. All showed some relation, which could be explained in a broadly similar way to that described above.

The scatter of sampling points on the canonical axes is also interesting. Figure 13.7 shows the scatter in the plane of the first pair of canonical axes. The symbols show which mapping classes the sites lay in. There is evidently a trend from group E to group D on the lower ground (concentrated in the lower left of the diagram), to group C scattered near the middle, to group B and finally group A (upper right) which occupy mainly high ground. The scatter of the sites on the second pair of axes similarly shows regions in which particular groups are most concentrated. Here at last we see the good sense of the mapping classification. Despite the complexity of the landscape and intimate mixture of soil types in it, the soil surveyor had intuitively identified relationships between soil and physiography.

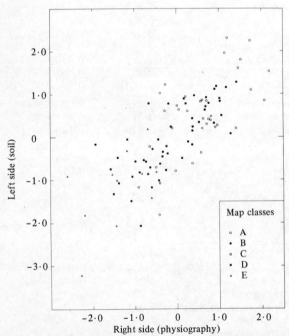

Figure 13.7. Scatter of Ginninderra sampling sites in plane of first pair of canonical axes. Symbols show the mapping classes in which sites lay (table 13.4).

13.4.2 Wungong

Another illustration of canonical correlation applied to soil survey is also drawn from Australia. It concerns the soil and its environment in the catchment of the Wungong Brook, a forested area of about 150 km² at the edge of the Great Plateau of Western Australia some 30 km southeast of Perth.

The country rock of the area is granite–gneiss into which are intruded many dolerite dykes. Beneath the plateau surface this rock is deeply weathered to a ferruginous and bauxitic laterite cap, and beneath that there is a mottled zone, a pallid zone, and then more or less fresh rock. The Wungong Brook and its tributaries dissect the plateau giving relief of about 80 m in their middle reaches and up to 160 m in the lower reaches of the main stream. Its valley sides, which range in gradient from 4° to 20°, occupy most of the area, and these bevel in turn the horizons of the weathering profile and the rock beneath. An idealised section is shown in figure 13.8. The upper metre of soil varies from loose yellow–brown sandy material that contains more or less lateritic gravel on the highest ground to friable, finer textured, and usually redder or browner soil on the lower slopes. The former soil is obviously the top of the old laterite profile. At least part of the latter was probably formed more recently from fresh rock or the deepest horizons of the weathered plateau mantle, but the extent to which this was mixed with material from higher ground could not be gauged easily. The association between soil and physiography seemed strong enough here for canonical correlation to lead to worthwhile prediction of soil character at many sites from a knowledge of their physiography; hence this study.

The catchment was surveyed at 30 m intervals on transects each extending from an interfluve crest to a valley bottom and as nearly as possible directly downslope. There were 347 sampling points at each of which the soil was examined with a 10 cm diameter auger to 1 m. The following data from these sites were used for analysis.

Soil: colour on the Munsell scales, clay content and gravel content at 0 cm–5 cm, 15 cm–30 cm, 45 cm–60 cm, and 75 cm–100 cm; depth at which clay content increases most markedly and the depth range over which the increase occurs.

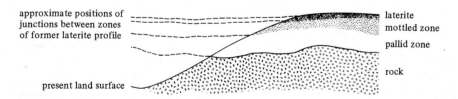

Figure 13.8. An idealised cross section through the dissected laterite plateau of Western Australia.

Physiography: absolute height, slope gradient and convexity, length of slope, distance to crest, distance to valley floor, height above valley floor, height of slope crest above site, and distance to nearest high-order (fourth or more) stream.

The first six canonical correlations appeared significant ($\alpha < 0.01$), and are listed in table 13.7. The first is substantially larger than the others, and we shall devote our attention to it.

Interpretation was fairly straightforward. Of the soil variates (left side), colour and clay content in the 75 cm–100 cm layer, and clay and gravel content at the surface made important contributions. On the right side (physiography) the vertical interval between the site and the slope crest carried most weight, though length of slope and height above valley floor also seemed important.

The most striking feature of the analysis is in the distribution of sampling points in the plane of the first canonical axes, figure 13.9. There are two fairly distinct clusters, the lower-left one fairly tight and roughly circular, the other, upper-right, elongated along the 45° bisector of the axes. The interpretation of the vector elements, shown on the axes, suggests that the lower-left group consists mainly of profiles on high ground over laterite, whereas the profiles in the upper-right group occurred on valley sides and contained less lateritic material. In fact, when the results were referred back to the field it transpired that the tighter group consisted entirely of sites above the local level of the laterite outcrop, whereas the looser group contained only those sites lying below. The two groups could be neatly separated in figure 13.9 by a line AA' at right angles to the line of correlation.

Evidently the outcrop of the laterite marks an important division between two systems of soil and physiography in the area. Whereas further east on the Plateau the laterite makes substantial contributions to the soil profile downslope from its outcrop (Mulcahy and Hingston, 1961; Mulcahy et al, 1972), in the Wungong catchment it appears not to. It seems that the laterite once eroded from its original position is transported

Table 13.7. First six canonical roots and correlations for the Wungong Brook catchment.

Order	Root	Percentage of trace	Cumulative percentage	Canonical correlation
1	0·7739	41·2	41·2	0·88
2	0·3497	18·6	59·8	0·59
3	0·2615	13·9	73·7	0·51
4	0·1645	8·8	82·5	0·41
5	0·1120	5·9	88·4	0·33
6	0·1046	5·6	94·0	0·32

out of the landscape with too little remaining on lower slopes to mask the effects of the fresher material exposed there.

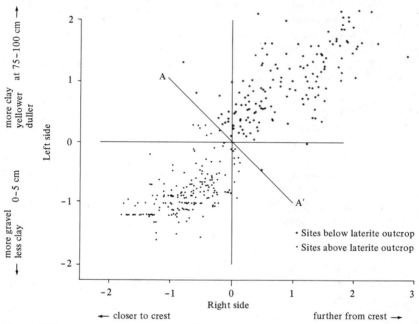

Figure 13.9. Scatter of Wungong sampling sites projected into the plane of the first pair of canonical axes. Symbols distinguish soil above and below local laterite outcrop.

13.5 Multiple discriminant analysis

Multiple discriminant analysis (MDA) is, of course, formally equivalent to canonical correlation in which one set of variables are dummies representing classes of a classification instead of measured quantities. In soil survey the context is different, and it is well to regard the applications of the two techniques separately. One of the main uses of MDA in general is to identify individuals as belonging to particular classes or to allocate them to the most appropriate classes. In land survey, MDA is being used increasingly to identify kinds of crop and soil from airborne sensors, and it has been used to identify constituents of soil from automatic chemical analysers. The technique has also been valuable for studying soil classification, and it is this use of MDA that is illustrated here.

13.5.1 Kelmscot, Oxfordshire

Burrough et al (1971) were concerned with the cost-effectiveness of soil survey, and compared soil maps made by different degrees of effort in three test areas in Oxfordshire. One of these areas, area 3, was at Kelmscot, home of William Morris. The data for that area have been reanalysed, and the results are given here.

Kelmscot lies on low terraced gravel in the upper Thames Valley. The soil ranges from shallow, stony brown loam over gravel to somewhat stony, grey, mottled clay loam and clay, also over gravel, but deeper. Three classes of soil (soil series) were recognised there, as follows.

(1) Badsey series; the shallowest, lightest in texture (loam), and well drained (brown throughout the profile).

(2) Carswell series; the deepest, heaviest in texture (often clay), and most strongly gleyed (indicating waterlogging).

(3) Kelmscot series; intermediate in character between Badsey and Carswell.

Attempts were made to map, by means of soil boundaries, these three classes of soil with three different degrees of effort in a test area 1400 m × 600 m. Two of the maps were made as though for publication at 1 : 63360 and 1 : 25000 by conventional ground survey. The third was made by interpreting 1 : 20000 air photographs as though for publication at 1 : 40000. They represented a progression in detail: the 1 : 63360 map was the least intricate, the 1 : 25000 map the most.

The area was then sampled at eighty-four points 100 m apart on a square grid and the following twenty-one characters were measured: colour hue, value, and chroma, degree of mottling, and sand and clay contents both in topsoil and subsoil; contents of organic matter, potassium, magnesium, and phosphorus, cation exchange capacity, structure size and degree of development, and consistence in the topsoil only; and depth of mottling. The data were then analysed for each map classification in turn, and Wilks's criterion, $L = |S^W|/|S^\pi|$, and $\mathrm{tr}(S^{W-1}S^B)$ were calculated to measure the effect of each classification. S^π, S^W, and S^B are respectively the total, within-groups, and between-groups matrices of sums of squares and products. The results are given in table 13.8. The classes of the 1 : 25000 map are most compact, those of the 1 : 63360 map least so, as judged by the values of L. The groups are best separated in the 1 : 25000 map and closest together (in the Mahalanobis sense) in the 1 : 63360 map, as judged by $\mathrm{tr}(S^{W-1}S^B)$.

The effects are perhaps best displayed as canonical variates, and figure 13.10 shows the three groups represented as 90% confidence circles centred on their mean canonical points for each classification. There is evidently a good deal of overlap in the 1 : 63360 classification.

Table 13.8. Wilks's criterion L, and $\mathrm{tr}(S^{W-1}S^B)$ for three map classifications and an improved classification of the Kelmscot area.

Classification	L	$\mathrm{tr}(S^{W-1}S^B)$
1 : 63360 map	0·470	0·94
1 : 40000 map	0·301	1·98
1 : 25000 map	0·176	3·04
Improved	0·056	7·44

The 1 : 40 000 air photograph interpretation classification is an improvement, but is achieved entirely by a better separation between the Carswell series and the other two. In the 1 : 25 000 map, Carswell and Badsey series are well separated, though both overlap somewhat with Kelmscot series.

There is also increasing interest in improving soil classifications by MDA along the lines proposed by Rubin (1967); that is, by reallocating individuals that lie nearer to some group centroid other than their own. The last line of table 13.8 and the lower-right-hand graph in figure 13.10 show the result of reallocation that starts from the 1 : 25 000 map classification. For these data there was evidently still room for considerable improvement.

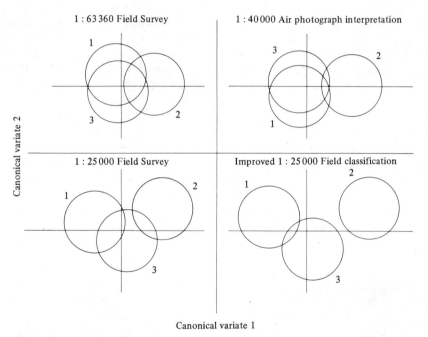

Figure 13.10. Differences among groups of four classifications of soil profiles at Kelmscot, displayed as 90% confidence circles in the plane of canonical axes.

13.5.2 Sabah

The last application of multivariate analysis described in this paper represents an intermediate stage in a soil survey. It concerns a reconnaissance survey made by Acres et al (1975) in the Keningau and Tambunan districts of Sabah, which cover some 4400 km² of rugged terrain. The country rock consists of highly contorted Tertiary greywackes (sandstones) and shales, more or less deeply weathered and planed during Pliocene times, and subsequently strongly dissected. The result is a complex soil pattern. Nevertheless, one of the aims of the soil survey was to try to express the soil variation by simple classification.

It was decided to use a classification of soil profiles based on the
FAO/UNESCO legend for the soil map of the world (FAO, 1968). Three
classes, orthic acrisol, gleyic acrisol, and recent alluvium were well-
represented [see FAO (1968) for definitions]. The first spanned a broad
spectrum, however, and it was thought desirable to divide it. Initially this
was done on physiographic grounds—thus soil derived by rock weathering
in situ was distinguished from that formed in Pliocene terrace deposits.
Table 13.9 lists the classes and their definitions. It soon became clear that
this classification was unsatisfactory. Classes 3 and 4 were easily identified,
but distinguishing between the two classes of orthic acrisol was difficult in
the field. The situation was therefore studied on a small sample.

Sixty-six sites were chosen randomly within the survey area. The soil
profile at each site was examined to $1 \cdot 2$ m, three horizons were recognised,
and each horizon was described separately in detail. Horizon thickness,
colour (Munsell hue, value, and chroma), structure size and degree of
development, quantity and size of pores, degree of development and
thickness of cutans (clay skins), and root quantity were recorded in
the field; pH, soluble phosphorus, organic carbon, total nitrogen,
exchangeable calcium, magnesium, and potassium, cation exchange
capacity, and clay, silt, and sand contents were measured in the laboratory.
The soil at each site was also allocated to its class in table 13.9 on the
field evidence.

The data were first standardised and transformed to principal components
to reduce computing in later analysis. The first nine components accounted
for 71% of the variance present in the sixty-seven variates, and these were
retained for discriminant analysis. The results of MDA applied to the
field classification are displayed as canonical variates in figure 13.11.
Wilks's criterion is $0 \cdot 050$. The percentage of the trace of $S^{W-1}S^B$ is
given for each axis. The third dimension accounts for almost none of the
between-groups variance. Figure 13.11(a) shows groups 3 and 4 clearly
separated, but almost complete overlap between groups 1 and 2.

Webster and Burrough (1974) then tried to improve matters by initially
allocating to the classification only those profiles that seemed easy to
identify. The MDA was repeated, and previously unidentified profiles

Table 13.9. Initial classification of the soil of Keningau and Tambunan, Sabah.

Class	Definition
1 Orthic acrisol	well drained, derived *in situ* either from sandstone or shale
2 Orthic acrisol	well drained, weathered alluvium (terrace deposits)
3 Gleyic acrisol[a]	poorly drained, weathered alluvium
4 Recent alluvium	river floodplain alluvium

[a] Two members of this class were later identified as gleyic podzols (see text).

were allocated to the groups to which they were nearest. Figure 13.11(b) shows the result; as before over 98% of the between-groups variance is expressed by the first two canonical variates. Wilks's criterion is 0·026.

There are several features of interest. First, the new classification is in general better than the first; Wilks's L is almost halved. Second, several profiles were so far from any group centroid that they were best regarded as representing other groups. They are ringed in figure 13.11(b). The two at the upper left were later recognised as a group of gleyic podzols. Third, and most important in the context of the study, the overlap between groups 1 and 2 is still substantial. Neither the original classification nor the modified means of using it were satisfactory for creating mapping strata from the information available. For a more routine survey either the surveyors would have to be content with a single group of orthic acrisols or they would need to find better means of discriminating among them.

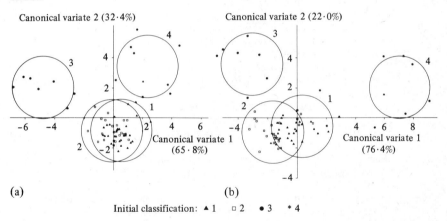

(a)   (b)

Initial classification: ▲ 1   □ 2   ● 3   * 4

Figure 13.11. (a) Soil profiles and 90% confidence circles for initial classification of Keningau and Tambunan districts, Sabah, in plane of first two canonical axes. (b) Scatter of Sabah soil profiles and 90% confidence circles after reallocation. The encircled points show sites that were considered to represent additional groups.

In the event the second alternative was pursued. Additional data were obtained, the classification was revised to produce a reasonable compromise between a numerically optimal result and one that would be practicable in routine survey. This was done by a combination of multivariate techniques, and is the subject of a lengthy paper on its own (Burrough and Webster, 1976).

Acknowledgements. The results described above derive from collaboration with several colleagues; namely, Dr P A Burrough (Kelmscot and Sabah), B E Butler (ACT), H M Churchward (Wungong), and Dr H E Cuanalo (West Oxfordshire). It is my pleasure to thank them all.

References

Acres B D, Bower R P, Burrough P A, Folland C J, Kalsi M S, Thomas P, Wright P S, 1975 "The soils of Sabah Land Resource Study" number 20, Ministry of Overseas Development, London

Beckett P H T, Webster R, 1965a "A classification system for terrain" Report number 872, Military Engineering Experimental Establishment, Christchurch, Hants

Beckett P H T, Webster R, 1965b "Field trials of a terrain classification—organisation and methods" Report number 873, Military Engineering Experimental Establishment, Christchurch, Hants

Burrough P A, Beckett P H T, Jarvis M G, 1971 "The relation between cost and utility in soil survey" (I-III) *Journal of Soil Science* **22** 359-394

Burrough P A, Webster R, 1976 "Improving a reconnaissance classification by multivariate methods" *Journal of Soil Science* **27** 554-571

Cuanalo de la C H E, Webster R, 1970 "A comparative study of numerical classification and ordination of soil profiles in a locality near Oxford" *Journal of Soil Science* **21** 340-352

FAO, 1968 "Definitions of units for the soil map of the world" World Soil Report 33, Food and Agriculture Association, Rome

Gower J C, 1966 "Some distance properties of latent root and vector methods used in multivariate analysis" *Biometrika* **53** 325-338

Gower J C, 1971 "A general coefficient of similarity and some of its properties" *Biometrics* **27** 857-871

Hernandez Avila A, 1976 *The Use of Canonical Correlation to Investigate the Relations Between Soil and Its Environment* M Sc thesis, Oxford University, Oxford, England

Hodgson J M (Ed.), 1974 "Soil survey field handbook" technical monograph 5, Soil Survey, Harpenden, Herts, England

Kaiser H F, 1958 "The varimax criterion for analytic rotation in factor analysis" *Psychometrika* **23** 187-200

Mulcahy M J, Churchward H M, Dimmock G M, 1972 "Landforms and soils on an uplifted peneplain in the Darling Range, Western Australia" *Australian Journal of Soil Research* **10** 1-14

Mulcahy M J, Hingston F J, 1961 "The development and distribution of soils of the York-Quairading area, Western Australia, in relation to the landscape evolution" CSIRO *Australia Soil Publication* 17

Rayner J H, 1966 "Classification of soils by numerical methods" *Journal of Soil Science* **17** 79-92

Rubin J, 1967 "Optimal classification into groups; an approach for solving the taxonomy problem" *Journal of Theoretical Biology* **15** 103-144

Webster R, 1977 "Canonical correlation in pedology: how useful?" *Journal of Soil Science* **28** 196-221

Webster R, Burrough P A, 1974 "Multiple discriminant analysis in soil survey" *Journal of Soil Science* **25** 120-134

Webster R, Butler B E, 1976 "Soil classification and survey studies at Ginninderra" *Australian Journal of Soil Research* **14** 1-24

Index